VIRAL DIAGNOSTICS

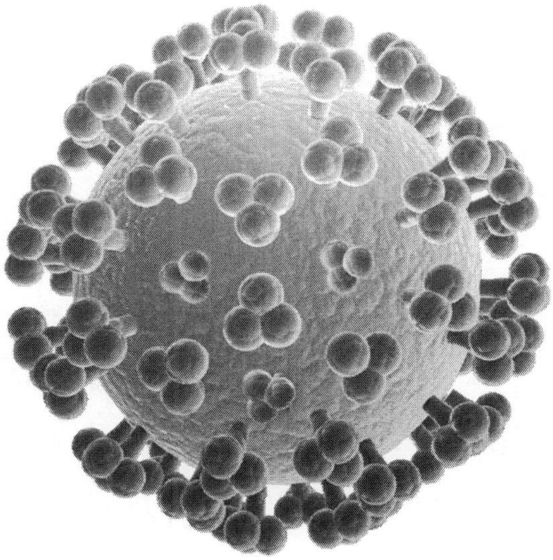

Pan Stanford Series on the High-Tech of Biotechnology

Robert S. Marks
Series Founding Editor

Avram and Stella Goldstein-Goren Department of Biotechnology Engineering
National Institute for Biotechnology Engineering
Ilse Katz Institute for Nanoscale Science & Technology
Ben Gurion University of the Negev
Israel

Titles in the Series

Published

Vol. 1

Nanoantenna: Plasmon-Enhanced Spectroscopies for Biotechnological Applications

Marc Lamy de la Chapelle and Annemarie Pucci, eds.

2013

978-981-4303-61-3 (Hardcover)
978-981-4303-62-0 (eBook)

Vol. 2

Viral Diagnostics: Advances and Applications

Robert S. Marks, Leslie Lobel, and Amadou Alpha Sall, eds.

2015

978-981-4364-43-0 (Hardcover)
978-981-4364-44-7 (eBook)

Vol. 3

Electrochemical Biosensors

Serge Cosnier, ed.

2015

978-981-4411-46-2 (Hardcover)
978-981-4411-47-9 (eBook)

Forthcoming

Vol. 4

Nanomaterials for Water Management: Signal Amplification for Biosensing from Nanostructures

Robert S. Marks and Ibrahim Abdulhalim, eds.

978-981-4463-47-8 (Hardcover)
978-981-4463-48-5 (eBook)

Vol. 5

Luminescent Biosensors

Gerald Thouand and Robert S. Marks, eds.

Vol. 6

Fibre-Optic Immunosensors and Biosensors

Robert S. Marks, ed.

Pan Stanford Series on the High-Tech of Biotechnology
Volume 2

VIRAL DIAGNOSTICS
Advances and Applications

edited by
Robert S. Marks
Leslie Lobel
Amadou Alpha Sall

PAN STANFORD PUBLISHING

Published by

Pan Stanford Publishing Pte. Ltd.
Penthouse Level, Suntec Tower 3
8 Temasek Boulevard
Singapore 038988

Email: editorial@panstanford.com
Web: www.panstanford.com

British Library Cataloguing-in-Publication Data
A catalogue record for this book is available from the British Library.

Viral Diagnostics: Advances and Applications

ISBN 978-981-4364-43-0 (Hardcover)
ISBN 978-981-4364-44-7 (eBook)

Printed in the USA

Viral Diagnostics: Advances and Applications *is dedicated to the newly opened Lee Kong Chian School of Medicine, Nanyang Technological University's joint medical school with Imperial College London in Singapore, with a vision to "Redefine Medicine and Transform Healthcare".*

We wish it a fruitful and successful journey.

The Editors

Contents

Foreword

Although the immediate genesis of this volume was the unique and very practical international Viralcheck Workshop held in Dakar, Senegal, in June 2008, it represents a signpost along the path begun by Leslie Lobel and Robert Marks years ago at the Ben-Gurion University of the Negev in Israel. The path through the jungle of infectious diseases does not end with the workshop or with this volume. As a result of their shared interest in infectious disease surveillance, diagnosis, and therapy, Lobel and Marks established individual scientific collaborations across the African continent, first simply as a means of collecting reagents to facilitate their development of advanced countermeasures back in the lab. They soon recognized the extraordinary skill and dedication of their collaborators in Africa and their desperate need for tools to fight disease. Together with their new colleagues, they defined unique challenges on the continent, including too few diagnostic tools, too little training and integration of new technologies, and too little continuity among infectious disease programs—many very good—already ongoing in Africa. Thus, at the Senegal workshop in 2008, the Bio-Africa Research Network (BARN) was born. Its purpose—"Research to directly address the local needs of Africans with respect to infectious diseases." Today, Lobel and Marks have strong scientific and human relationships in Senegal, Uganda, and South Africa and budding ones in Ethiopia, Kenya, Mali, Madagascar, Tanzania, and Ghana, obviously spanning both English- and French-speaking regions. This book serves as a record of these activities and underscores their value.

Here, the authors have brought together a set of chapters at once educational and interesting, technologically cutting edge, and practical. They begin with real-world examples of the impact, epidemiology, and control of naturally occurring infectious diseases, which extends from the popularized filovirus family of viruses to the timely influenzas and from field and laboratory diagnostics and global surveillance to the business side of medical countermeasures: research, development, and commercialization. Part one of the book provides the historic and contemporary underpinnings for the hard technical and field work that must follow, all with the goal of enhancing capacity in the developing world.

The two middle parts cover proven methods of diagnosis of infectious disease. First, classical antigen-antibody reactions that have been used for well-known agents (Ebola hemorrhagic fever virus, the Lyssaviridae, and the common viral respiratory killers) are described. Then, an in-depth and balanced description of new technologies is provided, some still not fielded: phage display, biochips, optical fiber-based systems, chemiluminescence, and DNA hybridization arrays. The chapters in these two sections clearly delineate the power of the biotech revolution's weapons and the point-of-care diagnostics that will have a huge impact on the battles against infectious disease. While the technologies are being developed in universities and start-up companies of the developed world, holding the workshop in the developing world underscored their global promise. Therefore, in the course of the book, the reader is transitioned from the challenges of confronting common and emerging infectious diseases in the developing world to the enormous promise of the biotech revolution and back to the laboratory where the hard work must be done.

Advanced Detection of Viral Pathogens is extraordinarily timely and near the working edge geographically and chronologically as the biotech revolution roars through the developed world and infectious diseases continue to burn through the developing world and beyond, killing 14 million humans per year globally. It would have been easy for Drs. Lobel and Marks to simply continue their very useful technical and clinical activities in the comfort of their university, but they had a vision. And they acted on their vision. I have admired their work at the bench, developing diagnostic tools and therapeutic antibodies. I have watched them articulate their vision to the good people who helped with early funding of their activities in Africa. I will always respect them for rolling up their sleeves, leaving the comfort of their university campus, building human relationships, and gaining the trust of their colleagues . . . all based on the common language of science and medicine and powered by the will to make a difference in the developing world. This volume is a lasting monument to the power of the synergies that can be found when good science and good human relationships are brought together by those who care enough to make the effort.

Dave Franz

Preface

Dear Reader,

From the beginning of recorded history it is evident that viruses evolved in our midst. Were they one of the unwanted gifts as Adam left the Garden of Eden? Our sudden nudity exposed our fragility and ever since we have had to survive them while continually seeking defenses.

The book you have in your hands illustrates a variety of challenges we bug-hunters try to overcome. Indeed, we have gathered here insights and expertise of an interdisciplinary group of researchers from all walks of life: virologists, physicians, immunologists, electrochemists, physicists, computer scientists, biotechnologists, epidemiologists, molecular biologists, all working in fields that cross scientific boundaries—working at the forefront of diagnostics.

Notwithstanding a dire moment as that of the recent Ebola outbreak in West Africa that saddens us all, we must more than ever peregrinate across oceans, and take the fight to them, as these invisible and exotic but deadly enemies know no barriers and constantly invade our lives. The coming epidemics and killer pandemics require us to continuously innovate in areas of prevention, detection and therapy. However, the present volume focuses on creating virus alerts or hand-held detectors to bring to the field during an epidemic, when possible, or tools for the healthcare worker to rapidly diagnose the patient's ailment. We describe the conventional cell cultures and molecular biology methodologies, while introducing state-of-the-art multidisciplinary biosensors. This book is not a manual, nor a set of protocols, nor a textbook, but our glimpse into the advances of virus diagnostic research. Included are a variety of topics that encompass world health issues, local folklore, such as associating outbreaks with witchcraft, problems in getting healthcare workers to the outbreak areas and insights in viral diagnostic pitfalls.

The concept of this book came about in Dakar, 2008, during a "Viralcheck" workshop the co-editors organized, where researchers from around the world came to test their respective diagnostic kits with samples provided by the Institute Pasteur in Dakar. The workshop introduced the participants to the difficulties involved in testing "real-life" samples, where discrepancies, data inconsistencies and inappropriateness for local use of tests were discovered. As I write this preface in my stay in Singapore at the CREATE laboratories where dozens of cases of dengue have occurred just meters away from us, my friends Amadou Alpha Sall and Leslie Lobel and our other colleagues are busy each in their own way to help fight this terrible wrath (Ebola) hitting humanity. I conclude that the timeliness of our book is mind-boggling. Finally, I would like to express my gratitude and thank all my co-authors, participants of Viralcheck, Dakar, Senegal, and our more than patient publisher, and of course, you the reader, in showing your interest in a topic that will never stop to concern all of us.

Robert S. Marks
Ben Gurion University of the Negev
October 2014

Chapter 1

Practical Experience with an Integrated Syndromic Surveillance System in the Medical, Veterinary, Nursing, and Emergency Response Communities

William Stanhope,[a] Tigi Ward,[b] R. Michael Ragain,[c] Gary Simpson,[d] and Alan Zelicoff[e]

[a] *St. Louis University, School of Public Health, St. Louis, USA*
[b] *City of Lubbock Health Department*
[c] *Texas Tech University Health Sciences Center, Department of Family Medicine*
[d] *New Mexico Department of Health*
[e] *School of Medicine, University of New Mexico, Albuquerque, NM, USA*
wm.stanhope@gmail.com, tward@mylubbock.us, zelicoff@slu.edu

1.1 More Than Half of All Human Infectious Agents are Zoonotic

Public health officials (PHOs) have long recognized the critical importance of comprehensive reporting of infectious disease in both animals and humans. It is estimated that more than 60% of 1,400 or so human infectious agents are, in fact, zoonotic in origin. Recent events, such as the arrival of West Nile virus (WNV) into

Viral Diagnostics: Advances and Applications
Edited by Robert S. Marks, Leslie Lobel, and Amadou Alpha Sall

ISBN 978-981-4364-43-0 (Hardcover), 978-981-4364-44-7 (eBook)
www.panstanford.com

North America and the subsequent spread of the virus to bird and mosquito populations across the continental United States, underscore the importance of early recognition of unusual signs and symptoms in wildlife, domestic animals, fowl, and humans to identify novel disease and its scope.

Unfortunately, as WNV and other surprising outbreaks (e.g., monkeypox in prairie dogs in Midwestern states in 2004) make clear, there is virtually no near-real-time communication between veterinary and human medical practitioners and precious little training in medical school in recognizing zoonotic disease in humans. Further, PHOs who would ostensibly evaluate clinical reports and provide advice, and initiate contact investigations in general, have no method for reliably and rapidly communicating with clinicians in their jurisdictions. For example, very few physicians subscribe to or read state public health offices' health alert networks (HANs); the number of veterinarians and emergency management and ambulance systems (emergency medical services [EMS]) who use local HANs is unknown but is likely to be a very small percentage of these professionals.

The current reportable disease paradigm that has long dominated public health reporting is inherently limited by (a) the predictable and indivisible duration in laboratory analysis of clinical samples; (b) failure to recognize signs and symptoms of classical infectious diseases that have now become rather rare—for example, measles, diphtheria, mumps, and tularemia; (c) failure to obtain and handle laboratory samples correctly; and (d) poor compliance on the part of physicians to report per se even when reportable conditions are diagnosed. Poor compliance is due to either a lack of awareness of the requirement to do so or the time-consuming process of notifying PHOs via telephone or fax.[1]

While there is no question that laboratories are valuable sources of clinical test results that are diagnostic of key reportable conditions—indeed, laboratories are, for all intents and purposes, the main source of infectious disease data for most PHOs—substantial time may be required for processing samples and generating actionable results. By definition, truly novel diseases or disease syndromes (such as the causative coronavirus of severe acute respiratory syndrome [SARS]) elude straightforward laboratory detection because of the absence of reagents for identifying the presence of the organism.

Finally, the threat of bioterrorism, along with the potential for rapid spread of naturally occurring communicable diseases of animals and humans, places an extraordinary premium on rapid diagnosis and situational awareness of population health. Many exercises conducted by the Departments of Defense, Homeland Security, and Health and Human Services lead inevitably to the conclusion that hours matter in a bioterrorism event or in some naturally occurring diseases (particularly in the agricultural large-animal population). Multiple disease propagation models utilizing agent-based and deterministic approaches as well confirm these lessons.

1.2 Syndrome Reporting Information Systems

Syndrome-based disease surveillance systems (SBDSS) are defined as data-gathering approaches focusing on clinical descriptors (signs and symptoms) instead of specific diseases. In theory, SBDSS, by virtue of their timeliness and volume of information flows, could assist in meeting these central public health responsibilities. In practice, however, the specific designs, underlying technical features, the scientific approach, and the ease of use are dramatically different across the dozens of SBDSS currently in existence, some of which have been implemented only in narrowly defined demographic settings or they have other limiting features. The promise is often not met in real-world use.

All SBDSS fall into two basic categories:[2]

- **Automated** or **passive** surveillance systems that seek to exploit existing data streams and employ various statistical algorithms to detect the presence of infectious disease. Some of the data sources that are tapped by these passive systems include:
  - Pharmacy sales (including over-the-counter medications)
  - Total volume of nurse hotline calls

  - Brief chief complaint summaries from emergency room logs and ambulance logs
  - School and work absenteeism

- **Active** or **clinical** surveillance systems that depend on selected reporting from physicians, veterinarians, EMS, and other health care providers on the basis of the clinical judgment when assessing the severity of illness among patients (whether animal or human). It is difficult to quantify the value of clinical judgment, but it is nonetheless a key component of daily parlance and processes of medical and veterinary practice.

It is also important to note that the overwhelming majority of SBDSS data-gathering features focus solely on human patients, despite the fact that in all significant outbreaks of novel diseases over the past decade or more in North America, animals were the primary source of the diseases.

A recent report has called into question the utility and timeliness of data mining approaches.[3] Data mining, though superficially attractive because it obviates the need for additional effort on the part of the clinician to enter data, appears to have an unacceptable signal-to-noise ratio. False positives are common, and because the data gathered is nonspecific (such as billing codes), a highly stochastic background incidence of syndromes is seen. Several data mining systems, and two clinician-driven reporting systems, have been utilized in the epidemiologically complicated, mixed urban, rural, and agricultural environment in west Texas and eastern New Mexico. These conclusions are consistent with other data mining experiences during 1999–2001.

Starting in 2003, and on the basis of the experience of Zelicoff and Simpson, inventors of the Rapid Syndrome Validation Program (RSVP)—a proof of principle experiment funded by the US Department of Energy but not currently in widespread use—the authors have developed and implemented the Syndrome Reporting Information System (SYRIS).[5] SYRIS, a fully integrated animal and human syndrome-based disease surveillance tool, has been in use by the City of Lubbock Health Department and Health Service Region I, a separate public health jurisdiction comprising the 40 counties around the city of Lubbock. In addition, SYRIS is in use by the state of California and the Centers for Disease Control's Border Infectious Disease Surveillance system (BIDSS). In total, more than one million people and countless animals are in the SYRIS coverage areas, with the participation of approximately 120 physicians and nurse-practitioners, 25 veterinarians, 25 school nurses, multiple EMS companies, and animal control officials. Wildlife rehabilitators who are often the first to recognize disease in bird and free-ranging mammalian species also report to SYRIS. Early experience with this system will be described and contrasted with other syndromic surveillance systems.

The basic design requirements for the SYRIS software are:

- Complete platform independence. SYRIS is written in JAVA and works on all Windows, Macintosh, LINUX, and other Java-enabled operating systems (including tablet personal computers [PCs]).
- Local PHOs serve as data analysts for all information reported from the clinical communities listed above. PHOs also register users via an administrative page in SYRIS.
- Rapid (less than 30 seconds) data entry based on syndromic classifications for animals or humans suspected of having symptoms and signs indicative—based on clinical judgment—of serious infectious disease.
- Immediate updating of geographic maps and temporal graphs so that all users can quickly visualize local-, regional-, and national-level reporting. Veterinarian, physician, school nurse, and EMS reports are color coded for easy interpretation (see illustration below). The specific details of individual case reports (e.g., reporting clinician, specific clinical findings, and approximate patient age) is under the control of local PHOs but can be easily shared as a spreadsheet file readable by almost all statistics software packages.
- Geographic information system (GIS)-based analysis tools for PHOs and clinicians complemented by statistical tools for rapidly determining unusual events.
- Free text advice and analysis from PHOs.

- Automated alarm criteria set on an ad hoc basis by PHOs such that any report meeting those criteria generates immediate notification of on-call PHOs via e-mail and/or digital pager.
- A highly secure database (SQL server) that meets all health level (HL)-7 and National Electronic Disease Surveillance System (NEDSS) requirements and that can easily import other data streams (e.g., laboratory data) as are available.
- Easy installation and automatic updating of software.
- On-line training via videos and manuals.
- Low cost—approximately 15 cents per capita for the annual licensing fee.

1.2.1 *Examples*

Data entry in SYRIS is via intelligent pulldown menus, which automatically change depending on the specific situation and radio buttons. There are nine veterinary syndromes and six human syndromes. An example of a veterinary submission is seen in Fig. 1.1.

Note that data fields include a "site descriptor" (zoo vs. feedlot vs. domestic/residential). Veterinary infectious disease experts who advised us in the design of SYRIS noted that this stratification was particularly important, given the unique diseases that are found in each of these subpopulations of animals. A wide variety of animal species are listed (and more can easily be added), and the veterinarian can enter their clinical impression, in addition to categorizing the case into a specific syndrome.

After entering the case by clicking Submit Report, the data is securely transferred to a central database. Within a few seconds the veterinarian receives the following screen (see Fig. 1.2).

The map (in this case centered on the area around Albuquerque, New Mexico) shows the case just reported (in zip code 87042, obscured in the previous figure by the topmost pull-down menu), as well as the case report on the time graph on the left (third graph from top "Incidence of Veterinary Neurological/Lameness"). Both the map and the graph report data as over a 30-day rolling window in time (this is an arbitrary parameter that can be set by the PHO). The color coding on the map reflects the syndromes

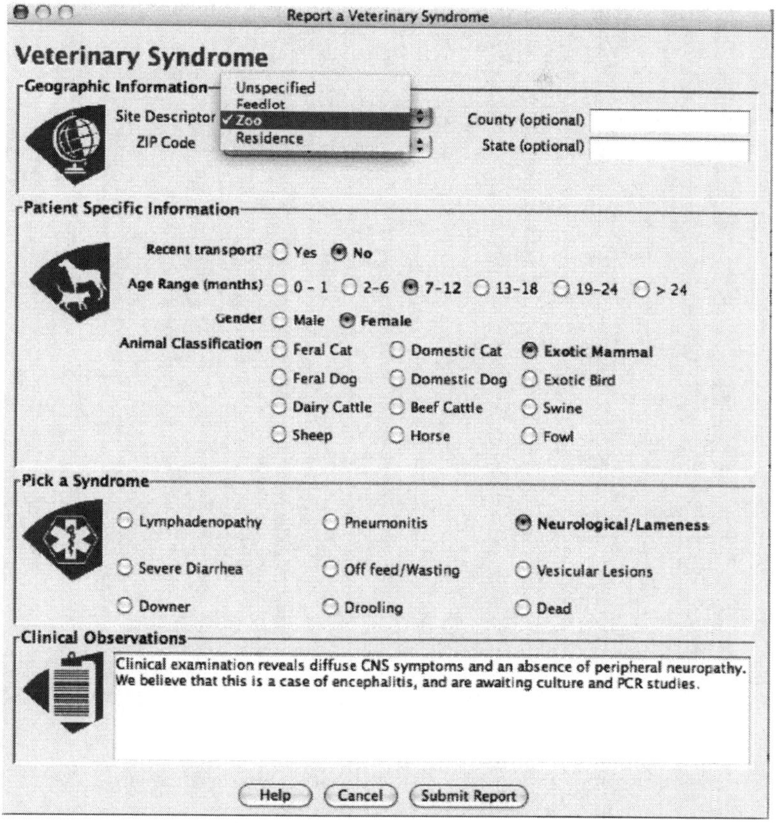

Figure 1.1

reported: horizontal stripes represent human syndromes; vertical stripes represent veterinary syndrome reports, such as the one just entered; and diagonal stripes indicate school nurse, EMS, or animal control syndrome reports. In any given public health jurisdiction (e.g., county or city health departments), all reports from users registered in that jurisdiction appear on the map.

Data querying is similarly straightforward. The map can be panned and zoomed in real time; maximum resolution is at the zip code level (which has been assessed as sufficiently precise for analysis purposes while at the same time protecting patient confidentiality), and entire regions can be visualized and queried.

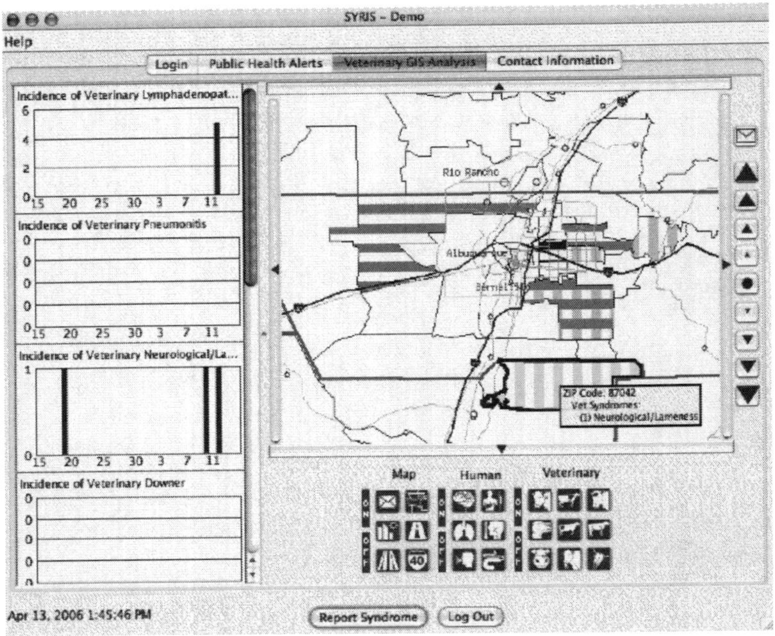

Figure 1.2

For example, by zooming out and right-clicking a state, SYRIS summarizes all case reports by syndrome from the database in a few seconds (see Fig. 1.3).

Finally, clinicians and PHOs can use the GIS tools beneath the map to quickly ascertain the overlap(s) of syndromes that may help narrow differential diagnoses. The blue icons beneath the map represent geographic features (such as major highways, political boundaries, and postal codes), each of the six human syndromes (fever with central nervous system [CNS] findings, acute hepatitis, adult respiratory distress syndrome, influenza-like illness, fever with skin rash, and severe diarrhea) and the nine veterinary syndromes (see Fig. 1.1 for a list of these categories). So, if users are concerned about encephalitis cases, they can turn off all syndromes except animal neurologic syndrome and the human syndrome of fever with CNS findings (see Fig. 1.4). In this case there is a geographical and temporal overlap of veterinary and human reports,

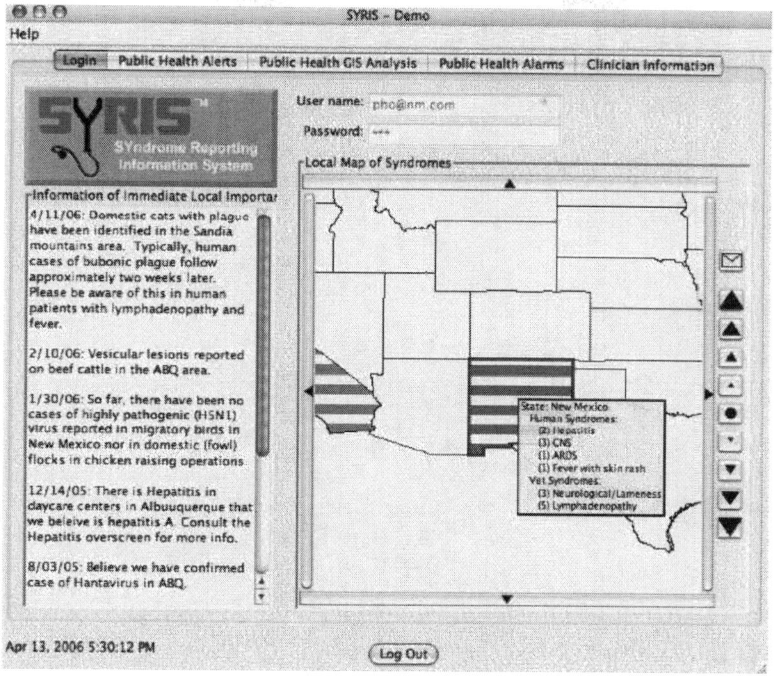

Figure 1.3

and as local PHOs would be aware, the specific areas involved are both urban and rural and are also known to be heavily infested with mosquitoes as the zip code boundaries span the Rio Grande in the center of Albuquerque (see Fig. 1.4).

PHOs can easily determine which clinicians have reported cases of interest that may warrant further discussion with the health care provider. When a PHO logs into SYRIS, he or she can perform the same GIS analysis. SYRIS organizes the data by clinician type (physicians, veterinarians, school nurses) and then by date within each category. Included in the database is the name and phone number of the reporting clinician. Thus, initiating a contact investigation is streamlined and allows the index of suspicion of PHOs to be rapidly assessed. PHOs can easily view the database in almost any spreadsheet or statistics program, facilitating further analysis.

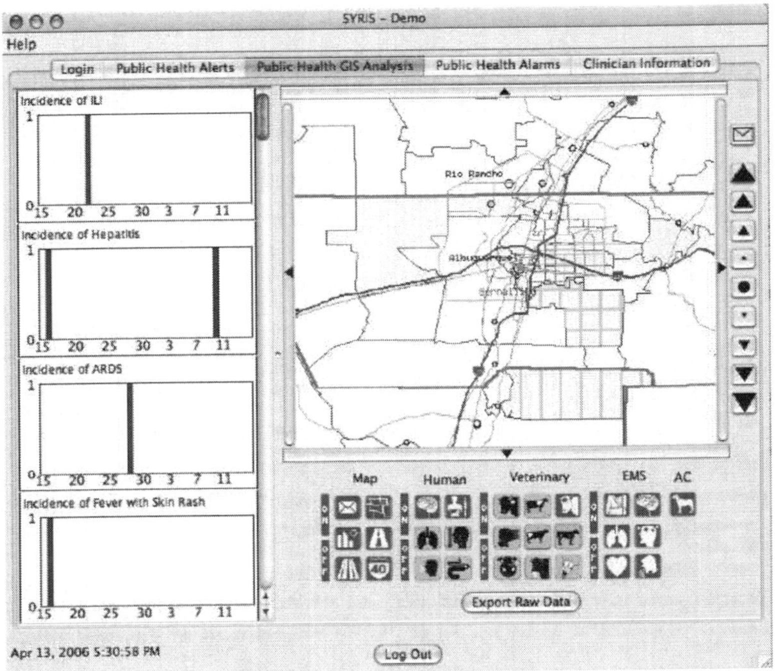

Figure 1.4

Further, PHOs can set specific criteria—syndromes, signs, and symptoms within those syndromes—for which they wish to be notified immediately, even if they are away from their computers. For example, if a PHO was interested in knowing about all cases of international travelers over the age of 17 with severe diarrhea, he or she simply goes to the Alarm tab, initiates a New Alarm, and selects criteria from intelligent pull-down menus (see Fig. 1.4).

This process makes operational the public health principle of "case definitions," which are, in practice, almost impossible to communicate to clinicians. Thus, when a clinician enters a case that meets the criteria of the case definition, SYRIS notifies the PHO with an e-mail and/or a digital page, specifying the name of the clinician, his or her telephone number, the syndrome reported, and any other clinical details reported. The PHO can then make the decision to contact or not contact the clinician. Should the criteria be too general such that too many alarms are generated, the PHO can at any time refine (define with higher specificity) the criteria of interest.

1.2.2 *Some Successes with SYRIS*

SYRIS has become the primary public health tool for routine, daily communication among clinicians, animal control officials, and PHOs in the city of Lubbock and the surrounding 40 counties of northwest Texas. In addition to providing regular, timely summaries of advice to clinicians, SYRIS was also employed during the Katrina disaster by physicians pressed into service to staff two medical clinics at the evacuee center, established for approximately 850 people in the mostly abandoned Reese Air Force Base approximately 20 miles west of Lubbock. Because the volunteer physicians worked in shifts and the evacuees were at high risk for unusual infectious disease (e.g., *Vibrio vulnificus* cellulitis, meningococcal disease, and enteric infections) with which most physicians had little experience, PHOs and doctors installed SYRIS in two laptop computers within a few minutes of the arrival of evacuees. Local physicians volunteered to staff the intake sessions for evacuees and then staffed an ongoing clinic over a two-week period. Many of these physicians ordinarily worked in private practice or as faculty members at Texas Tech School of Medicine. None of these doctors had any formal training with the SYRIS program or syndromic surveillance. The system was so intuitive in its design that little training was required for these physicians to be able to use it. Ward provided a single-page handout to guide physicians. Data collection was very simple, and therefore Ward and her team were able to collect syndromic surveillance data successfully.[4]

The data from SYRIS allowed for planning and preventing the outbreak of communicable disease within other evacuees and to the Lubbock community. The display of data from the SYRIS system is easy to read and allowed physicians working with the evacuees to assess and react to important outbreak information in a timely manner. In the system, a single screen summarizes syndromic data for easy interpretation.

The system also aided in the care of individual patients with symptoms to help guide diagnosis. SYRIS was added to the health director's station in Lubbock's Emergency Operations Center (EOC) when the EOC was activated for the Katrina response. The director of the health department was able to monitor SYRIS from the EOC bunker 20 miles away and determine if resources would be allocated to meet the health needs of the evacuees. On the basis of information observed in SYRIS, the director was able to identify, with reasonable certainty, the most prevalent health complaints of the evacuees. Ordering unnecessary supplies and unneeded resources was thereby reduced significantly.

Because of the concern about unusual disease and associated misdiagnosis and incorrect treatment, the local community wanted to be prepared for outbreaks of communicable diseases should they occur either at the evacuee housing site or in the surrounding area. SYRIS addressed both problems easily. For example, on the main SYRIS login page, continuously updated advice was provided by PHOs and the SYRIS map displayed the incidence and location of a variety of infectious disease syndromes (see Fig. 1.5).

At intake, numerous cases of diarrhea, skin rash, and headache, with or without fever (fever with CNS findings), were reported but rapidly waned, reassuring physicians, evacuees, and residents that there was little likelihood of a communicable disease outbreak. Few, if any, unnecessary diagnostic tests were ordered, and there were no complications from missed diagnoses or inappropriate treatment. PHOs at multiple levels were apprised of clinical assessments, dramatically reducing costs and waste of limited resources. Because of its intuitive interface and online training, clinicians were able to train themselves in the use of the system with minimal assistance from busy PHOs.

1.2.3 *Actionable Information: The SYRIS Advantage*

PHOs have long recognized the desirability of high-specificity, high-sensitivity data characterizing the population health of animals and humans in near real time to provide actionable information to identify and interrupt infectious disease outbreaks. SYRIS has met this essential function while at the same time providing an additional unexpected benefit—strengthening the relationships between public health and clinical practice professionals. In west

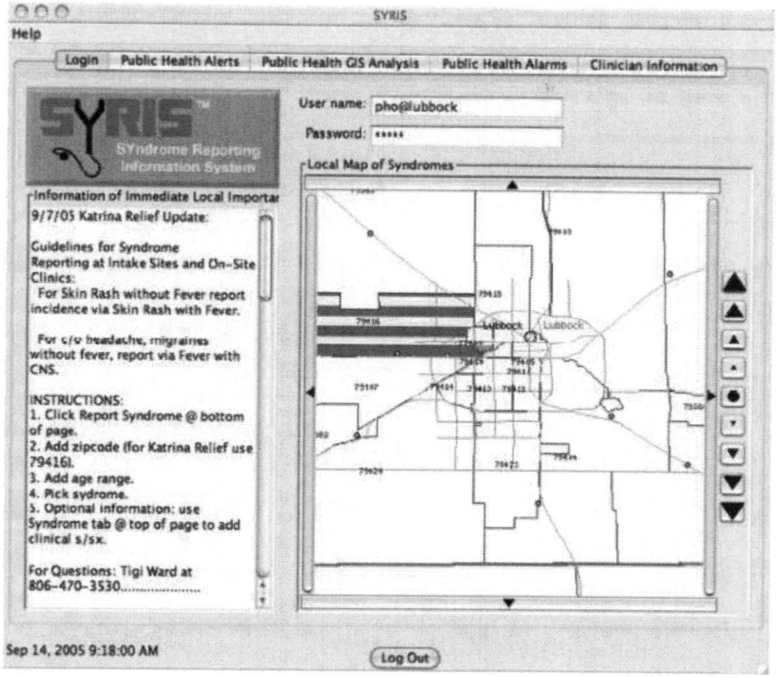

Figure 1.5

Texas and parts of New Mexico, multiple data mining and clinician-driven syndrome surveillance systems have been evaluated, but to date, no system has met SYRIS's ease of use, low cost, low false-positive rate, and broadly based clinician acceptance. The authors are in the process of a detailed chart review to measure the sensitivity and specificity of SYRIS over the wide range of diseases covered by the clinical syndromes,[4] but on the basis of interviews with physicians and the ever larger number of clinicians voluntarily participating in SYRIS, the authors have the impression is that compliance with reporting from physicians and school nurses is high.

Unlike some other electronic syndrome-based systems, there have been no violations of patient confidentiality after many hundreds of case reports, no false-positive results (defined here as cases that led to the expenditure of scarce public health resources on insignificant events), nor false negatives (defined as cases of classical reportable disease that were identified by laboratory examination but of which the Department of Health had no awareness from clinical syndromic data entry from clinicians). Physicians, in particular, utilize SYRIS because it provides them with a real-time assessment of infectious disease epidemiology, thus fostering efficient use of laboratory tests and appropriate antibiotic prescribing. In short, SYRIS makes it possible to practice evidence-based medicine across all age groups who present signs and symptoms of acute, multisystem infectious disease.

Near-term augmentations in SYRIS include real-time environmental monitoring via National Oceanic and Atmospheric Administration (NOAA) satellites, dust model prediction 48 hours in advance to anticipate environmental respiratory disease in humans and animals, and a neural network backbone to detect anomalies using nonlinear methods.

1.3 Conflict of Interest Disclosure

All authors declare that they have no financial or ownership interest in ARES Corporation or the SYRIS software system, either singly or jointly.

References

1. Doyle, T. J., M. K. Glynn, and S. L. Groseclose. Completeness of notifiable infectious disease reporting in the United States: an analytical literature review. 2002. *Am. J. Epidemiol.*, **155**, 866–874.
2. Bravata, D. M., K. M. McDonald, W. M. Smith, C. Rydzak, H. Szeto, D. L. Buckeridge, C. Haberland, and D. K. Owens. Systematic review: surveillance systems for early detection of bioterrorism-related diseases. 2004. *Ann. Intern. Med.*, **140**, 910–922.
3. Pavlin, J. A. Medical surveillance for biological terrorism agents. 2005. *Human Ecol. Risk Assess.*, **11**, 525–537.

4. Ward, T., T. Camden, A. Zelicoff, G. Simpson, and W. Stanhope. The sensitivity, specificity and physician compliance in a clinician-driven syndromic monitoring system. In preparation.
5. SYRIS is owned by ARES Corporation, Burlingame, California. http://www.arescorporation.com/software/syris/

Chapter 2

Environmental Surveillance for Polioviruses in Israel: Bioerror, Bioterror, or just Mother Nature

Lester M. Shulman,[a,d,e] **Yossi Manor,**[b,e] **Danit Sofer,**[c,e] **and Ella Mendelson**[d,e]

[a]*Environmental Virology Laboratory*
[b]*National Center for Viruses in the Environment*
[c]*National Center for Polio and Enteroviruses*
[d]*Department of Epidemiology and Preventive Medicine, School of Public Health, Sackler Faculty of Medicine, Tel Aviv University, Tel Aviv, Israel*
[e]*Central Virology Laboratory, Public Health Services, Israel Ministry of Health, Chaim Sheba Medical Center, Tel Hashomer, Israel*
lester.shulman@sheba.health.gov.il, yossi.manor@sheba.health.gov.il, danitso@sheba.health.gov.il, ellamen@sheba.health.gov.il

2.1 The Silent Presence or Circulation of Polioviruses in Poliomyelitis-Free Communities

The Environmental Virology Laboratory is one of two laboratories that form the Israeli Public Health Services Central Virology Laboratory. This chapter is based on the cooperative effort of three groups within the Environmental Virology Laboratory, namely,

Viral Diagnostics: Advances and Applications
Edited by Robert S. Marks, Leslie Lobel, and Amadou Alpha Sall

ISBN 978-981-4364-43-0 (Hardcover), 978-981-4364-44-7 (eBook)
www.panstanford.com

the National Center for Viruses in the Environment, the National Center for Polio and Non-polio Enteroviruses, and the molecular epidemiology group. This chapter describes the routine, monthly environmental surveillance program for the detection of the silent presence and/or circulation of polioviruses within our communities and how molecular epidemiological analysis can answer whether the presence of a particular poliovirus isolate was due to bioerror, bioterror, or just Mother Nature.

The terms "silent presence" and "silent circulation" refer to the presence and circulation, respectively, of polioviruses within a community before the occurrence of paralytic poliomyelitis in individual members of the community. The silence occurs due to the fact that the majority (greater than 90%) of primary infections of naive immunocompetent individuals with neurovirulent polioviruses are asymptomatic. The remaining infections are symptomatic and can take one of three forms: abortive poliomyelitis—a nonspecific febrile illness with headache, malaise, sore throat, and/or vomiting; nonparalytic poliomyelitis or aseptic meningitis—a febrile illness accompanied with back pain and muscle spasms; and paralytic poliomyelitis—an illness characterized by asymmetrical acute flaccid paralysis (AFP) of the limbs, pharynx, vocal chords, and/or respiratory paralysis. These overt diseases occur in less than 5%, 1–2%, and 0.1–0.2%, respectively, of infected individuals.

2.2 Global Eradication of Poliomyelitis

With the development of inactivated poliovirus vaccine (IPV) and live oral poliovirus vaccine (OPV) in the 1950s, paralytic poliomyelitis became a vaccine-preventable disease. Both vaccines are trivalent, that is, they contain one representative for each of the three poliovirus serotypes. The World Health Organization (WHO) Global Poliomyelitis Eradication Initiative of 1988, inspired by the success of the smallpox eradication program, had as its goal, the eradication of wild poliovirus by 2000 through universal vaccination of all children in order to break the chains of transmission. The eradication initiative also envisioned a discontinuance of polio vaccinations three years after the last case of poliomyelitis due to the circulating wild poliovirus. This would free the approximately

1.5 billion dollar annual expense for global vaccination for other health programs. To date, universal vaccination with OPV, and to a lesser extent IPV, or combined schedules of OPV and IPV, successfully eliminated transmission of wild type 2 poliovirus by 1995 and reduced the number of countries with endemic transmission of wild type 1 and/or type 3 poliovirus to six countries by 2004.

Three problems have contributed to the delay in attaining eradication: vaccine failure, failure to vaccinate, and reversion of live, attenuated oral polio vaccine strains to neurovirulent strains that in rare circumstances have caused limited outbreaks of poliomyelitis. The combination of a very high number of births, a high force of endemic wild poliovirus transmission, and widespread competitive interference of OPV reproduction by concurrent enteroviral infections has contributed to vaccine failure, for example, wild poliovirus has induced AFP in children who received greater than 10 OPV doses in the Uttar Pradesh province of India.[10,20] Cessation of vaccination for religious/political reasons in Nigeria in 2003[11] is an example of failure to vaccinate. The cessation led to a rapid spread of wild poliovirus with associated cases of AFP among naive individuals in Nigeria and more than 15 previously poliomyelitis-free countries.[1]

Finally, live attenuated OPV strains like wild polioviruses rapidly accumulate nucleotide substitutions[7,14] and their genomes can undergo recombination with other polioviruses and clade C, nonpolio enteroviruses during replication.[3,9,33] Some of these substitutions can cause the progeny of vaccine virus to revert to a neurovirulent phenotype (as few as two to three nucleotide substitutions may differentiate between the two phenotypes[19]), and this reversion to neurovirulence may even occur during infection of the primary vaccine.[8] On rare occasions, these vaccine-derived polioviruses (VDPVs) have caused limited outbreaks of poliomyelitis (see reviews[2,13]).

Even when eradication is obtained, there is a very high probability for reemergence of wild poliovirus.[29]

Potential sources for reemergence are bioerror, bioterror, and Mother Nature. Bioerror refers to the accidental release of poliovirus from a vaccine production or storage facility, from known laboratory poliovirus stocks, from stored poliovirus contaminated clinical material obtained when wild poliovirus circulated endemically, or even from stocks of poliovirus recreated[6,32] by genetic engineering.

Bioterror refers to the intentional release of poliovirus from any of these sources. Its effectiveness increases in proportion to the growing cohort of naive individuals if universal vaccination were to be discontinued. The contribution of Mother Nature could include release of poliovirus preserved in permafrost, genetic mutation, and phenotypic reversion of live attenuated vaccine or emergence of a fourth serotype of poliovirus from a strain of enterovirus that has mutated so that it binds to the same human cell receptor as polioviruses. In the current preeradication era, we tend to classify vaccine failure as a bioerror and unintentional introduction from an endemic region into a poliomyelitis-free region as primarily due to Mother Nature.

2.3 The Need for Routine Surveillance

For eradication to succeed, it is not enough to employ universal vaccination to break the chains of transmission of the wild virus. The eradication program must also include a surveillance arm to ensure that these chains of transmission have, in fact, been broken. The major pathway for transmission of poliovirus is fecal-oral. As a consequence of initial infection of cells in the oropharynx and intestine, high titers of polioviruses are excreted in stools. The primary surveillance method recommended by the WHO requires laboratory examination of stool samples to confirm whether poliovirus was the cause in each case of AFP. However, the very low percentage of infections that cause AFP means that for each case due to poliovirus, there is a large segment of the population already silently infected. Since most of the excreted poliovirus (with the major exception of virus in soiled diapers) ends up in sewage, another method for detecting silent circulation of polio is to search for and characterize all polioviruses recovered from sewage.

2.4 The Program of Environmental Surveillance for Poliovirus in Israel

The remainder of this chapter is a description of the development and application of the routine environmental surveillance program for Israel, the Gaza district, and the West Bank for detecting reintroduction of poliovirus through bioerror, bioterror, or Mother Nature. It is important to point out that Israel, the West Bank, and the Gaza district have been poliomyelitis free from 1989 to date, and all polioviruses isolated between 1989 and today have been identified from environmental samples in the absence of any clinical cases of poliomyelitis.

Israel was one of the first countries to implement nationwide polio immunization, starting with IPV in 1957, followed by OPV in 1961, and a combined inactivated/oral polio vaccine IPV/OPV schedule in 1989.[28] In 2005, Israel switched to exclusive use of IPV. This reduced the incidence of poliomyelitis from greater than 140 per 100,000 in 1950 to sporadic cases between 1983 and 1986. In 1988, there was a cluster of poliomyelitis cases in Israel and the West Bank[26] that will be referred to as the 1988 outbreak. Countrywide administration of OPV to all individuals less than 40 years of age succeeded in breaking the chain of poliovirus transmission of the outbreak, and the last diagnosed case of poliomyelitis in Israel occurred on October 3, 1988.

A comprehensive regional sewage surveillance plan was initiated in 1989 to detect poliovirus circulation before the appearance of AFP cases.[18] It involved monthly surveillance of between 20 and 30 sentinel communities in Israel, the West Bank, and the Gaza district. Sites were chosen after careful consideration of a number of factors, including identifying communities where disease occurred in the past, or where residents were likely to travel or come in contact with people from external poliomyelitis endemic regions, the socioeconomic status of the residents, the type of sewage system, and the ease of access. A mobile field laboratory (see Fig. 2.1) with portable automatic sewage samplers provides extra flexibility in choosing sites and allows for rapid response to specific situations.

The number of sentinel communities was reduced to 15 in 2000. The first consideration was to develop a method for high-throughput virus isolation from sewage that was suitable for a poliomyelitis-free country with high documented vaccination coverage. The method needed to detect wild and vaccine-derived poliovirus in the presence of high backgrounds of OPV and nonpolio enteroviruses. Such a system would also be appropriate for countries with vaccination

Figure 2.1 The Mobile Environmental Virology Laboratory. The mobile laboratory includes an electric generator, centrifuges, a pump and filtration system for sampling large volumes of water, refrigerators, incubators, and a mobile automatic system for collecting composite sewage samples, (a pool of samples from a given site taken at regular intervals over a 24-hour period). One of the composite samplers is shown in the forefront of the picture.

programs based solely on IPV. Protocol details and their validation appear elsewhere.[16,17] Molecular characterization of viral isolates currently provides the most detailed and easily manipulatable data for quantitative analyses of the relationship between viral isolates. As currently recommended,[5] poliovirus isolates are classified as vaccine, vaccine-derived, or wild-type poliovirus on the basis of the percentage nucleotide sequence homology between the isolate's capsid protein, VP1, and that of the corresponding OPV serotype. An isolate with a VP1 nucleotide divergence of 0% to 1% is classified as vaccine virus. When divergence ranges between 1% and 15%, it is classified as a VDPV. Finally, isolates with a greater than 15% difference are considered wild-type polioviruses. As knowledge continues to accumulate on vaccine evolution, isolates with a greater than 15% VP1 nucleotide difference are still considered VDPVs, when a genetic linkage to a known VDPV with less than 15% divergence can be demonstrated.

2.5 Polioviruses Isolated from Environmental Samples in Israel

Between 1989 and 2008, molecular characterization of all sewage isolates from Israel, the Gaza district, and the West Bank has provided evidence for all three classifications of poliovirus. Most isolates (greater than 98%) were vaccine strains with minimal divergence (less than 0.5%). This was expected because of the extensive use of OPV in Israel between 1989 and 2005 and in the West Bank and Gaza district from 1989 to date. The next most frequent type was wild poliovirus. Wild type 1 poliovirus was isolated from two sewage samples in Gaza in 1991; from 11 samples in Israel, Gaza district, and the West Bank between 1994–1995; and from 1, 1, and 2 samples, respectively, from the Gaza district in 1996, 1999, and 2002. Wild type 3 was isolated from one environmental sample from the Gaza district in 1990. The third and least frequent type consisted of highly diverged VDPVs that had diverged from their respective Sabin 2 type by 8% to greater than 15%.[24,25]

2.6 Molecular Analysis Yields Epidemiological Information

During the course of our environmental surveillance program, molecular analysis has shifted from restriction fragment-length polymorphism of reverse transcriptase polymerase chain reaction (RT-PCR)-amplified viral templates,[30] through sequence analyses of 250 nucleotides of a region spanning the 3′ end of VP1 and the 5′ end of the 2A gene, to analysis of the entire VP1 gene (906 nucleotides for type 1, 903 nucleotides for type 2, and 900 nucleotides for type 3).[22] For isolates of interest, we may also sequence 450 nucleotides of the 5′ untranslated region,[26] approximately 2,240 nucleotides encoding all four viral capsid proteins, the 1,500 nucleotides of the viral polymerase 3D gene, or the entire 7,400-nucleotide genome.

Phylogenetic relationships among isolates from Israel and the world can then be inferred using free open source, shareware, or commercial sequence analysis programs. The chance of finding nucleotide differences between two isolates increases in proportion to the length of the portion of the genome sequenced. A corollary of this is that it becomes easier to detect smaller differences between isolates as the length of the sequences used for comparison is increased. The data for the sequence comparisons comes from the Israeli Laboratory of Environmental Virology local data bank of Israeli, the Gaza district, and West Bank environmental isolate sequences; public data banks (EMBL, GenBank, DDBJ); and specialized data banks such as those maintained by the Poliovirus Laboratory at the Respiratory and Enteric Virus Branch of the Division of Viral and Rickettsial Disease at the Centers for Disease Control (CDC) in Atlanta, Georgia, USA, and from the National Public Health Laboratory, KTL, in Finland. The large repertoire of sequences in these data banks provides the necessary information for identifying the source of imported strains (see examples below).

The first level of molecular epidemiology is based on the generation of these rather straightforward phylogenetic relationships. In general, an isolate's VP1 sequence is used to determine the viral genotype (equivalent to molecular serotyping), whether it originated from vaccine or wild virus, and its relationship to viruses already isolated from the same region. For example, comparative VP1 analysis of isolates from clinical cases in 1987 and from the 1988 outbreak and concurrent environmental samples from 1988 linked the 1987 and 1988 isolates to a single epidemiological event and provided information on the type and rates of nucleotide substitutions that occur during outbreaks of short duration.[16,22]

This information allowed us to formulate a profile of substitutions that would be consistent with those that accumulate during short outbreaks of polio evolving from a single progenitor virus called POESP[16,22] for inferring whether two viruses isolated within a given time interval from the same region were linked by endemic transmission or were separately introduced from different external reservoirs. Confidence in POESP has been enhanced by the fact its parameters also fit a 10-year outbreak[9] of poliomyelitis in South

America. The most important aspect of the model is based on the rapid more or less constant rate of nucleotide substitutions during poliovirus replication. These nucleotide substitutions are believed to occur randomly and at the same rate throughout the noncoding and coding regions of the viral genome. However, most observed substitutions occur at the third codon positions and most of these are synonymous, for example, do not code for a different amino acid, due to redundancy of the genetic code and strong selection against deleterious amino acid substitutions. For example, greater than 90% of the substitutions were synonymous when compared to the vaccine parent of our most highly diverged (greater than 15%) VDPV isolate. Similarly, the need for conservation of structure (stems and loops) in the 5′ untranslated region generally restricts observed substitutions to those that maintain function. The confidence for ruling out a direct chain of transmission increases the higher the substitution rate is above POESP expectations, especially when the interval between isolations is short.

The second level of phylogenetic analysis involves locating the potential source of individual isolates. The accumulation of large databases of poliovirus sequences allowed the development of a second analytic tool to aid in differentiating between separate introductions and endemic transmission. This tool, called relative similarity (RS) analysis, is based on determination of whether the nucleotide sequences of two local isolates are more similar to each other than either or both are to isolates from external reservoirs.[16]

Two examples of POESP and RS analyses will be presented. In the first, two wild type 1 polioviruses isolated from the Gaza district four months apart in 2002 were shown to be separate introductions.[16] Furthermore RS indicated two separate regions in the southern governorates of Egypt as the most probable source for the introductions and indirectly provided the Egyptian Health Ministry with information that there might still have been wild virus circulation in those areas. In the second, a large wave of wild polioviruses isolated from the Gaza district sewage in 1994–1995 was correlated with the large influx of Palestinians from Egypt into the Gaza district after the signing of the Oslo Peace Accord in 1994.[18] POESP and RS analysis of this relatively large silent outbreak indicated multiple separate introductions with the

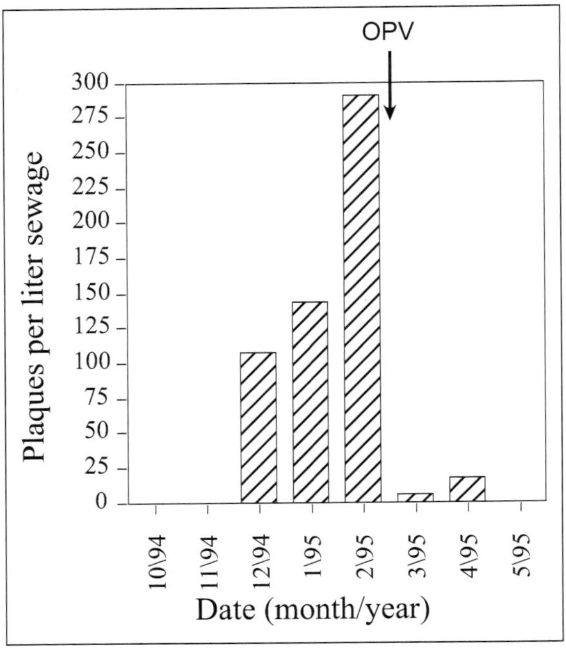

Figure 2.2 Sewage surveillance. Detection of silent circulation of wild poliovirus (in the absence of AFP) and its disappearance after intervention. In December 1994 wild poliovirus was isolated from environmental surveillance in Beit Lahiya and other communities in the Gaza district. Data is shown for Beit Lahiya. The number of plaques of wild poliovirus per liter of sewage increased over the next two months until subregional immunization with OPV [black arrow] broke the chains of transmission of wild poliovirus. The last wild poliovirus was isolated in April 1995.

highest similarity to isolates in Egypt. More importantly, it also provided evidence for local endemic transmission of one lineage (Shulman et al., unpublished) that required active intervention,[18] that is, initiation of a round of subregional immunization. In this case environmental analysis not only identified penetration of virus but also was provided documentation of the efficacy of response (see Fig. 2.2).[18] The ability to detect silent wild poliovirus circulation and then monitor efficacy of intervention is a major justification for maintaining a regular sewage surveillance program.

2.7 The Contribution of Routine Surveillance toward Understanding One Potential Route for Reemergence of Neurovirulent Polioviruses

Routine environmental analysis has increased our understanding of the potential reemergence of phenotypically wild virus from live attenuated vaccine. There are two major pathways for evolution of vaccine-derived poliovirus. The more common pathway is during chains of person-to-person transmission starting from a single dose of live vaccine. VDPVs arising in this manner are called circulating VDPVs or cVDPVs. Individuals in the chain of transmission have usually not been previously exposed to wild or vaccine polioviruses; however, we have shown that even highly immunized individuals can excrete virus for up to three weeks after the challenge.[27] The less common pathway is by the establishment of a persistent poliovirus infection in immunodeficient individuals either by direct administration of live vaccine or as a consequence of contact with someone who received a dose of live vaccine. VDPVs that arise by persistent infection of immunodeficient individuals are designated iVDPVs. Efforts to cure these persistently infected individuals have so far proven ineffective.[15]

VDPVs, such as those isolated from the environment whose pathways of origin are unknown or ambiguous, have been designated as aVDPVs. In 1988, the first Israeli aVDPV was isolated at the inlet of a sewage treatment plant servicing 1.6 million individuals in the greater Tel Aviv area using the automatic sampling system located at the entry to the treatment plant.[23] Its VP1 had diverged from Sabin 2VP1 by 8%. Its isolation raised a number of questions, only some of which have been answered. The most important questions and responses are outlined below:

1. **Q.** What was the likelihood of finding evidence of a single excreter or even of a few excreters in a population of 1.6 million? **A.** Models such as those of Ranta et al.[21] and empirical observations are providing some information, but more fieldwork needs to be done.
2. **Q.** Would more aVDPVs be isolated and over how long a time interval?

A. Between 1998 and 2008, another 32 type 2 aVDPVs were isolated from 23 additional sewage samples in Israel. The longest series of genotypically related isolates were periodically isolated over a span of 10 years. In general, moving surveillance sites to upstream sewage branches was much more effective in procuring new aVDPV isolates than increasing the number of samplings within a given time interval at any given site.

3. **Q.** Are there pockets of under immunized individuals who might have been the reservoir for the aVDPVs?

   A. Sero-prevalence vaccine coverage studies on potentially high-risk populations, such as documented and undocumented foreign workers,[4] and routine annual studies on different sectors of the Israeli population have so far not identified any such pockets.

4. **Q.** How many separate epidemiological events giving rise to aVDPVs might occur in a population the size of Israel, approximately 7,282,000 in 2008?

   A. Greater than one-third of the Israeli population is covered by our sewage surveillance sites. Phylogenetic analysis of the 32 aVDPVs indicated that there were at least two separate epidemiological clusters arising from separate administrations of Sabin 2 poliovaccine.[25] Until the end of 2012, sewage samples from the greater Tel Aviv area have continued to sporadically contain type 2 aVDPVs from both of these clusters. Two related, highly diverged, type 1 aVDPVs were isolated from sewage collected in Haifa in 2009 and 2012, providing evidence for a third epidemiological cluster. It is not known if this excreter is a resident or a periodic visitor to Haifa.

5. **Q.** How many people were excreting the virus, who were they, and where were they located?

   A. See Shulman et al.[25] for evidence supporting excretion by two persistently infected immunodeficient individuals or small cohorts in contact with such individuals. Surveillance sites, which were gradually moved to upstream branches in the sewage system, have currently restricted one foci to a branch serving 50,000 people and the other to a branch serving 25,000 people.

6. **Q.** If there are at least two foci of long-term excretions of aVDPV in the Israeli population, how many are there in the world?

A. The Israeli surveillance program is currently the most comprehensive, longest-running surveillance program. However, similar programs in Europe, Japan, and elsewhere have also yielded aVDPVs (reviewed by Kew et al.[13] and in the MMWR[2]).

7. **Q.** What are the phenotypic characteristics of these aVDPVs?
   A. Maximum nucleotide divergence has exceeded 15% for the first group. Five isolates were tested, and all were highly neurovirulent in a transgenic mouse model system.[25] In silico translation of the capsid proteins indicated an average of 12.1 +/− 1.5 amino acid substitutions in or near neutralizing antigenic sites for one of the foci and 5.0 +/− 1.0 for the other, suggesting alteration in the antigenic properties of these isolates.
8. **Q.** How well is the Israeli public protected against these potentially antigenically altered isolates?
   A. Vaccine coverage in Israel exceeds 95%. However, the geometric mean titers of neutralizing antibodies in the general population was threefold lower against the aVDPVs than against vaccine strains for all cohorts studied between the ages of 1.5 and 50 years.[25] While all serum samples from the cohort of 20- to 50-year-olds still had sufficient titers to neutralize the vaccine, a small number (7%) no longer neutralized the aVDPVs, implying for the first time that there may be a group of poorly protected or unprotected individuals even in a population with high documented vaccination coverage with high efficacy.

The wild polioviruses and aVDPVs isolated between 1989 and 2008 from the Israeli sewage surveillance program fall into the category of emergence or reemergence of potentially neurovirulent poliovirus into a poliomyelitis-free region through the intervention of Mother Nature. A sensitive limit of detection as shown by our ability to repeatedly detect infection of one chronically infected individual or a few infected contacts in large populations[25] and to detect and follow silent introductions and circulation of wild poliovirus[16,18] provide confidence that had there been emergence or reemergence through bioerror or bioterror, it would also have been detected.

2.8 Monitoring Silent Poliovirus Infections: The Contribution of Sewage Surveillance and Molecular Epidemiology

Sewage surveillance coupled with molecular analysis is the method of choice for monitoring silent poliovirus infections within a large population in the absence of AFP from poliovirus.[12,16,31] It requires development of a comprehensive plan for close coordination between public health laboratories, epidemiology departments, district health offices and hospitals, the CDC, public works engineering departments, and international agencies, including the WHO and the CDC. Manual or automatic sampling can be performed under difficult field conditions and the analysis carried out in well-equipped central laboratories. The surveillance plan can be easily modified to suit new conditions, allow increased sampling frequencies when needed, and allow alterations in sampling locations. The ability to detect silent wild poliovirus circulation and then monitor efficacy of intervention is a major justification for maintaining a regular sewage surveillance program. As the global effort to eradicate poliomyelitis approaches its goal, molecular epidemiological methods and environmental sampling assume greater importance for documenting that circulation of wild or highly diverged vaccine poliovirus has ceased. The public health implications and intervention strategies are different depending on whether isolation of two or more wild or highly diverged viruses within a short span of time represents local circulation or separate introductions into an area of high vaccine coverage. Advanced sequence analyses of environmental isolates may even provide epidemiological information about poliovirus circulation in the population being monitored, such as whether there is silent endemic transmission or separate introductions[16] or whether aVDPVs are more likely to have originated from persistently infected immunodeficient individuals or through person-to-person transmission.[25] An extensive global sequence database allows extraction of additional epidemiological information about the source (bioerror, bioterror, or Mother Nature) of local isolates and by inference provides surveillance information about poliovirus circulation in external reservoirs of viral infection. The information gleaned from the environmental surveillance program in Israel has contributed to formulation of the WHO guidelines for establishing environmental surveillance[31] and towards the understanding of some of the difficulties associated with cessation of vaccination programs when eradication of poliomyelitis has been achieved. Our laboratory personnel, under WHO sponsorship, have assisted with the initiation of environmental surveillance in other countries. And finally, our environmental surveillance plan could also be expanded to include surveillance for additional pathogens.

References

1. Resurgence of wild poliovirus type 1 transmission and consequences of importation: 21 countries, 2002–2005. 2006. *MMWR*, **55**, 145–150.
2. Centers for Disease Control and Prevention. Update on vaccine-derived polioviruses: worldwide, January 2006–August 2007. 2007. *MMWR*. **56**, 996–1001.
3. Agol, V. I. Recombination and other genomic rearrangements in picornaviruses. 1997. *Sem. Virol.*, **8**, 77–78.
4. Calderon-Margalit, R., Sofer, D., Gefen, D., Lewis, M., Shulman, L., Mendelson, E., Swartz, T. A., and Shohat, T. Immune status to poliovirus among immigrant workers in Israel. 2005. *Prev. Med.*, **40**, 685–689.
5. CDC. Laboratory surveillance for wild poliovirus and vaccine-derived poliovirus, 2000–2001. 2002. *MMWR*, **51**, 369–371.
6. Cello, J., Paul, A. V., and Wimmer, E. Chemical synthesis of poliovirus cDNA: generation of infectious virus in the absence of natural template. 2002. *Science*, **297**, 1016–1018.
7. Domingo, E., Martinez-Salas, E., Sobrino, F., de la Torre, J. C., Portela, A., Ortin, J., Lopez-Galindez, C., Perez-Brena, P., Villanueva, N., Najera, R., et al. The quasispecies (extremely heterogeneous) nature of viral RNA genome populations: biological relevance: a review. 1985. *Gene*, **40**, 1–8.
8. Georgescu, M. M., Delpeyroux, F., Tardy-Panit, M., Balanant, J., Combiescu, M., Combiescu, A. A., Guillot, S., and Crainic, R. High diversity of poliovirus strains isolated from the central nervous system from patients with vaccine-associated paralytic poliomyelitis. 1994. *J. Virol.*, **68**, 8089–8101.
9. Guillot, S., Caro, V., Cuervo, N., Korotkova, E., Combiescu, M., Persu, A., Aubert-Combiescu, A., Delpeyroux, F., and Crainic, R. Natural genetic

exchanges between vaccine and wild poliovirus strains in humans. 2000. *J. Virol.,* **74**, 8434–8443.

10. John, T. J. Problems with oral poliovaccine in India. 1972. *Indian Pediatr.,* **9**, 252–256.
11. Kapp, C. Surge in polio spreads alarm in northern Nigeria. Rumours about vaccine safety in Muslim-run states threaten WHO's eradication programme. 2003. *Lancet,* **362**, 1631–1632.
12. Kew, O. M., Mulders, M. N., Lipskaya, G. Y., de Silva, E., and Pallansch, M. A. Molecular epidemiology of polioviruses. 1995. *Sem. Virol.,* **6**, 401–405.
13. Kew, O. M., Sutter, R. W., de Gourville, E. M., Dowdle, W. R., and Pallansch, M. A. Vaccine-derived polioviruses and the endgame strategy for global polio eradication (*). 2005. *Annu. Rev. Microbiol.,* **59**, 587–635.
14. Kinnunen, L., Huovilainen, A., Poyry, T., and Hovi, T. Rapid molecular evolution of wild type 3 poliovirus during infection in individual hosts. 1990. *J. Gen. Virol.,* **71**(Pt 2), 317–324.
15. MacLennan, C., Dunn, G., Huissoon, A. P., Kumararatne, D. S., Martin, J., O'Leary, P., Thompson, R. A., Osman, H., Wood, P., Minor, P., Wood, D. J., and Pillay, D. Failure to clear persistent vaccine-derived neurovirulent poliovirus infection in an immunodeficient man. 2004. *Lancet,* **363**, 1509–1513.
16. Manor, Y., Blomqvist, S., Sofer, D., Alfandari, J., Halmut, T., Abramovitz, B., Mendelson, E., and Shulman, L. M. Advanced environmental surveillance and molecular analyses indicate separate importations rather than endemic circulation of wild type 1 poliovirus in Gaza district in 2002. 2007. *Appl. Environ. Microbiol.,* **73**, 5954–5958.
17. Manor, Y., Handsher, R., Halmut, T., Neuman, M., Abramovitz, B., Mates, A., and Mendelson, E. A double-selective tissue culture system for isolation of wild-type poliovirus from sewage applied in a long-term environmental surveillance. 1999. *Appl. Environ. Microbiol.,* **65**, 1794–1797.
18. Manor, Y., Handsher, R., Halmut, T., Neuman, M., Bobrov, A., Rudich, H., Vonsover, A., Shulman, L., Kew, O., and Mendelson, E. Detection of poliovirus circulation by environmental surveillance in the absence of clinical cases in Israel and the Palestinian authority. 1999. *J. Clin. Microbiol.,* **37**, 1670–1675.
19. McGoldrick, A., Macadam, A. J., Dunn, G., Rowe, A., Burlison, J., Minor, P. D., Meredith, J., Evans, D. J., and Almond, J. W. Role of mutations G-480 and C-6203 in the attenuation phenotype of Sabin type 1 poliovirus. 1995. *J. Virol.,* **69**, 7601–7615.

20. Paul, Y., and Priya. Polio eradication in India: some observations. 2004. *Vaccine,* **22**, 4144–4148.

21. Ranta, J., Hovi, T., and Arjas, E. Poliovirus surveillance by examining sewage water specimens: studies on detection probability using simulation models. 2001. *Risk Anal.,* **21**, 1087–1096.

22. Shulman, L. M., Handsher, R., Yang, C. F., Yang, S. J., Manor, J., Vonsover, A., Grossman, Z., Pallansch, M., Mendelson, E., and Kew, O. M. Resolution of the pathways of poliovirus type 1 transmission during an outbreak. 2000. *J. Clin. Microbiol.,* **38**, 945–952.

23. Shulman, L. M., Manor, Y., Handsher, R., Delpeyroux, F., Halmut, T., Silberstein, I., Alfandari, J., Quay, J., Robinov, J., Kew, O. M., Crainic, R., and Mendelson, E. Presented at the Annual Meeting of the ISM, Beersheva, Israel, 2001.

24. Shulman, L. M., Manor, Y., Handsher, R., Delpeyroux, F., McDonough, M. J., Halmut, T., Silberstein, I., Alfandari, J., Quay, J., Fisher, T., Robinov, J., Kew, O. M., Crainic, R., and Mendelson, E. Molecular and antigenic characterization of a highly evolved derivative of the type 2 oral poliovaccine strain isolated from sewage in Israel. 2000. *J. Clin. Microbiol.,* **38**, 3729–3734.

25. Shulman, L. M., Manor, Y., Sofer, D., Handsher, R., Swartz, T., Delpeyroux, F., and Mendelson, E. Neurovirulent vaccine-derived polioviruses in sewage from highly immune populations. 2006. *PLoS ONE,* **1**, e69.

26. Slater, P. E., Orenstein, W. A., Morag, A., Avni, A., Handsher, R., Green, M. S., Costin, C., Yarrow, A., Rishpon, S., Havkin, O., et al. Poliomyelitis outbreak in Israel in 1988: a report with two commentaries. 1990. *Lancet,* **335**, 1192–1195, discussion 1196–1198.

27. Swartz, T. A., Green, M. S., Handscher, R., Sofer, D., Cohen-Dar, M., Shohat, T., Habib, S., Barak, E., Dror, Z., Somekh, E., Peled-Leviathan, T., Yulzari, R., Libling, A., Mendelson, E., and Shulman, L. M. Intestinal immunity following a combined enhanced inactivated polio vaccine/oral polio vaccine programme in Israel. 2008. *Vaccine,* **26**, 1083–1090.

28. Swartz, T. A., Handsher, R., Manor, Y., Stoeckel, P., Barkay, A., Mendelson, E., and Leventhal, A. Immune response to an intercalated enhanced inactivated polio vaccine/oral polio vaccine programme in Israel: impact on the control of poliomyelitis. 1998. *Vaccine,* **16**, 2090–2095.

29. Tebbens, R. J., Pallansch, M. A., Kew, O. M., Caceres, V. M., Jafari, H., Cochi, S. L., Sutter, R. W., Aylward, R. B., and Thompson, K. M. Risks of paralytic disease due to wild or vaccine-derived poliovirus after eradication. 2006. *Risk Anal.,* **26**, 1471–1505.

30. Vonsover, A., Handsher, R., Neuman, M., Guillot, S., Balanant, J., Rudich, H., Mendelson, E., Swartz, T., and Crainic, R. Molecular epidemiology of type 1 polioviruses isolated in Israel and defined by restriction fragment length polymorphism assay. 1993. *J. Infect. Dis.*, **167**, 199–203.
31. World Health Organization, Department of Vaccines and Biologicals. 2003. *Guidelines for Environmental Surveillance of Poliovirus Circulation.*
32. Wimmer, E. The test-tube synthesis of a chemical called poliovirus. The simple synthesis of a virus has far-reaching societal implications. 2006. *EMBO Rep.*, 7(Spec. no.), S3–S9.
33. Wimmer, E., Hellen, C. U., and Cao, X. Genetics of poliovirus. 1993. *Annu. Rev. Genet.*, **27**, 353–436.

Chapter 3

Filoviruses: Deadly Pathogens and Potential Bioweapons

Michael Schümann[a,*] **and Elke Mühlberger**[a,b,c]
[a] *Institute of Virology, Philipps-University Marburg, Marburg, Germany*
[b] *Department of Microbiology, Boston University School of Medicine, 72 East Concord Street, Boston, MA, USA*
[c] *National Emerging Infectious Diseases Laboratories, Boston, MA, USA*
michael.schuemann@cslbehring.com, muehlber@bu.edu

3.1 Emergence of Marburg and Ebola Viruses

More than 40 years ago, filoviruses were first isolated and initially characterized following accidental infections of laboratory workers in Marburg and Frankfurt, Germany, and Belgrade, Yugoslavia, who were handling African green monkeys imported from Africa for polio vaccine production. The virus caused a severe disease with high fever and inner bleeding and in spite of immediate hospital care, 7 of 25 primary infected persons died during this outbreak.[1] According to the place of its isolation, the virus was named Marburg virus (MARV).

*Present address: CSL Behring GmbH, Marburg, Germany

Viral Diagnostics: Advances and Applications
Edited by Robert S. Marks, Leslie Lobel, and Amadou Alpha Sall

ISBN 978-981-4364-43-0 (Hardcover), 978-981-4364-44-7 (eBook)
www.panstanford.com

This incident remained exclusive, until nine years later epidemics among 284 patients in Sudan and 318 patients in the Democratic Republic of Congo, formerly named Zaire, occurred with similar symptoms causing 431 deaths.[2] The causative virus was found to be closely related to MARV and in allusion to the nearby river was termed Ebola virus (EBOV). In accordance with the filamentous shape of the virion, the viruses were subsumed in the family of *Filoviridae*.[1]

Despite the severity of the disease and the high mortality among its victims, filoviruses were regarded for a long time as exotic pathogens, with their occurrence restricted to Central Africa and the Philippines, and no major health impact was suspected outside the endemic areas. Knowledge on their biology and pathogenicity remained limited, and the development of therapeutics promised low profits. However, the increase of global tourism, and even more the increased awareness of (bio)terrorist action after September 11 and the anthrax attacks in late 2001, have reignited research endeavors to elucidate the biology of the filoviruses and develop effective therapeutic strategies.[3]

3.2 The Virus and the Disease

Filoviruses possess a single-stranded, nonsegmented RNA genome of negative polarity. Due to this genome organization, filoviruses are classified along with the paramyxo-, rhabdo-, and bornaviruses in the order *Mononegavirales*. The filovirus family is subdivided into two genera: *Ebolavirus* with five distinct species (*Zaire ebolavirus* [ZEBOV], *Sudan ebolavirus* [SEBOV], *Tai Forest ebolavirus* [TEBOV], *Bundibugyo ebolavirus* [BEBOV], and *Reston ebolavirus* [REBOV]) and *Marburgvirus* with a single species, *Marburg marburgvirus*.

The most pathogenic strain in humans with a case fatality rate of about 80% is ZEBOV, followed by SEBOV and BEBOV with case fatality rates of about 50% and 25%, respectively. There are only two reported nonfatal cases of TEBOV infection. Some individuals with a known exposure to REBOV sero-converted but did not develop any signs of a disease, suggesting that REBOV is not pathogenic for humans. Since the first-reported MARV outbreaks in Germany and Yugoslavia, it was believed for a long time that MARV was less pathogenic compared to EBOV. However, recent outbreaks of MARV in the Democratic Republic of the Congo in 1998–2000 and in Angola in 2004 were associated with fatality rates above 80%, indicating that MARV is as virulent as ZEBOV.[1,2]

3.3 Filovirus Biology

The thread-shaped filovirus particles consist of the RNA genome, seven structural proteins, and a membranous envelope (see Fig. 3.1). The genome organization of the different filoviruses is similar. The negative-sense RNA genome is about 19 kb in length and contains seven viral genes in line. Each gene is composed of 3′ and 5′ nontranslated regions and the respective open reading frame. The genes are flanked by highly conserved transcription start and stop signals that are used as capping and polyadenylation sites, respectively. The genome ends consist of short nontranscribed

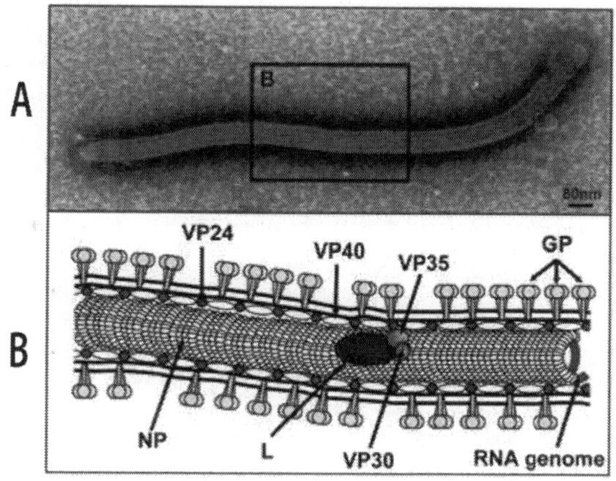

Figure 3.1 Filovirus structure. (A) Electron micrograph of an EBOV particle (courtesy of L. Kolesnikova, University of Marburg, Germany). (B) Scheme presentation of the EBOV structure. The RNA genome is encapsidated by the nucleocapsid proteins NP, VP35, VP30, and L. VP40 and VP24 are matrix proteins. GP trimers are inserted into the viral membrane. (Reproduced by permission of Future Medicine Ltd.)

regions containing *cis*-acting signals important for replication, transcription initiation, and encapsidation of the genomic RNA.[4]

Among the seven structural proteins, filoviruses possess only a single surface protein, the type I transmembrane glycoprotein GP that mediates attachment to target cells and virus entry. Notably, EBOV GP is an important cytopathic factor inducing cellular damage, cell rounding, detachment, and membrane permeabilization.[5]

Once the virus has entered the cell, replication and transcription occur in the cytoplasm of the infected cell. The nucleoprotein NP, the viral proteins VP35 and VP30, and the L protein are tightly associated with the RNA genome, forming the nucleocapsid. NP enwraps the viral genome; VP35 and L constitute the RNA-dependent RNA polymerase, which is responsible for replication of the genome and transcription of the single filoviral genes; and EBOV VP30 serves as a transcription activator.[4] Besides its function as a polymerase cofactor, VP35 acts as an inhibitor of antiviral pathways. VP35 antagonizes the induction of interferon (IFN)-β by interfering with the activation of IFN regulatory factor 3 and blocks the activation of protein kinase activated by double-stranded RNA (PKR).[6]

Filovirus genomes encode two matrix proteins, VP40 (the functional equivalent of the matrix proteins M of other nonsegmented negative-stranded RNA viruses) and the minor matrix protein VP24, unique for filoviruses. VP40 is critical for viral budding by interacting with cellular proteins involved in vesicle formation and facilitates virus release.[7] The minor matrix protein VP24 is involved in nucleocapsid formation and assembly. EBOV VP24 and MARV VP40 have also been shown to counteract the type I IFN response.[8,75]

3.4 Pathogenesis and Clinical Presentation

Information on the pathogenesis of filoviruses in humans is rare but so far matches the more extensive data collected from laboratory infections of nonhuman primates (NHPs).[9] Macrophage and dendritic cells are the primary target cells at the site of virus entry and mediate the rapid transport of the virus to regional lymph nodes. With ongoing replication, free viral particles and/or infected monocytes are then disseminated through the lymphatic system and the blood stream to infect other macrophage and dendritic cells at secondary sites of infection, which typically include the lymph nodes, spleen, and liver. After further replication, the virus spreads to surrounding parenchymal cells, where the resulting tissue injury causes the signs of multifocal organ injury. Although lymphocytes are not infected by filoviruses, they undergo massive apoptosis during later stages of infection, presumed due to the lack of co-stimulatory signaling from the dendritic cells the virus has hindered to fully maturate.[72]

The infected macrophage and dendritic cells produce large amounts of procoagulant proteins, like tissue factor (TF), cytokines, chemokines, and free oxygen radicals. The increased levels of TF evoke the signs of disseminated intravascular coagulation (DIC) and mutually enhance pro-inflammatory responses in the vascular endothelium that finally lead to progressive tissue injury and vascular leakage, while the cytopathic effects caused by the virus replication itself appear to be of minor importance. The large amounts of pro-inflammatory cytokines, which overwhelmingly accumulate late in infection, are presumed to evoke sepsis-like symptoms that finally lead to death from multiorgan failure and shock.[9,10]

The incubation time of EBOV infection ranges from 2 to 21 days, and the onset of the disease is typically abrupt. The initial signs of infection are of a rather general nature and may resemble influenza, malaria, typhoid fever, sepsis, other hemorrhagic diseases, or even noninfectious syndromes.[11,12] Thus, a symptomatic differential diagnosis at this stage of infection must remain presumptive and requires confirmation by laboratory results.

The early symptoms include acute fever, chills, myalgia, headache, arthralgia, and anorexia, followed by nausea, vomiting, abdominal pain, hypotension, tachypnea, relative bradycardia, conjunctivitis, conjunctival injection, pharyngitis, and diarrhea, which may already be bloody.[11] From days 5–7 fulminant hemorrhagic diathesis can occur, yielding the first symptoms of increased specificity to filovirus infection. Hemorrhagic signs include epistaxis, hematuria, hematemesis, petechiae, melena, and mucous membrane and conjunctival hemorrhage. Intriguingly, the eponymous signs of hemorrhage occur in less than half of the patients, further complicating a reliable bedside diagnosis. Erythematous rash, spreading from the face to the trunk, may develop with confluent, nonitchy maculopapular lesions.[12]

Laboratory findings of infected patients typically indicate leukopenia, atypical lymphocytes, thrombocytopenia, elevated transaminase levels, hyperproteinemia, proteinuria, hematuria, and prolonged bleeding, prothrombine, and activated partial thromboplastine time.[11]

Additionally, in some cases symptoms of the central nervous system, such as delirium, somnolescence, or convulsions, may occur. Patients with fatal disease progression die due to multiorgan failure and hypovolemic shock around day 12 after the onset of symptoms.

Despite all the severity of the disease, some patients are able to successfully establish an immune response to the virus within the first week and begin to recover 7 to 10 days after appearance of the first symptoms. The rash disappears around day 12, and desquamation occurs at days 12–14. The period of slow and prolonged convalescence may involve complications such as weakness, fatigue, recurrent hepatitis, uveitis, orchitis, and other clinical sequelae.[11,12]

3.5 The Bioweapon Potential of Filoviruses

The National Institute of Allergy and Infectious Diseases (NIAID) categorized microorganisms with biowarfare potential into three classes according to their potential to cause mass casualties and social disruption. Category A agents include bacteria as well as viruses (see Table 3.1). The following characteristics have been assigned to the pathogens listed in this category:

(1) easy dissemination and transmission from person to person;
(2) high mortality, with potential for major public health impact;
(3) cause of public panic and social disruption; and
(4) requirement of special action for public health preparedness.

The following section comments on how filoviruses meet each of these criteria.

Table 3.1 Category A priority pathogens (and diseases) as listed by the NIAID

Bacteria	
Bacillus anthracis (anthrax)	
Clostridium botulinum toxin (botulism)	
Yersinia pestis (plague)	
Francisella tularensis (tularemia)	
Viruses	
Variola virus (smallpox) and other related pox viruses	
Viral hemorrhagic fevers:	
Arenaviruses	Lymphocytic choriomeningitis virus, Junin virus Machupo virus, Guanarito virus, Lassa virus
Bunyaviruses	Hantaviruses, Rift Valley fever virus
Flaviruses	Dengue virus
Filoviruses	Ebola viruses, Marburg virus

3.5.1 *Dissemination and Transmission*

The transmission of filoviruses during outbreaks has so far been ascribed to contact with infected persons or exposure to contaminated equipment. Especially during the late stages of the disease, body fluids, including saliva, blood, vomit, and stool, have been shown to contain high titers of infectious virus.[13] Thus, particularly hospital personnel, caring relatives, or participants of ritual burial ceremonies, which frequently included the cleansing of the dead body, exhibited the highest risk of becoming secondarily infected.[14] To address this problem, infection control guidelines have been established by the World Health Organization (WHO) and the Centers for Disease Control (CDC), which include the isolation and surveillance of infectious patients, barrier nursing techniques, and safe transport and burial of the deceased.[15,16]

Strikingly, convalescent patients have been shown to secrete infectious virus through breast milk or semen months after the acute phase of illness, representing potential sources of transmission even a long time after the peaking outbreak. Educational efforts should prove effective to counter this risk.

Under experimental conditions, filoviruses have been shown to remain infectious on contaminated surfaces for several days;[17] thus fomites may impose a serious source of transmission. However, considering daylight exposure, filoviruses are estimated to belong to the most UV-sensitive viruses as assessed in a calculational study by the US Army.[18] Still, disinfection and decontamination routines are requisite infection control measures in an outbreak setting.

Although there has been no evidence of aerosol dissemination of filoviruses during natural outbreaks, it cannot be reliably excluded either. In fact, it is held responsible for a number of unexplained transmissions during an EBOV outbreak in Kikwit, for which physical contact to infectious patients is excluded.[19]

Laboratory findings have demonstrated that EBOV is able to efficiently and lethally infect rhesus monkeys by the airborne route.[20] The concept of naturally occurring aerosol transmission is further strengthened by a report of unintentional laboratory infections with ZEBOV of two of three control monkeys in a biocontainment laboratory. Here as well, physical contact with infected monkeys could be excluded.[21]

Applied to the scenario of an intentional filovirus release, the mode of person-to-person transmission is of importance, especially for the design of infection control measures, and definitely requires further investigation. However, the question of feasibility of efficient initial virus dissemination through aerosols is of far greater relevance when evaluating the bioweapon potential of filoviruses and regrettably is much easier to answer.

In violation of the Biological and Toxin Weapons Convention of 1972, which was signed by more than 170 nations, the former Soviet Union maintained a covert bioweapon program of enormous dimensions, involving at least 50,000 people working in more than 50 research and production facilities.[22,23] Among the aims of this endeavor were the improvement of microbial delivery, enhancement of virulence and viability, increased survivability of aerosols, and the establishment of large-scale industrial production processes. Shortly after the first filovirus incidences in Marburg, Frankfurt, and Belgrade in 1967, the Soviet military was able to obtain the virus and immediately added it to its bioweapon research program. According to Kanatjan Alibekov (alias Ken Alibek), a former Soviet biological warfare expert, field studies performed in 1991 proved the potency of aerosolized MARV to efficiently infect monkeys, and the newly established bioweapon entered industrial production until the collapse of the Soviet Union and the declared end of the bioweapon program in 1993.[23]

Despite international endeavors to render the research and production facilities civilian, serious concerns remain regarding the location and accessibility of the huge bioweapon arsenals and even more regarding the whereabouts of the thousands of people who worked in the program and the documents detailing every aspect of production and deployment.[22,23] The possibility that rogue nations and terrorist groups have used this opportunity to gain access to sophisticated bioweaponry knowledge must be suspected and considered when preparing for bioterrorist attacks.

3.5.2 *Mortality and Impact on Public Health*

As past epidemics of filovirus hemorrhagic fever (FHF) have shown, filoviruses are the most deadly pathogens known. The EBOV species Zaire as well as the Angola isolate of MARV caused case fatality rates as high as 90% during recent outbreaks in the Republic of Congo and Angola,[1,2] despite international on-site help in patient care, infection control, and public surveillance. No effective therapy is available to date, and thus the treatment of FHF patients remains merely supportive.[24] Even though Western health care facilities will provide superior measures of patient care as compared to their African counterparts, it remains speculative whether this circumstance alone will suffice to significantly reduce FHF mortality.

The group of hemorrhagic fever viruses is expected to be the most infectious among all category A biowarfare agents. The US military estimates 1–10 particles in aerosolized form to be sufficient for productive infection. WHO estimates for the casualties produced by attacks with various category A agents in aerosolized form range from 400 to 95,000 deaths and up to 125,000 incapacitated people among a population of 500,000. The windward reach of the aerosol may exceed 20 km.[25] In consideration of their successful use as weapons in 1991, MARV particles can be expected to equally fulfill these parameters.

With respect to a deliberate filovirus release, the possibility of aerosol dissemination provides the chance to conduct a covert attack, unless the terrorists should choose to inform the public in advance. The release of the virus may thus remain unrecognized until the first victims develop clinical symptoms. Due to the long incubation period of filoviruses, the first cases are likely to occur timely and geographically scattered, concealing the initial point of virus release, and unspecific early symptoms will complicate an appropriate diagnosis. Thus, immediate and locally concentrated mass casualties are rather unlikely to appear and it will depend on the vigilance of the first responding clinicians to get on the track of filoviral infections in time. Importantly, a rapid diagnosis of FHF will be fundamental to initiate effective countermeasures and prevent the further spread of the virus.[26]

Once the disease is identified, the management of FHF patients will occur according to well-prepared guidelines set up by WHO and the CDC. The infection control measures require the placement of the patients in private rooms with restricted access and appropriate surveillance. Health care personnel are obliged to use adequate personal protection equipment, including gloves, gowns, face shields with eye protection, and face masks when pulmonary complications with coughing could favor the formation of virus-containing aerosols. The proper removal, disinfection, and disposal of protective wear after leaving the patient is also mandatory.[15,16,24] These barrier nursing precautions will, however, with increasing numbers of patients, quickly impose logistical problems to the health care facility, not only due to the limited availability of protective or medical equipment and care personnel, but even more due to the increasing requirement for space to isolate the patients.[27]

In addition to the potentially large number of factually infected patients, an abundance of worried individuals, possibly with symptoms resembling those of early FHF, will rush to the hospitals to seek medical advice and diagnostic checkup. Equally, these people will contribute to eventually exhaust the capacity of the local health care systems and must be included in considerations about health care preparedness for bioterrorist events.[27]

3.5.3 *Public Panic and Social Disruption*

Among the different infectious agents that have been classified as category A pathogens, filoviruses probably have the most horrifying reputation. This makes them an attractive tool for bioterrorists, because the appearance of only a few cases will already suffice to induce panic and disrupt social life. The public perception of filoviruses may, in part, result from their deadly appearance with high case fatality rates, but additionally, other contributing factors are likely.

Natural outbreaks have so far been restricted to remote locations in Central Africa, concealing the actual appearance of the disease from the public and potentially enhancing a mystification of the virus. The strongest and most widespread impression that people associate with filovirus infections is the severe hemorrhages that develop late in the disease. Although hemorrhages occur in less than 50% of the infected individuals, filoviruses are popularly described as "liquefying" their victims.[9] This picture is well illustrated by movies like *Outbreak* and various examples of popular literature, which mostly represent the only information that people have gathered on filoviruses. Therefore, public information and education should extend to exotic pathogens with bioweapon potential to reduce their fear-inducing potential based on myth and ignorance.

Interestingly, pathogens of past or present epidemic relevance in our hemisphere, such as plague, influenza, and human immunodeficiency virus (HIV), do not share the thrilling reputation of EBOV, although the casualties these pathogens caused and still cause far exceed the number of FHF victims. This disproportion may result, in part, from a familiarization with these pathogens, but more likely, the availability of therapeutics and/or vaccines against these pathogens has trivialized the public perception (a fatal misjudgement, especially in the case of HIV). Therefore, attempts to develop effective treatment strategies for FHF should be regarded not only as providing a cure for a small number of victims but also as an important measure to obliterate the terrifying public reputation of filoviruses and to devaluate their attractiveness as bioterrorist tools.

3.5.4 *Public Health Preparedness*

Of course, there was and there is no actual need to specifically prepare for an FHF outbreak in the Western world. But the terrorist attack on the World Trade Center and the deliberate spread of *Bacillus anthracis* spores through the US mail system in 2001 have revived general considerations about the preparedness for assaults with biological weaponry. In contrast to a warfare scenario, a terrorist event will most likely occur covert and unannounced, imposing on the public health systems the challenges described before. Most importantly, a bioterrorist attack has to be recognized as such in order to take appropriate countermeasures. Thus, the specific training of all clinicians, regardless of their specialty, to recognize select agents of bioterrorism, the adaptation of diagnostic routines, and the extension of public health alert networks to build up an efficient frontline surveillance system to recognize in time any kind of bioterrorist attack have been proposed by the CDC.[28]

The US government initiated a large biodefense research program, which is maintained by the NIAID.[3]

The program aims at funding basic as well as applied research endeavors on category A–C agents with respect to the study of their biology and host interaction; the development of vaccination and treatment strategies, including model systems to prove their efficacy; and the establishment of enhanced diagnostic tests. Even though the program has not been immediately successful in providing efficient therapy to FHF, it has enhanced the research endeavors in filovirology and increased the willingness of medical and pharmaceutical companies to participate in this process. The following sections describe the current status of vaccination, treatment, and diagnostic research.

3.5.4.1 Vaccination

The first attempts of filovirus vaccination employed whole virus particles of ZEBOV or MARV inactivated either by treatment with heat, formalin, or γ-irradiation. The application of inactivated material to guinea pigs was able to demonstrate protective efficacies from 50–100% after a lethal challenge with the virus. In NHPs, however, only weak protection was conferred, and the few animals that survived the virus challenge did not exhibit any obvious correlation between the observed adaptive immune response and survival.

In addition to the poor immunogenicity of inactivated filovirus particles, safety concerns would surely have been a major constraint against this approach, since the factual inactivation or attenuation of vaccination virus is difficult to prove. Thus, a recent approach of vaccine design circumvents these concerns by exploiting a unique biochemical property of the filovirus matrix protein VP40. Upon overexpression alone or along with other virus proteins such as NP and GP, VP40 assembles in virus-like particles (VLPs) that morphologically mimic the natural virion. VLPs normally bud from the transfected cell membrane and are even capable of attaching to and entering host cells just like the entire virion. Yet, the risk of reversion to infectious virus can be excluded.[30]

VLPs of ZEBOV (eVLP) or MARV (mVLP) have been shown to fully protect mice or guinea pigs when challenged with the corresponding virus. A generation of hybrid VLPs showed that only expression of the homologous GP conferred protection to each virus. However, co-vaccination with a mixture of eVLP and mVLP led to full protection of guinea pigs against ZEBOV and MARV, demonstrating that simultaneous vaccination to different filoviruses is possible. The latest studies evaluating the efficacy of VLPs in the protection of NHPs yielded the first promising results for both eVLP and mVLP,[31,32] but further studies in NHPs are needed to demonstrate if VLPs will be able to catch up with the other promising vaccine candidates described later.

DNA vaccines provide a different strategy to induce protective immunity from filoviruses. Instead of administering complex virus particles, immunogenic proteins are transiently expressed from DNA plasmids. A big advantage of this approach, which is able to stimulate both humoral and cell-mediated immune responses, is the fast construction and manipulation of the expression vectors and cheap and easy production, even in large quantities.[29] DNA vaccines expressing the GP of either ZEBOV or MARV have shown good protection in rodent models. For the MARV vaccines, boosting with baculovirus-derived GP was shown to be efficient in guinea pigs. However, a similar attempt for ZEBOV vaccination failed.[29] In NHPs, the DNA vaccine to MARV was able to elicit reproducible adaptive immune responses, but only two-thirds of the animals were protected against a lethal challenge of homologous virus. The combination of DNA vaccination with subsequent adenovirus vaccine boosts in NHPs has shown the most convincing results so far, but the contribution of the DNA vaccine was only of supportive nature and dispensable for complete protection of the animals.[33]

Recent endeavors aimed to increase the efficacy of DNA vaccines by altering the glycosylation pattern of the GP transgene[34] and to demonstrate the safety and efficacy of immune stimulation in humans in a phase I clinical trial. Future studies will show whether DNA vaccination is a successful strategy for filoviruses.

The most promising advances in the development of vaccine candidates for filoviruses, however, have been achieved with the use of heterologous viral vectors to deliver immunogenic filoviral proteins.

Probably the oldest and best characterized viral vector system is derived from adenovirus serotype 5 (Ad5), which is widely used in gene therapeutic as well as vaccination approaches. The first successful vaccination attempt with adenoviral vectors expressing the EBOV GP in NHPs was carried out in combination with a DNA vaccination prime. However, complete protection was also achieved when a rAd-GP/NP vaccine was administered alone.[33] A recent study demonstrates that GP vaccination alone is sufficient to confer full protection in the NHP model even with reduced toxicity when using vaccination-optimized GP mutants.[35]

Adenovirus vaccines for MARV GP and bivalent vaccines for ZEBOV and SEBOV GP exhibited pan-species protection for each MARV and EBOV in rodent models,[29,36] and mucosal administration was shown to be effective.[37] Clinical trials will be necessary to demonstrate the safety and immunogenicity of the adenoviral vaccines in humans.

Adenoviruses are easy to manipulate, reliably transduce a wide variety of target cells, and can be easily grown to high titers, and the commonly used vector strains are nonpathogenic to humans. However, the immunogenic potential may be hampered in humans by preexisting immunity, since up to 60% of the Western population carries antiadenovirus antibodies.[29]

Recent improvements thus aimed to use nonhuman adenovirus vectors for EBOV vaccination. Chimpanzee adenovirus vectors were shown to overcome preexisting Ad5 immunity and elicit a substantial immune response in the guinea pig model,[38] while a vector derived from simian adenovirus 21 protected mice from a lethal virus challenge and was immunogenic in rhesus macaques.[39]

Another point of concern to adenovirus-based vaccinations is the relatively high dosage needed for vaccination (10^{10} pfu in NHPs[35]), since the commonly used adenoviral vectors are replication incompetent. The adverse effects resulting from the administration of large amounts of adenovirus material are heavily discussed, since in 1999 one patient died in a gene therapy approach with adenovirus vectors.[40]

Replication competent vector systems may overcome the need for high dosage administration, and thus, different alternatives have been evaluated, including Venezuelan equine encephalitis virus (VEEV), vaccinia virus, human parainfluenza virus (HPIV) type 3, and vesicular stomatitis virus (VSV). While VEEV and vaccinia virus approaches were less successful, HPIV and VSV vaccines showed very promising results.[29]

HPIV type 3 is a common pediatric pathogen of the airway epithelium and thus provides the advantage of respiratory administration. Constructs expressing GP alone or in combination with NP exhibited strong immunogenicity and full protection against a ZEBOV challenge after single-dose nasal inhalation in guinea pigs and full protection after only two administrations in rhesus monkeys.[41] However, the natural prevalence of this virus in the population may raise similar concerns as for adenovirus regarding vaccine efficacy in humans due to preexisting immunity.

The most promising results so far have been achieved with VSV constructs having their glycoprotein replaced by the filoviral counterpart. Vaccines for both ZEBOV and MARV conferred full protection to cynomolgus monkeys already after single administration,[42] and the GP vaccine derived from the MARV isolate Musoke even revealed cross protection against other relevant MARV isolates. Most importantly, the MARV vaccine even exhibited postexposure efficacy when administered up to 24 hours after the virus challenge,[43,68] and similar results could be demonstrated recently with a vaccine vector against SEBOV.[44] For the ZEBOV vaccine, however, postexposure efficacy is only 50%.[45]

3.5.4.2 Treatment

Besides the promising vaccination approaches with VSV, effective postexposure medication for filovirus infections is not available. In the search for efficient therapy, most desirably with postexposure applicability, numerous strategies to block or at least delay virus replication have been evaluated, while other approaches focused on relieving the deleterious disease symptoms to increase the chance of spontaneous convalescence.

On the basis of the successful treatment of many other viral infections, early attempts to treat FHF included administration of IFN-α or ribavirin. While triggering the innate immune response with IFN-α obviously showed promising results in mice, IFN treatment of cynomolgus macaques infected with ZEBOV could only delay death by several days.[46,47] Ribavirin, a nucleoside analogue, which efficiently blocks the polymerase of various RNA viruses, was shown to be ineffective in the treatment of FHF[46,48] and is therefore contraindicated when FHF is diagnosed.[49]

Passive immunization has been evaluated as a measure to reduce acute virus load and infectivity. First attempts employing equine antibodies against ZEBOV in monkeys showed little to no effect, but the survival of seven of eight patients who received blood transfusions from convalescent patients in Kikwit in 1995 revived the approach of passive antibody treatment.[46,47] ZEBOV-neutralizing antibodies showed promising results even when administered up to four days postexposure in mice, but a human monoclonal antibody, which had previously been shown to protect guinea pigs, proved ineffective in NHPs, not only in terms of survival, but also in reducing viremia.[50] Finally, a last attempt to understand the treatment success of the Kikwit patients failed, when whole-blood transfusions from convalescent EBOV-immune monkeys were unable to protect ZEBOV-infected naïve animals.[51]

In addition to these classical attempts to treat filovirus infection, other ingenious approaches, which target single steps of the viral life cycle, were developed.

The processing of the viral glycoprotein was believed to be important for efficient cell-to-cell transmission of the virus, and thus inhibitors of furin and furin-like proteases were evaluated in tissue culture.[52] However, these inhibitors did not exhibit any effect on viral titers, and similar results have been obtained in reverse genetics studies with viruses lacking the respective processing site in tissue culture and NHPs.[53] GP processing by cathepsins has recently been demonstrated, and respective inhibitors were able to diminish virus replication in cell culture.[54,73] Their therapeutic value in vivo remains to be evaluated.

Another promising approach for the development of anti-filoviral drugs is to block filovirus entry.[72,70] The recent discovery of the endosomal filovirus receptor, Niemann-Pick C1 protein, has opened up new strategies to inhibit one of the earliest steps in filovirus infection, the binding of GP to its receptor.[67,66]

Attempts to interfere with viral gene expression have utilized diverse antisense approaches. Phosphorodiamidate morpholino oligomers (PMO), stabilized RNA antisense molecules specifically interfering with the translation of viral mRNA, have been found to have pre- and postexposure efficacy in rodents and nonhuman primates.[55,56] Small interfering RNAs (siRNA) have been shown to interfere with MARV replication in cell culture when directed against NP, VP35, or VP30, and siRNA targeting ZEBOV L completely protected guinea pigs when administered shortly after the virus challenge.[47] A mixture of siRNAs targeting the viral genes L, VP24, and VP35 conferred complete protection to ZEBOV-infected macaques when administered several times postexposure.[69] Along with the PMO studies, these data suggest that targeting viral mRNAs might be a powerful strategy for postexposure treatment of filovirus infection.

Another way to interfere with the translation of viral mRNA is through the blockage of (S)-adenosyl-homocysteine (SAH) hydrolase, which is critical for the methylation of viral mRNA. SAH hydrolase inhibitors, like 3-deazaneplanocin A (c^3-Npc A), showed good efficacy in vitro and also in controlling viral replication in infected mice, even when administered only once a day one or two postinfection. Further investigations revealed that the efficacy of this approach largely relies on the induction of IFN-α, an effect that is unfortunately only achieved in mice, since NHPs failed to show a comparable response.[47] Currently, substances exhibiting these properties in primates are searched for.

Yet another group of treatment approaches aims at curing the deleterious symptoms of filovirus infection, including the exaggerated cytokine response, lymphocyte apoptosis, and coagulopathy.

To reduce the levels of harmful cytokines such as tumor necrosis factor (TNF)-α or interleukin (IL)-1, neutralizing antibodies for these cytokines, desferoxamine (desferal), a substance preventing TNFα–dependent induction of endothelial cell adhesion molecules, or recombinant IL-1 receptor antagonist (IL1RA) have been administered to guinea pigs infected with MARV. While desferal showed only little effect, IL1RA and anti-TNF serum were able to reduce the amount of circulating TNFα and led to survival of half of the animals.[46]

Finally, there have also been attempts to antagonize the severe effects of the DIC syndrome by application of recombinant nematode anticoagulant protein c2 (rNAPc2), a potent inhibitor of TF-induced coagulation. In ZEBOV-infected rhesus macaques, an attenuation of the coagulation and proinflammatory response was observed, which led to a prolonged survival time and an overall increase of survival by 33%.[47] In another experiment with rhesus macaques infected with MARV, however, rNAPc2 could not delay disease progression, presumably because DIC is less important for the pathogenesis of this virus.[57]

Recombinant human-activated protein C (rhAPC), the major physiological antagonist of the coagulation factors Va and VIIIa, has already exhibited some efficacy in the treatment of sepsis.[58] Consequently, it was tested for its efficacy against ZEBOV-induced coagulopathy in rhesus macaques. Similarly to rNAPc2 treatment, rhAPC was able to increase survival time by 33% and overall survival by 20%.[59] Although these experiments demonstrate significant effects, the modulation of virus-induced coagulation alone is unlikely to alter the disastrous course of filovirus disease. However, in combination with other treatments it may help to improve the prognosis of FHF patients.

In summary, neither classical nor sophisticated, neither direct nor indirect approaches did originate an effective treatment strategy for FHF. The increasing knowledge on the biology of filoviruses, however, may continuously yield new targets for therapeutic intervention, which may eventually lead to the development of efficient medication for FHF.

3.5.4.3 Diagnostics

The early symptoms of a filovirus infection are too common and unspecific to allow an accurate bedside diagnosis during the first days of the disease. However, in the past outbreaks of FHF, the rapid diagnosis and subsequent isolation of infected patients have been fundamental to prevent virus transmission and control the spread of the disease.

The most basic yet quite reliable method to diagnose FHF is the direct isolation and subsequent characterization of the virus. Filoviruses can be propagated in a large variety of cell lines, and the distinctive appearance of the virion allows easy electron microscopic identification. Immunological detection methods yield another easy means of characterization. However, diagnostic samples must be processed under biosafety level 4 conditions (see Fig. 3.2), and laboratories of that standard are rare and often remote from outbreak locations. The processing of the specimens is time consuming and expensive, and the transport bears the risk of accidental virus release or may be detrimental to the sample quality.[60]

Enzyme-linked immunosorbent assays (ELISAs) or reverse-transcriptase polymerase chain reaction (RT-PCR) methods represent affordable and reliable diagnostic alternatives, which are currently used on-site in FHF outbreak situations.

Serological diagnosis includes both IgM- and IgG-ELISA and is particularly suitable for patients in later stages of the disease or for monitoring the patients' immune status. IgM levels rise on days 2 to 9 following onset of symptoms and disappear one to five months after. IgG antibodies appear after one to three weeks and persist

Figure 3.2 The handling of filoviruses requires biosafety level 4 laboratory conditions. (Photograph by Uwe Brock).

for years.[11] Since fatal cases usually succumb to infection before the onset of the antibody response, the detection of antibodies is merely suitable to monitor benign cases of the disease.[60]

The antigen-capture ELISA allows the detection of virus at earlier stages of the illness. Viral antigen becomes detectable in the patients' blood 3 to 6 days after the onset of symptoms, yet it vanishes again 7 to 16 days later.[11] Antigen-capture ELISAs apply antibodies raised against authentic or recombinant viral proteins, including MARV NP, REBOV NP, and ZEBOV NP, GP, or VP40. Depending on the specificity of the antibodies, antigen-capture ELISAs are even useful for the differential diagnosis of FHF.[60]

Nonetheless, the RT-PCR methodology is the most sensitive device to detect filoviruses in patient samples. As few as 10–50 RNA copies per reaction suffice to yield positive results,[61,62] and viral genomes can be detected up to 48 hours prior to successful antigen capture.[63] RT-PCR analyses are cheap and easy to conduct, and the technical enhancement of PCR machines by now allows routine applications in outbreak settings.[60,61] Quantitative RT-PCR yielding information on the amount of template RNA, which mirrors a patient's virus load, may help to predict the outcome of the disease.[63] RT-PCR diagnostic assays are available for a large number of infectious agents and easily allow differential diagnoses in evolving outbreaks. The development of assays employing consensus primers, primer mixtures,[64] mass tag labeling,[65] or the coupling of amplification and sequencing[62] yields an opportunity for the rapid differential diagnosis of patient samples and easy characterization of new virus variants. Other recent advancements include the establishment of a high-throughput RNA purification procedure that is applicable to outbreak locations.[61]

Despite all advantages of RT-PCR techniques in terms of specificity and sensitivity, there is an increasing risk with the length of an outbreak to produce erroneous results due to the contamination of diagnostic samples. False-positive results lead to the unnecessary isolation of healthy patients and expose them to an increased risk of infection in isolation wards. False-negative results prevent infected patients from being isolated in time and potentiate the risk that these persons will further spread the virus to the public. To keep both misdiagnoses at a minimum, antigen-capture ELISA and RT-PCR techniques are employed concomitantly in today's outbreak situations.[60]

3.6 Future Perspectives

With the increasing awareness of bioweapons as potential tools for terrorists, large biodefense programs were launched to enhance the preparedness of public health for bioterrorist events.

Endeavors range from the improvement of local health communication and surveillance systems to the support of basic research on select agents, development of effective vaccination and treatment strategies, and enhanced diagnostic tests. The latter will be of particular importance in order to control any deliberate outbreak. Although reliable systems are available to eventually diagnose filoviruses, the long incubation period of filoviruses prevents their detection before the onset of the early disease symptoms, a time period that may be too long to effectively counteract the spread of infected people or infectious material. Innovative diagnostic strategies, for example, biosensors, will be necessary to rapidly locate contaminated persons and objects, define the dimensions of a deliberate virus release, and initiate quarantine and infection control measures.

Most important will be the development of an efficient medication for FHF. Effective postexposure treatment will be of immediate importance to counteract the deleterious symptoms and reduce mortality of the disease, thereby reducing the terrifying reputation of the virus in the public, and finally to devaluate the attractiveness of filoviruses as bioterrorist agents.

References

1. World Health Organization. *Marburg Haemorrhagic Fever: Factsheet.* http://www.who.int/mediacentre/ factsheets/fs_marburg/en/index.html.
2. World Health Organization. *Ebola Haemorrhagic Fever: Factsheet No. 103.* http://www.who.int/media- centre/factsheets/fs103/en/index.html.
3. National Institute of Allergy and Infectious Diseases. *NIAID Strategic Plan for Biodefense Research.* 2002. http://www3.niaid.nih.gov/topics/BiodefenseRelated/ Biodefense/PDF/strategic_plan.pdf.
4. Mühlberger, E. (2007). Filovirus replication and transcription. *Future Virol.*, **2**(2), 205–215.
5. Yang, Z. Y., Duckers, H. J., Sullivan, N. J., Sanchez, A., Nabel, E. G., and Nabel, G. J. (2000). Identification of the Ebola virus glycoprotein as the main viral determinant of vascular cell cytotoxicity and injury. *Nat. Med.*, **6**(8), 886–889.
6. Ramanan, P., Shabman, R. S., Brown, C. S., Amarasinghe, G. K., Basler, C. F., and Leung, D. W. (2011). Filoviral immune evasion mechanisms. *Viruses*, **3**, 1634–1649.
7. Hartlieb, B., and Weissenhorn, W. (2006). Filovirus assembly and budding. *Virology*, **344**(1), 64–70.
8. Reid, S. P., Valmas, C., Martinez, O., Sanchez, F. M., and Basler, C. F. (2007). Ebola virus VP24 proteins inhibit the interaction of NPI-1 subfamily karyopherin alpha proteins with activated STAT1. *J. Virol.*, **81**(24), 13469–13477.
9. Mahanty, S., and Bray, M. (2004). Pathogenesis of filoviral haemorrhagic fevers. *Lancet Infect. Dis.*, **4**(8), 487–498.
10. Hensley, L. E., Jones, S. M., Feldmann, H., Jahrling, P. B., and Geisbert, T. W. (2005). Ebola and Marburg viruses: pathogenesis and development of countermeasures. *Curr. Mol. Med.*, **5**(8), 761–772.

11. Casillas, A. M., Nyamathi, A. M., Sosa, A., Wilder, C. L., and Sands, H. (2003). A current review of Ebola virus: pathogenesis, clinical presentation, and diagnostic assessment. *Biol. Res. Nurs.*, **4**(4), 268–275.
12. World Health Organization. *WHO Recommended Guidelines for Epidemic Preparedness and Response: Ebola Haemorrhagic Fever (EHF).* http://www.who.int/csr/resources/publications/ebola/whoemcdis977E.pdf.
13. Bausch, D. G., Towner, J. S., Dowell, S. F., Kaducu, F., Lukwiya, M., Sanchez, A., Nichol, S. T., Ksiazek, T. G., and Rollin, P. E. (2007). Assessment of the risk of Ebola virus transmission from bodily fluids and fomites. *J. Infect. Dis.*, **196**(Suppl 2), S142–S147.
14. Salvaggio, M. R., and Baddley, J. W. (2004).Other viral bioweapons: Ebola and Marburg hemorrhagic fever. *Dermatol. Clin.*, **22**(3), 291–302.
15. World Health Organization. *Interim Infection Control Recommendations for Care of Patients with Suspected or Confirmed Filovirus (Ebola, Marburg) Haemorrhagic Fever.* 2008. http://www.who.int/csr/bioriskreduction/interim_recommendations_filovirus.pdf.
16. Centers for Disease Control. *Interim Guidance for Managing Patients with Suspected Viral Hemorrhagic Fever in U.S. Hospitals.* 2005. http://www.cdc.gov/ncidod/dhqp/pdf/bbp/VHFinterimGuidance05_19_05.pdf
17. Piercy, T. J., Smither, S. J., Steward, J. A., Eastaugh, L., and Lever, M. S. (2010). The survival of filoviruses in liquids, on solid substrates and in a dynamic aerosol. *J. Appl. Microbiol.*, **109**, 1531–1539.
18. Lytle, C. D., and Sagripanti, J. L. (2005). Predicted inactivation of viruses of relevance to biodefense by solar radiation. *J. Virol.*, **79**(22), 14244–14252.
19. Roels, T. H., Bloom, A. S., Buffington, J., Muhungu, G. L., Mac Kenzie, W. R., Khan, A. S., Ndambi, R., Noah, D. L., Rolka, H. R., Peters, C. J., and Ksiazek, T. G. (1999). Ebola hemorrhagic fever, Kikwit, Democratic Republic of the Congo, 1995: risk factors for patients without a reported exposure. *J. Infect. Dis.*, **179**(Suppl 1), S92–S97.
20. Johnson, E., Jaax, N., White, J., and Jahrling, P. (1995). Lethal experimental infections of rhesus monkeys by aerosolized Ebola virus. *Int. J. Exp. Pathol.*, **76**(4), 227–236.
21. Jaax, N., Jahrling, P., Geisbert, T., Geisbert, J., Steele, K., McKee, K., Nagley, D., Johnson, E., Jaax, G., and Peters, C. (1995). Transmission of Ebola virus (Zaire strain) to uninfected control monkeys in a biocontainment laboratory. *Lancet*, **346**(8991–8992), 1669–1671.

22. Davis, C. J. (1999). Nuclear blindness: an overview of the biological weapons programs of the former Soviet Union and Iraq. *Emerg. Infect. Dis.*, **5**(4), 509–512.
23. Alibek, K., and Handelman, S. *Biohazard: The Chilling True Story of the Largest Covert Biological Weapons Program in the World–Told from Inside by the Man Who Ran It Paperback*, April 11, 2000.
24. Bossi, P., Tegnell, A., Baka, A., van Loock, F., Hendriks, J., Werner, A., Maidhof, H., and Gouvras, G. (2004). Bichat guidelines for the clinical management of haemorrhagic fever viruses and bioterrorism-related haemorrhagic fever viruses. *Eur. Surveill.*, **9**(12), E11–E12.
25. Hilleman, M. R. (2002). Overview: cause and prevention in biowarfare and bioterrorism. *Vaccine*, **20**(25–26), 3055–3067.
26. Cunha, B. A. (2002). Anthrax, tularemia, plague, Ebola or smallpox as agents of bioterrorism: recognition in the emergency room. *Clin. Microbiol. Infect.*, **8**(8), 489–503.
27. Koenig, K. L., Kahn, C. A., and Schultz, C. H. (2006). Medical strategies to handle mass casualties from the use of biological weapons. *Clin. Lab. Med.*, **26**(2), 313–327.
28. Gerberding, J. L., Hughes, J. M., and Koplan, J. P. (2002). Bioterrorism preparedness and response: clinicians and public health agencies as essential partners. *JAMA*, **287**(7), 898–900.
29. Geisbert, T. W., Bausch, D. G., and Feldmann, H. (2010a). Prospects for immunisation against Marburg and Ebola viruses. *Rev. Med. Virol.*, **20**, 344–357.
30. Warfield, K. L., Swenson, D. L., Demmin, G., and Bavari, S. (2005). Filovirus-like particles as vaccines and discovery tools. *Expert Rev. Vaccines*, **4**(3), 429–440.
31. Swenson, D. L., Warfield, K. L., Larsen, T., Alves, D. A., Coberley, S. S., and Bavari, S. (2008). Monovalent virus-like particle vaccine protects guinea pigs and nonhuman primates against infection with multiple Marburg viruses. *Expert Rev. Vaccines*, **7**(4), 417–429.
32. Warfield, K. L., Swenson, D. L., Olinger, G. G., Kalina, W. V., Aman, M. J., and Bavari, S. (2007). Ebola virus-like particle-based vaccine protects nonhuman primates against lethal Ebola virus challenge. *J. Infect. Dis.*, **196**(Suppl 2), S430–S437.
33. Sullivan, N. J., Geisbert, T. W., Geisbert, J. B., Xu, L., Yang, Z. Y., Roederer, M., Koup, R. A., Jahrling, P. B., and Nabel, G. J. (2003). Accelerated vaccination for Ebola virus haemorrhagic fever in non-human primates. *Nature*, **424**(6949), 681–684.

34. Dowling, W., Thompson, E., Badger, C., Mellquist, J. L., Garrison, A. R., Smith, J. M., Paragas, J., Hogan, R. J., and Schmaljohn, C. (2007). Influences of glycosylation on antigenicity, immunogenicity, and protective efficacy of Ebola virus GP DNA vaccines. *J. Virol.*, **81**(4), 1821–1837.

35. Sullivan, N. J., Geisbert, T. W., Geisbert, J. B., Shedlock, D. J., Xu, L., Lamoreaux, L., Custers, J. H., Popernack, P. M., Yang, Z. Y., Pau, M. G., Roederer, M., Koup, R. A., Goudsmit, J., Jahrling, P. B., and Nabel, G. J. (2006). Immune protection of nonhuman primates against Ebola virus with single low-dose adenovirus vectors encoding modified GPs. *PLoS Med.*, **3**(6), e177.

36. Wang, D., Raja, N. U., Trubey, C. M., Juompan, L. Y., Luo, M., Woraratanadharm, J., Deitz, S. B., Yu, H., Swain, B. M., Moore, K. M., Pratt, W. D., Hart, M. K., and Dong, J. Y. (2006). Development of a cAdVax-based bivalent ebola virus vaccine that induces immune responses against both the Sudan and Zaire species of Ebola virus. *J. Virol.*, **80**(6), 2738–2746.

37. Patel, A., Zhang, Y., Croyle, M., Tran, K., Gray, M., Strong, J., Feldmann, H., Wilson, J. M., and Kobinger, G. P. (2007). Mucosal delivery of adenovirus-based vaccine protects against Ebola virus infection in mice. *J. Infect. Dis.*, **196**(Suppl 2), S413–S420.

38. Kobinger, G. P., Feldmann, H., Zhi, Y., Schumer, G., Gao, G., Feldmann, F., Jones, S., and Wilson, J. M. (2006). Chimpanzee adenovirus vaccine protects against Zaire Ebola virus. *Virology*, **346**(2), 394–401.

39. Roy, S., Zhi, Y., Kobinger, G. P., Figueredo, J., Calcedo, R., Miller, J. R., Feldmann, H., Wilson, J. M. (2006). Generation of an adenoviral vaccine vector based on simian adenovirus 21. *J. Gen. Virol.*, **87**(Pt 9), 2477–2485.

40. Marshall, E. (1999). Gene therapy death prompts review of adenovirus vector. *Science*, **286**(5448), 2244–2245.

41. Bukreyev, A., Rollin, P. E., Tate, M. K., Yang, L., Zaki, S. R., Shieh, W. J., Murphy, B. R., Collins, P. L., and Sanchez, A. (2007). Successful topical respiratory tract immunization of primates against Ebola virus. *J. Virol.*, **81**(12), 6379–6388.

42. Jones, S. M., Feldmann, H., Ströher, U., Geisbert, J. B., Fernando, L., Grolla, A., Klenk, H. D., Sullivan, N. J., Volchkov, V. E., Fritz, E. A., Daddario, K. M., Hensley, L. E., Jahrling, P. B., and Geisbert, T. W. (2005). Live attenuated recombinant vaccine protects nonhuman primates against Ebola and Marburg viruses. *Nat. Med.*, **11**(7), 786–790.

43. Daddario-DiCaprio, K. M., Geisbert, T. W., Ströher, U., Geisbert, J. B., Grolla, A., Fritz, E. A., Fernando, L., Kagan, E., Jahrling, P. B., Hensley, L. E., Jones, S. M., and Feldmann, H. (2006). Postexposure protection against Marburg haemorrhagic fever with recombinant vesicular stomatitis virus vectors in non-human primates: an efficacy assessment. *Lancet*, **367**(9520), 1399–1404.
44. Geisbert, T. W., Daddario-DiCaprio, K. M., Williams, K. J., Geisbert, J. B., Leung, A., Feldmann, F., Hensley, L. E., Feldmann, H., and Jones, S. M. (2008). Recombinant vesicular stomatitis virus vector mediates postexposure protection against Sudan ebola hemorrhagic fever in nonhuman primates. *J. Virol.*, **82**(11), 5664–5668.
45. Feldmann, H., Jones, S. M., Daddario-DiCaprio, K. M., Geisbert, J. B., Ströher, U., Grolla, A., Bray, M., Fritz, E. A., Fernando, L., Feldmann, F., Hensley, L. E., and Geisbert, T. W. (2007). Effective post-exposure treatment of Ebola infection. *PLoS Pathog.*, **3**(1), e2.
46. Bray, M., and Paragas, J. (2002). Experimental therapy of filovirus infections. *Antiviral Res.*, **54**(1), 1–17.
47. Stroher, U., and Feldmann, H. (2006). Progress towards the treatment of Ebola haemorrhagic fever. *Expert Opin. Investig. Drugs*, **15**(12), 1523–1535.
48. Paragas, J., and Geisbert, T. W. 2006). Development of treatment strategies to combat Ebola and Marburg viruses. *Expert Rev. Anti Infect. Ther.*, **4**(1), 67–76.
49. Borio, L., Inglesby, T., Peters, C. J., Schmaljohn, A. L., Hughes, J. M., Jahrling, P. B., Ksiazek, T., Johnson, K. M., Meyerhoff, A., O'Toole, T., Ascher, M. S., Bartlett, J., Breman, J. G., Eitzen, E. M. Jr., Hamburg, M., Hauer, J., Henderson, D. A., Johnson, R. T., Kwik, G., Layton, M., Lillibridge, S., Nabel, G. J., Osterholm, M. T., Perl, T. M., Russell, P., and Tonat, K.; Working Group on Civilian Biodefense (2002). Hemorrhagic fever viruses as biological weapons: medical and public health management. *JAMA*, **287**(18), 2391–2405.
50. Oswald, W. B., Geisbert, T. W., Davis, K. J., Geisbert, J. B., Sullivan, N. J., Jahrling, P. B., Parren, P. W., and Burton, D. R. (2007). Neutralizing antibody fails to impact the course of Ebola virus infection in monkeys. *PLoS Pathog.*, **3**(1), e9.
51. Jahrling, P. B., Geisbert, J. B., Swearengen, J. R., Larsen, T., and Geisbert, T. W. (2007). Ebola hemorrhagic fever: evaluation of passive immunotherapy in nonhuman primates. *J. Infect. Dis.*, **196**(Suppl 2), S400–S403.

52. Ströher, U., Willihnganz, L., Jean, F., and Feldmann, H. (2007). Blockage of filoviral glycoprotein processing by use of a protein-based inhibitor. *J. Infect. Dis.*, **196**(Suppl 2), S271–S275.

53. Neumann, G., Geisbert, T. W., Ebihara, H., Geisbert, J. B., Daddario-DiCaprio, K. M., Feldmann, H., and Kawaoka, Y. (2007). Proteolytic processing of the Ebola virus glycoprotein is not critical for Ebola virus replication in nonhuman primates. *J. Virol.*, **81**(6), 2995–2998.

54. Chandran, K., Sullivan, N. J., Felbor, U., Whelan, S.P., Cunningham, J. M. (2005). Endosomal proteolysis of the Ebola virus glycoprotein is necessary for infection. *Science*, **308**(5728), 1643–1645.

55. Enterlein, S., Warfield, K. L., Swenson, D. L., Stein, D. A., Smith, J. L., Gamble, C. S., Kroeker, A. D., Iversen, P. L., Bavari, S., and Mühlberger, E. (2006). VP35 knockdown inhibits Ebola virus amplifi cation and protects against lethal infection in mice. *Antimicrob. Agents Chemother.*, **50**(3), 984–993.

56. Warren, T. K., Warfield, K. L., Wells, J., Swenson, D. L., Donner, K. S., Van Tongeren, S. A., Garza, N. L., Dong, L., Mourich, D. V., Crumley, S., Nichols, D. K., Iversen, P. L., and Bavari, S. (2010). Advanced antisense therapies for postexposure protection against lethal filovirus infections. *Nat. Med.*, **16**, 991–994.

57. Geisbert, T. W., Daddario-DiCaprio, K. M., Geisbert, J. B., Young, H. A., Formenty, P., Fritz, E. A., Larsen, T., and Hensley, L. E. (2007). Marburg virus Angola infection of rhesus macaques: pathogenesis and treatment with recombinant nematode anticoagulant protein c2. *J. Infect. Dis.*, **196**(Suppl 2), S372–S381.

58. Bernard, G. R., Vincent, J. L., Laterre, P. F., LaRosa, S. P., Dhainaut, J. F., Lopez-Rodriguez, A., Steingrub, J. S., Garber, G. E., Helterbrand, J. D., Ely, E. W., and Fisher, C. J. Jr.; Recombinant human protein C Worldwide Evaluation in Severe Sepsis (PROWESS) study group (2001). Efficacy and safety of recombinant human activated protein C for severe sepsis. *N. Engl. J. Med.*, **344**(10), 699–709.

59. Hensley, L. E., Stevens, E. L., Yan, S. B., Geisbert, J. B., Macias, W. L., Larsen, T., Daddario-DiCaprio, K. M., Cassell, G. H., Jahrling, P. B., and Geisbert, T. W. (2007). Recombinant human activated protein C for the postexposure treatment of Ebola hemorrhagic fever. *J. Infect. Dis.*, **196**(Suppl 2), S390–S399.

60. Saijo, M., Niikura, M., Ikegami, T., Kurane, I., Kurata, T., and Morikawa, S. (2006). Laboratory diagnostic systems for Ebola and Marburg hemorrhagic fevers developed with recombinant proteins. *Clin. Vaccine Immunol.*, **13**(4), 444–451.

61. Towner, J. S., Sealy, T. K., Ksiazek, T. G., and Nichol, S. T. (2007). High-throughput molecular detection of hemorrhagic fever virus threats with applications for outbreak settings. *J. Infect. Dis.*, **196**(Suppl 2), S205–S212.
62. Zhai, J., Palacios, G., Towner, J. S., Jabado, O., Kapoor, V., Venter, M., Grolla, A., Briese, T., Paweska, J., Swanepoel, R., Feldmann, H., Nichol, S. T., and Ian Lipkin, W. (2007). Rapid molecular strategy for filovirus detection and characterization. *J. Clin. Microbiol.*, **45**(1), 224–226.
63. Towner, J. S., Rollin, P. E., Bausch, D. G., Sanchez, A., Crary, S. M., Vincent, M., Lee, W. F., Spiropoulou, C. F., Ksiazek, T. G., Lukwiya, M., Kaducu, F., Downing, R., and Nichol, S. T. (2004). Rapid diagnosis of Ebola hemorrhagic fever by reverse transcription-PCR in an outbreak setting and assessment of patient viral load as a predictor of outcome. *J. Virol.*, **78**(8), 4330–4341.
64. Drosten, C., Göttig, S., Schilling, S., Asper, M., Panning, M., Schmitz, H., and Günther, S. (2002). Rapid detection and quantification of RNA of Ebola and Marburg viruses, Lassa virus, Crimean-Congo hemorrhagic fever virus, Rift Valley fever virus, dengue virus, and yellow fever virus by real-time reverse transcription-PCR. *J. Clin. Microbiol.*, **40**(7), 2323–2330.
65. Palacios, G. Briese, T., Kapoor, V., Jabado, O., Liu, Z., Venter, M., Zhai, J., Renwick, N., Grolla, A., Geisbert, T. W., Drosten, C., Towner, J. S., Ju, J., Paweska, J., Nichol, S. T., Swanepoel, R., Feldmann, H., Jahrling, P. B., and Ian Lipkin, W. (2006). MassTag polymerase chain reaction for differential diagnosis of viral hemorrhagic fevers. *Emerg. Infect. Dis.*, **12**(4), 692–695.
66. Carette, J. E., Raaben, M., Wong, A. C., Herbert, A. S., Obernosterer, G., Mulherkar, N., Kuehne, A. I., Kranzusch, P. J., Griffin, A. M., Ruthel, G., Dal Cin, P., Dye, J. M., Whelan, S. P., Chandran, K., and Brummelkamp, T. R. (2011). Ebola virus entry requires the cholesterol transporter Niemann-Pick C1. *Nature*, **477**, 340–343.
67. Cote, M., Misasi, J., Ren, T., Bruchez, A., Lee, K., Filone, C. M., Hensley, L., Li, Q., Ory, D., Chandran, K., and Cunningham, J. (2011). Small molecule inhibitors reveal Niemann-Pick C1 is essential for Ebola virus infection. *Nature*, **477**, 344–348.
68. Geisbert, T. W., Hensley, L. E., Geisbert, J. B., Leung, A., Johnson, J. C., Grolla, A., and Feldmann, H. (2010b). Postexposure treatment of Marburg virus infection. *Emerg Infect. Dis.*, **16**, 1119–1122.
69. Geisbert, T. W., Lee, A. C., Robbins, M., Geisbert, J. B., Honko, A. N., Sood, V., Johnson, J. C., de Jong, S., Tavakoli, I., Judge, A., Hensley, L. E., and

Maclachlan, I. (2010c). Postexposure protection of non-human primates against a lethal Ebola virus challenge with RNA interference: a proof-of-concept study. *Lancet*, **375**, 1896–1905.

70. Huang, I. C., Bailey, C. C., Weyer, J. L., Radoshitzky, S. R., Becker, M. M., Chiang, J. J., Brass, A. L., Ahmed, A. A., Chi, X., Dong, L., Longobardi, L. E., Boltz, D., Kuhn, J. H., Elledge, S. J., Bavari, S., Denison, M. R., Choe, H., and Farzan, M. (2011). Distinct patterns of IFITM-mediated restriction of filoviruses, SARS coronavirus, and influenza A virus. *PLoS Pathog.*, **7**, e1001258.
71. Olejnik, J., Ryabchikova, E., Corley, R. B., and Mühlberger, E. (2011). Intracellular events and cell fate in filovirus infection. *Viruses*, **3**, 1501–1531.
72. Radoshitzky, S. R., Warfield, K. L., Chi, X., Dong, L., Kota, K., Bradfute, S. B., Gearhart, J. D., Retterer, C., Kranzusch, P. J., Misasi, J. N., Hogenbirk, M. A., Wahl-Jensen, V., Volchkov, V. E., Cunningham, J. M., Jahrling, P. B., Aman, M. J., Bavari, S., Farzan, M., and Kuhn, J. H. (2011). Ebolavirus delta-peptide immunoadhesins inhibit marburgvirus and ebolavirus cell entry. *J. Virol.*, **85**, 8502–8513.
73. Shah, P. P., Wang, T., Kaletsky, R. L., Myers, M. C., Purvis, J. E., Jing, H., Huryn, D. M., Greenbaum, D. C., Smith, A. B., 3rd, Bates, P., and Diamond, S. L. (2010). A small-molecule oxocarbazate inhibitor of human cathepsin L blocks severe acute respiratory syndrome and ebola pseudotype virus infection into human embryonic kidney 293T cells. *Mol. Pharmacol.*, **78**, 319–324.
74. Valmas, C., Grosch, M. N., Schümann, M., Olejnik, J., Martinez, O., Best, S. M., Krähling, V., Basler, C. F., and Mühlberger, E. (2010). Marburg virus evades interferon responses by a mechanism distinct from ebola virus. *PLoS Pathog.*, **6**, e1000721.

Chapter 4

Bridging Diagnostics Research, Development, and Commercialization: Diagnostics for the Developing World

Rosanna W. Peeling
London School of Hygiene and Tropical Medicine, Keppel Street, London, UK
rosanna.peeling@lshtm.ac.uk

4.1 Lack of Access to Diagnostics as a Contributor to the Burden of Infectious Diseases

In the developing world infectious diseases account for the majority of deaths in adults and more than half of all deaths in children (see Table 4.1).[1] Women and children bear a disproportional share of the disease burden. Human immunodeficiency virus (HIV) and tuberculosis together are estimated to kill more than 3.5 million people a year.[2] Each year over two million people die of malaria, about four million of acute respiratory infections, and nearly three million of enteric infections. High-quality diagnostic tests for many infectious diseases are commercially available, but they are neither accessible nor affordable to patients in most of the developing

Viral Diagnostics: Advances and Applications
Edited by Robert S. Marks, Leslie Lobel, and Amadou Alpha Sall

ISBN 978-981-4364-43-0 (Hardcover), 978-981-4364-44-7 (eBook)
www.panstanford.com

Table 4.1 Top five causes of deaths in sub-Saharan Africa, South Asia, and Europe 2001 (from Kent et al. 2006)

Region	Sub-Saharan Africa	South Asia	Europe and Central Asia
1	HIV/AIDS	Ischemic heart disease	Ischemic heart disease
2	Malaria	Lower respiratory infections	Cerebrovascular disease
3	Lower respiratory infections	Perinatal conditions	Lung cancer
4	Diarrhoeal diseases	Cerebrovascular disease	Chronic Obstructive Pulmonary Disease
5	Perinatal conditions	Diarrhoeal diseases	Self-inflicted injuries

Abbreviations: HIV, human immunodeficiency virus; AIDS, acquired immunodeficiency syndrome.

world.[3,4] The lack of access to good quality diagnostic tests for infectious diseases to guide management decisions is a major contributor to the enormous burden of ill health in the developing world.

4.2 Role of Diagnostic Tests

Accurate diagnostic tests play a key role in patient management and the prevention and control of most infectious diseases (see Table 4.2). Early diagnosis and treatment reduce the risk not only of developing longterm complications in the patient but also for diseases such as tuberculosis, sexually transmitted infections (STIs), and HIV. Prompt treatment also reduces further transmission to other members of the community.

Table 4.2 Role of diagnostic tests in patient management and disease control

Case management	Screening for asymptomatic infections Diagnosis and/or staging of infections when clinical presentation is non-specific
Disease control	Surveillance Outbreak investigations Monitoring the effectiveness of interventions
Elimination	Certifying elimination
Treatment efficacy	Detecting and monitoring drug resistance

For diseases where clinical presentation is specific, management decisions can be made without the use of a diagnostic laboratory. But for the majority of infectious diseases, an accurate diagnosis requires the use of diagnostic tests.

4.3 Diagnostic Landscape in the Developing World

In many developing countries, laboratory services are often limited to major urban centers. *The 2004 World Development Report* cites the lack of accessibility as one of the major reasons why services fail. Table 4.3 shows the distances the poorest fifth of the population in selected countries have to travel to access the nearest health service.[5]

In settings where access to diagnostic laboratory services is limited, the World Health Organization (WHO) recommends the use of a syndromic approach to clinical management, where patients presenting with a particular syndrome are treated for all the major causes of the syndrome. Algorithms for syndromic management have been developed for STIs and for common childhood diseases.[6,7] Although these algorithms are inexpensive and easy to use, especially in primary health care settings, a major disadvantage

Table 4.3 Distance to nearest medical facility (selected from *The 2004 World Development Report*, p. 22)[6]

Country	Distance (km)
Bangladesh	0.9
Benin	7.5
Bolivia	11.8
Central African Republic	14.7
Chad	22.9
Haiti	8.0
India	2.5
Madagascar	15.5
Mali	13.6
Niger	26.9
Senegal	12.8
Zimbabwe	8.6

is overtreatment and the potential for development of antibiotic resistance. A syndromic approach does not allow contact tracing to prevent onward transmission and cannot be used to monitor specific disease trends and effectiveness of new interventions.

Even when laboratory services are available, there are often problems with the quality of the services due to a lack of resources, supply logistics, and trained personnel.[8] These problems lead to physicians not trusting laboratory results, which, in turn, leads to further neglect of laboratory services.

4.4 Lack of International and National Regulatory Standards for Approval of Diagnostics

National regulatory processes for drugs provide safeguards for the safety and effectiveness of drugs used in a country. Most countries have a process for reviewing the evidence from drug trials to support the introduction of new drugs, and this has done much to improve the quality of drugs used in developing countries. Unfortunately, apart from tests used for blood banking, regulatory standards are often lacking for diagnostic tests, especially those targeting diseases that are uncommon in industrialized countries.[9]As a result, diagnostic tests are often sold in most of the developing world without any formal evaluation of their performance and effectiveness.

4.5 The Ideal Diagnostic Tool

The UNICEF/UNDP/World Bank/WHO Special Programme for Research and Training in Tropical Diseases (WHO/TDR) has developed an acronym, ASSURED, for the ideal characteristics in a test to be useful for developingcountry settings (see Table 4.4).[10,11]

Few tests fulfill all these criteria. It is important for national authorities to develop an algorithm or criteria for the selection of the most appropriate test for a particular setting, population, or indication. Although the criteria upon which procurement decisions are made may vary, selections are generally based on:

Table 4.4 The ideal rapid test: ASSURED criteria

A = Affordable
S = Sensitive
S = Specific
U = User-friendly (simple to perform in a few steps with minimal training)
R = Robust and rapid (results available in less than 30 min)
E = Equipment free
D = Deliverable to those who need them

Test performance: Test sensitivity and specificity and positive and negative predictive values are important considerations. High sensitivity is important for a screening test in diseases such as HIV and syphilis, where there are serious consequences associated with a missed diagnosis. Poor specificity may matter less if overtreatment rarely results in adverse side effects, as in the treatment for syphilis, but may be a serious disadvantage if the treatment is highly toxic, for example, the case with drugs used to treat trypanosomiasis (sleeping sickness).

Ease of use: The number of processing steps will influence where the test can be carried out. It is important to consider the ease of specimen collection, that is, whether the test can use whole blood or noninvasive specimens. The need for accurate timing will influence the amount of training required.

Conditions of use: Tests that can be carried out in hot or humid conditions will be useful for primary health care or field settings. Heat stability of a test and reagents are important considerations for where the test can be used.

Conditions of storage: Storage temperatures are specified in product inserts for most tests. Unfortunately the results of most tests for use in the developing world are only guaranteed up to 30°C when the temperature in most clinics is above 30°C.

Shelf life: A longer shelf life reduces the pressure on the supply chain and the probability of wastage of expired tests.

Price: A control program may decide to select a cheaper test with slightly lower performance characteristics so that more patients can be diagnosed.

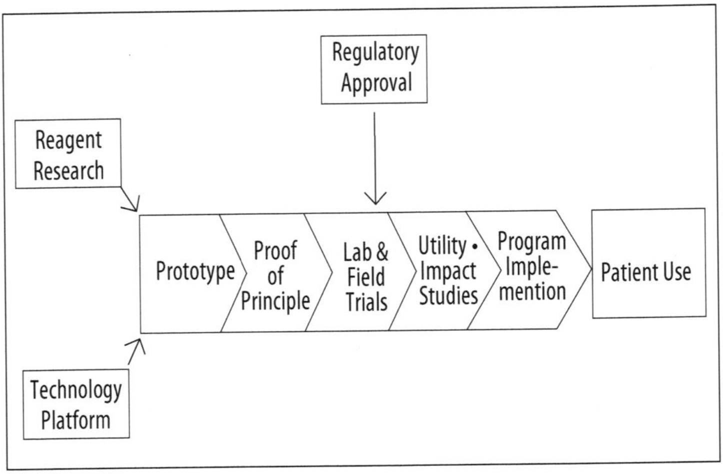

Figure 4.1 Bench to bedside pathway of diagnostics development and evaluation.

4.6 Development of Diagnostic Tests

The development of a diagnostic test starts with the identification of a diagnostic target, which is then put into an appropriate technology platform and optimized to produce a prototype. Proof-of-principle studies are then conducted to establish that the test detects the intended condition or infection. The test then undergoes further evaluations, first using "convenience" samples or archived specimens, followed by field evaluations in populations of intended use. These trial results are used in obtaining data for regulatory approval so that the test can be marketed and sold in a country. For postapproval marketing purposes, companies may also fund studies to demonstrate the utility and potential impact of a diagnostic test to advocate its widespread use.

4.7 Challenges in the Availability of Quality-Assured Diagnostic Tests in the Developing World

The pathway from target discovery to development of a diagnostic product (Fig. 4.1), and finally getting into the hands of providers, is a process that takes on an average of 5–10 years with an investment ranging from $2–10 million.[11]

For products with a viable commercial market, this is driven, funded, and managed largely by the private sector, drawing on appropriate expertise, as needed. For diagnostics of public health importance in the developing world, there has been little interest in investing in research for a pipeline of products that would be appropriate and useful, due to a perceived lack of return on investment. There have been efforts made in recent years to remedy this situation, but there is little coordination and leadership. The diagnostic landscape is fragmented with many gaps along this pathway.[11]

There are many examples of the inequity in investment in diagnostics for the developed and the developing world. One example is the West Nile virus, a virus of the family *Flaviviridae* and part of the Japanese encephalitis (JE) antigenic complex of viruses, which was first isolated in 1937 in the West Nile district of Uganda from a woman presenting with fever during a research study on yellow fever. It is the cause of severe human meningoencephalitis. Serosurveys later showed that the prevalence of antibodies to the West Nile virus ranged from 1.4% in the Congo to 46.4% in the White Nile region in Sudan. It was subsequently also found in Egypt and India. But diagnostics for this virus became widely available only after it surfaced in the U.S. and Canada in 1999, accompanied by a high level of media coverage. This coverage followed the spread of the virus across the continent and had the effect of increasing funding for research on this virus and related arthropod-borne viruses.

Many diseases that are serious public health concerns in the developing world remain problems because of a lack of user-friendly diagnostic tests that allow health care workers in the front line or primary health care settings to detect them before they become outbreaks. These include measles and the causes of viral hemorrhagic fevers, such as Ebola virus, Lassa fever, and Crimean Congo viruses. It is estimated that as many as 500,000 cases of Lassa fever occur in West Africa every year, but diagnostic tools are not readily available. There needs to be a concerted international effort to bring more equity to the availability of appropriate diagnostic

Table 4.5 Diagnostic research and development: Challenges and possible solutions

R&D phase	Challenges	Possible solutions
Research for new targets and technology platforms	- Test developers not aware of the diagnostic need and product specifications - Lack of funding for discovery research: large diagnostic companies not interested; small companies do not have sufficient research funding or expertise - Lack of access to reagents and well characterized specimens to facilitate test development	- convene expert consultations in disease endemic countries to define diagnostic needs and product specifications - synthesize existing knowledge through systematic reviews - advocate for more funding and public-private sector collaboration - establish banks of well-characterized specimens
Test development and validation	- Lack of convergence of researchers who have disease expertise and targets with those who have complementary technology platforms - Intellectual property issues acting as a barrier to innovation - Lack of access to good clinical trial sites to validate the performance and operational characteristics of a new test	- create networks for diagnostic innovation - Global Strategy and Plan of Action in response to the Intergovernmental Working Group on Public Health, Innovation and Intellectual Property will provide a framework to improve the situation - build capacity for diagnostic trials in developing countries
Commercialization of a test	- Lack of access to quality manufacturing - Lack of transparency in the regulation of diagnostics in most countries - Lack of standards on evaluations of test performance by country regulatory authorities	- advocate for regional harmonization of regulatory approval as most companies do not have funds to conduct trials in every country; or use WHO pre-qualified tests - TDR Nature Reviews Microbiology series on the design and conduct of diagnostic evaluations[12]
Scaling up the use of a new test	- Governments do not have clearly defined processes for test selection and developing guidelines for use - Lack of funding to purchase diagnostics of good quality and/or corrupt procurement practice - Poor supply chain management leading to frequent stockouts of diagnostics and/or drugs - Doctors have little confidence in results of diagnostics[8]	- need to create diagnostic policy platform - tests of acceptable performance are offered through WHO Bulk Procurement at negotiated pricing - build capacity for stock management and procurement - improved education of health providers on the performance and utility of diagnostic tests

tools to many areas of the developing world. It is important that these infections be recognized as soon as possible when they emerge, before they become widespread.

Table 4.5 shows some of the major challenges and possible solutions to resolve this situation.

4.8 Opportunities for a Better Future

In recent years, there has been increasing realization that the lack of diagnostics is a bottleneck for improving global health. More funding made available for research from donors and technical agencies could increase improved diagnostics for developing country settings.

4.8.1 *Technological Advances*

The genomes of many neglected infectious diseases of public health importance in developing countries have been sequenced. Investments in diagnostics target research will lead to the discovery of novel diagnostic targets, which can complement recent advances in rapid detection technologies driven and funded largely through antibioterrorism activities.

4.8.2 *More Funding and More Players*

There are now more funding and players in the field. Their efforts need to be coordinated and harmonized to avoid duplication and waste. The Bill & Melinda Gates Foundation organized a Global Health Diagnostics Forum and published a supplement with the Nature Publishing Group on the use of mathematical models to estimate the potential impact of improved diagnostics for the developing world and has funded the development of diagnostic tests and diagnostic platforms.[3,13]

The National Institute of Biomedical Imaging and Bioengineering (NIBIB) in the US National Institute of Health has funded a number of centers in the United States to develop point of care diagnostics. A number of public-private partnerships such as the Foundation for

Innovative New Diagnostics (FIND), the Program for Appropriate Technology in Health (PATH), and the Infectious Diseases Research Institute (IDRI) are now investing in appropriate technologies for diagnostics, which include tuberculosis, malaria, STIs, and other neglected tropical diseases.

The UNICEF/UNDP/World Bank/WHO/TDR, housed in WHO, has developed a diagnostics program aimed at promoting and facilitating research on the development, evaluation, and introduction of diagnostic tests appropriate for use in developing country settings. Its specific objectives include:

1. defining diagnostic needs for diseases of poverty and setting standards for diagnostics quality;
2. facilitating research for test development;
3. assessing and assuring diagnostic performance and quality; and
4. increasing access to diagnostics in the developing world, taking into account socioeconomic factors and issues of gender equity.

This program is involved in facilitating research on the development, evaluation, and deployment of diagnostic tools for two global elimination initiatives (malaria and congenital syphilis) and the elimination of visceral leishmaniasis from the Indian subcontinent. Its program of work has been developed based on WHO's comparative advantage of global convening power to define diagnostic needs, guidelines and standards for diagnostic performance, and appropriate use by working closely with its network of regional WHO offices and country disease control programs. Tests with acceptable performance and operational characteristics are made available to UN member states at negotiated pricing through the WHO Bulk Procurement Scheme.

To date, the TDR has convened global consultations to define diagnostic needs for tuberculosis, malaria, STIs, visceral leishmaniasis, schistosomiasis, dengue fever, and human African trypanosomiasis. The TDR has conducted systematic reviews to identify knowledge and research gaps and facilitated changes in tuberculosis diagnostic policy. To address the lack of quality standards in diagnostic evaluations, the TDR has assembled a Diagnostics Evaluation Expert Panel. In collaboration with the Nature Publishing Group, supplements on evaluating diagnostics have been published for malaria, STIs, and visceral leishmaniasis. Through the projects in this program, the TDR has trained scientists in research institutions and public health laboratories in more than 4 countries to conduct diagnostic research and evaluations. Tests with acceptable performance for malaria, syphilis, dengue, and visceral leishmaniasis are now available to UN member states at negotiated pricing through the WHO Bulk Procurement Scheme, allowing countries increased access to quality-assured diagnostics.

4.8.3 *Increased Efforts at Capacity Building*

Global health initiatives such as the Global Fund for AIDS, tuberculosis, and malaria, the President's Emergency Program for AIDS Relief (PEPFAR), and the President's Malaria Initiative (PMI), in addition to funding improved care and treatment, have made significant progress in terms of improving the capacity of developing countries to offer health care, including quality diagnostic services. Other nongovernmental organizations and universities have also increased their efforts at building capacity for diagnostic testing and quality management of laboratories.[14–16] TDR has also initiated funding of diagnostics fellowships.

4.9 Bridging Research, Product Development, and Commercialization

There is much optimism and unprecedented opportunities for advances in the research and development (R&D) of improved diagnostic tools for diseases prevalent in the developing world. There needs to be a concerted effort to nurture R&D in which developing countries play a pivotal role. This requires coordination of public and private sector efforts that will result in a robust pipeline of candidates and products along the diagnostic developmental pathway. Much of the developing world has the capacity to develop and produce diagnostics, but standards for quality have to be developed and enforced by national regulatory authorities. Bridging research, development, and commercialization is the key to reaching the goal of accessible, quality-assured diagnostic tools for the developing world.

References

1. Kent, M. M., and S. Yin. Controlling infectious diseases. 2006. *Population Bull.*, **61**, 3–9, *www.prb.org*.
2. WHO fact sheets at www.who.int.
3. Hay Burgess, D. C., Wasserman, J., and Dahl, C. A. Global health diagnostics. 2006. *Nature*, **S1**(23 November 2006), 1–2, doi:10.1038/nature05440.
4. Mabey, D., Peeling, R. W., Ustianowski, A., and Perkins, M. Diagnostics for the developing world. 2004. *Nat. Rev. Microbiol.*, **2**, 231–240.
5. World Bank. *World Development Report 2004: Making Services Work for Poor People.* New York: Oxford University Press for the World Bank.
6. WHO Guidelines on the Management of Sexually Transmitted Infections. 2003.
7. WHO Guidelines on the Integrated Management of Childhood Illness. 2004.
8. Petti, C. A., Polage, C. R., Quinn, T. C., Ronald, A. R., and Sande, M. A. Laboratory medicine in Africa: a barrier to effective health care. 2006. *Clin. Infect. Dis.*, **42**, 377–382.
9. Cunningham, J., et al. Regulation of in vitro diagnostics: a global perspective in diagnostics for tuberculosis. *Global Demand and Market Potential. Appendix*, **1**, 116–120.
10. Peeling, R. W., Mabey, D., and Holmes, K. K. Diagnostics for sexually transmitted infections: the way forward. 2006. *Sex. Transm. Infect.*, **82**(Suppl V), v1–v6.
11. Kettler, H., White, K., and Hawkes, S. Mapping the landscape of diagnostics for sexually transmitted infections. 2004. *WHO/TDR publication.*
12. Peeling, R. W., Smith, P. G., and Bossuyt, P. M. A guide for diagnostic evaluations. 2006. *Nat. Rev. Microbiol.*, **4**(9 Suppl), S2–S6.
13. Urdea, M., Penny, L. A., Olmsted, S. S., et al. Requirements for high impact diagnostics in the developing world. Nature supplement: determining the global health impact of improved diagnostic technologies for the developing world. 2006. *Nature*, **S1**, 73–79.

14. Martin, R., Hearn, T. L., Ridderhof, J. C., and Demby, A. Implementation of a quality systems approach for laboratory practice in resource-constrained countries. 2005. *AIDS,* **19**(Suppl 2), S59–S65.

15. Cheesborough, M. *District Laboratory Practice in Tropical Countries*. 2nd edition. 2005. Cambridge University Press.

16. Bates, I., and Maitland, K. Are laboratory services coming of age in sub-Saharan Africa? 2006. *Clin. Infect. Dis.,* **42**, 383–384.

Chapter 5

Oropouche Fever: An Overview of the Epidemiological and Molecular Aspects in the Brazilian Amazon Region

Pedro F. C. Vasconcelos and Marcio R. T. Nunes
Department of Arbovirology and Hemorrhagic Fevers, Instituto Evandro Chagas, Brazilian Ministry of Health, Av. Almirante Barroso, 490, Marco 66090-023, Belém, Pará, Brazil
pedrovasconcelos@iec.pa.gov.br, marcionunes@iec.pa.ov.br

5.1 Oropouche Outbreaks

The Oropouche virus (OROV; *Bunyaviridae, Orthobunyavirus*) is one of the most important arboviruses that infect humans in Northern South America, particularly in the Amazon region, causing an arboviral disease termed Oropouche (ORO) fever.[26] Between 1961 and 2007, several outbreaks were reported in different urban areas of Pará (Belém, Santa Isabel, Castanhal, Santarém, Oriximiná, Serra Pelada, Bragantina zone—Igarapé Açu, Maracanã, and Magalhães Barata municipalities), Amazonas (Manaus, Novo Airão, and Barcelos), Acre (Xapuri), Amapá (Mazagão), Maranhão (Porto Franco), Tocantins (Tocantinópolis), and Rondônia (Ariquemes and Oro

Viral Diagnostics: Advances and Applications
Edited by Robert S. Marks, Leslie Lobel, and Amadou Alpha Sall

ISBN 978-981-4364-43-0 (Hardcover), 978-981-4364-44-7 (eBook)
www.panstanford.com

Preto D'Oeste) states.[3,5,13,20,26,31,32,34] Molecular studies recently conducted have demonstrated the circulation of at least three major OROV lineages in the Brazilian Amazon region (genotypes I, II, and III). The genotypes I and II are more frequently detected in the western and eastern Amazon, respectively. The genotype III, previously recognized only in Panama, was recently described in the southeast region of Brazil. The association of molecular and epidemiological data has contributed substantially to the genetic characterization of OROV strains, isolated during different outbreaks in the past four decades.[3,15,29] In addition, it provided a better understanding of its molecular epidemiology with respect to emergence of new lineages and the dynamics of the evolution of these arboviruses in the Americas, principally in the Amazon region. In the Brazilian Amazon region, the ORO fever is considered the most frequent arboviral disease that affects humans, after dengue, and is characterized by epidemics of acute febrile illness accompanied mainly by headache, arthralgia, myalgia, photophobia, and other systemic manifestations. More rarely, some patients can present a clinical picture of aseptic meningitis or even meningoencephalitis. Interestingly, the symptoms of ORO fever commonly reappear a few days after the end of the first febrile episode. However the symptoms usually are less severe. Patients generally recovered completely without sequels, even in the more severe cases.[25]

No fatal cases due to ORO fever have been reported to date. One of the most relevant characteristics of the OROV is its capacity to cause outbreaks in urban areas, as observed in several reports in the Brazilian Amazon region since the early 60s.[3,20,27] The OROV was first isolated in 1955 from the blood of an acute febrile patient, a resident of the Vega de Oropouche village, Trinidad, and also from a pool of *Coquillettidia venezuelensis* mosquitoes.[1] In Brazil, the virus was isolated for the first time in 196 from the blood of a sloth (*Bradypus trydactilus*) captured in a forested area beside the BelemBrasilia highway and also from a pool where the *Ochlerotatus serratus* mosquitoes were captured in the same area.[17]

In 1961, the disease was reported in the municipality of Belem, Pará State, northern region of Brazil, an occasion in which the virus demonstrated its epidemic potential causing a huge outbreak, with an estimate of 11,00 people affected by the disease.[17] In the following years, the OROV caused outbreaks in urban areas of different northern and northeastern Brazilian states, more specifically in Pará, Amapá, Amazonas, Acre, Tocantins, Maranhão, and Rondônia.[27] Outside Brazil, febrile outbreaks caused by the OROV were reported in Panama in 1989,[27] as well as in the Peruvian Amazon region during the years 1992 and 1994.[7,35]

This chapter aims to present a comprehensive review regarding the epidemiological and molecular aspects of the OROV, emphasizing its geographic distribution, dynamics of the outbreaks occurring between 1961 and 2006, and its genotypic dispersion in Brazil.

5.2 The Oropouche Virus

ORO fever is caused by the OROV, which is a member of the family *Bunyaviridae*, genus *Orthobunyavirus*.[9] Morphologically, the virus is presented as a spherical particle, with a diameter ranging from 90 to 10 nm. Since the OROV is an enveloped virus it is sensitive to organic solvents (chloroform and ether) and detergents (sodium deoxycholate).[12] The virion symmetry is helicoidal, and the genetic material consists of ribonucleic acid (RNA).[4]

The viral genome, like other orthobunyaviruses, constitutes three singlestranded, negativesense RNA segments, namely, small RNA (SRNA), medium RNA (MRNA), and large RNA (LRNA).[9] These genome segments are responsible for encoding six proteins: three structural, one nucleocapsid (N) encoded in the segment SRNA, and the remaining two glycoproteins, Gn and Gc, encoded by the MRNA. The nonstructural proteins NSs, NSm, and L (viral polymerase) are encoded in the SRNA, MRNA, and LRNA, respectively.[9] As observed in several arbo- viruses, the OROV is able to induce the agglutination of goose erythrocytes due to the presence of the hemagglutinin protein. This protein can be easily recovered from sera of infected hamsters (*Mesocricetus auratus*) by using a sucrose-acetone method (A.P.A Travassos da Rosa, personal communication, 1969).

Regarding its pathogenesis, inoculations of the OROV into two- to three-dayold mice (*Mus musculus*) by intracerebral (i.c.) and intraperitoneal (i.p.) routes, as well as into adult hamsters by i.c., i.p., and subcutaneous routes, produced lethal infection in these animals.

Furthermore, the virus replicates in different tissue culture systems, including VERO, BHK-21, and chick embryonic cells (fibroblasts), yielding intense cytopathic effects.[25]

Laboratory studies conducted, during the outbreaks reported in the Brazilian Amazon, by researchers of the Department of Arbovirology and Hemorrhagic Fevers at the Instituto Evandro Chagas (IEC) demonstrated the crucial importance of the biting midge, *Culicoides paraensis* (family *Ceratopogonidae*), as the main vector of the OROV in urban and suburban areas.[19,22] This small insect, commonly named "maruim" in the Amazon region, is largely found in tropical and subtropical areas of the Americas. The insect is active predominantly during the day, more specifically in the afternoon, the period during which hematophagy is reported to be most intense.[11,12,18]

The virus is maintained in nature basically by two distinct ecological cycles, an urban and a forest one.[20] In the urban cycle, denominated also as an epidemic one, the OROV is transmitted among infected and healthsusceptible individuals by the *Culicoides paraensis* bite during the hematophagous process. Areas with accumulation of organic material in decomposition, such as decaying banana stalks and cacao hulls, constitute important breeding sites for *Culicoides paraensis*.[14] Studies using hamsters have demonstrated the transmission of the OROV using the mosquito vector *Culex quinquefasciatus*, which is commonly found in urban areas of the Amazon. However, the transmission was only possible when high levels of viremia in hamsters was observed, levels rarely found in infected humans. On the basis of this data, *Culex* mosquitoes could not be associated with OROV transmission during urban outbreaks.[20,28]

Regarding the forest cycle, evidence has suggested that among vertebrates, sloths (*Bradypus tridactylus*), monkeys, and wild birds (*Columbina talpacoti*) can be host to the OROV.[17,27] Although the virus was isolated once from a pool of *Ochlerotatus serratus* in Brazil and from a pool of *Coquillettidia venezuelensis* mosquitoes in Trinidad,[20] until recently, no studies have been carried out to assess the involvement of the *Culicoides paraensis* in the maintenance of the OROV in the forest cycle, and therefore, the sylvatic vector remains unknown. Probably, the link between the two epidemiological cycles is made by man when becoming infected in the forest (enzootic areas) and then carrying the virus to urban areas during the viremic period. In this case, man is a potential source of infection to other *Culicoides paraensis*. Furthermore, OROV replicates efficiently in the tissues of *Culicoides paraensis*, and after an extrinsic period for virus incubation, during the hematophagous processes, the culicoides can infect new susceptible individuals, initiating an urban transmission cycle, which culminates in the emergence of outbreaks.[27]

5.3 Geographic Distribution

Until now, ORO fever cases, as well as virus isolation, have only been reported and described in Brazil, Panama, Peru, and Trinidad. In Brazil, since the first isolation in 1960, the virus has been responsible for several outbreaks, geographically restricted to Para State, reaching the different municipalities of the Belem Metropolitan region (Belem, Ananindeua, Benfica, Caraparu, Castanhal and Santa Isabel do Para), northeast (Abaetetuba, Augusto Correa, Baião, Bragança, Capanema, Curuça, Tomé-Açu, Vigia, and Viseu), southeast (Itupiranga), Baixo Amazonas (Belterra, Mojui dos Campos, and Santarem), and Marajo (Portel). During this period, only the southeast region of Para State did not report cases of ORO fever[8,10,17,18] (see Fig. 5.1).

Between 1981 and 1996, cases of ORO fever were reported in both Para State (Oriximiná municipality; Baixo Amazonas region; and in Altamira, southwestern region) and other Amazonian states in the municipalities of Manaus Novo Airão and Barcelos (Amazonas), Mazagão (Amapa), Xapuri (Acre), Ariquemes, and Ouro Preto D'Oeste (Rondônia), Porto Franco (Maranhão), and Tocantinopolis (Tocantins)[5,13,20,26,31,32] (see Fig. 5.1).

In the years of 2003 and 2004, outbreaks of ORO fever were reported in the municipalities of Parauapebas (southeast Para) and Porto de Moz (Baixo Amazonas region of Para State), respectively.[3] In 2006, the OROV caused outbreaks in the municipalities of Maracanã, Igarapé-Açu, Magalhães Barata, and Viseu, all of them geographically located in the Bragantina region, northeastern Para,

Figure 5.1 Outbreaks of ORO fever reported in Central and South America between 1961 and 2007, Detection of anti-OROV IgM antibodies and OROV genome in serum samples of patients clinically suspected of ORO fever. (Courtesy of F. P. Pinheiro.)

which broke an epidemic silence of 26 years in the region[34] (see Fig. 5.1).

The geographic dispersion of the virus is probably related to the movement of human populations during the viremic phase, through areas where proper conditions for viral maintenance (susceptible populations, susceptible and infected arthropods, and environment) are observed.[27] Seroepidemiological studies conducted in Brazil and in other South and Central American countries between 1961 and 2007 estimate that a total of 357,00 people had been infected by the virus. However, this data appears to be an underestimate, since the incidence of this arboviral disease during important outbreaks reported in Belem (Para State) in 1968, Porto Franco (Maranhao State), and Tocantinopolis (Tocantins State) were not included. Thus, the more accurate estimate was about half a million people infected in the Amazon region and in the Americas during the past 48 years since the first isolation of the virus reported in 1960.[10,20,21,23,24]

Outbreaks of ORO fever have also been reported outside Brazil, in Panama and Peru. The outbreak that occurred in Panama was reported in 1989 in Benjuco, a small village 5 km west of Panama City.[27]

In Peru, the disease was clinically reported through laboratories in 1992 when the virus caused outbreaks in Iquitos.[7,35] Two additional outbreaks were reported in 1994 in the cities of Puerto Maldonato and Madre de Dios, both located within the Peruvian Amazon region.[35]

In 2005 seroepidemiological and molecular studies conducted in serum samples collected from febrile patients in Jujuy, Argentina, demonstrated the presence of both anti-OROV IgM antibodies and the OROV genome, suggesting the circulation of the virus in the region (Nunes and Fabri, personal communication, 2005 (see Fig. 5.1).

5.4 Molecular Biology of the OROV

More recently, with the implementation of molecular techniques, 55 strains of the OROV (see Table 5.1), isolated from different sources (humans, arthropods, and wild animals), periods of time (from 1955 to 2007), and geographic regions of the South and Central Americas, such as Brazil, Peru, Panama, and Trinidad,[3,15,29] had their N gene completely sequenced. By using phylogenetic methods, at least three genotypes have been described (I, II, and III) circulating in the Americas.[29] In Trinidad, only genotype I was found, while in Peru genotype II has been described. In Brazil, both genotypes I and II have been described with active circulation; however, genotype

I is more frequently isolated in the western Amazon region and genotype II in the eastern Amazon region.[29] In 2000, an OROV strain was isolated from a novel vertebrate host (*Callithrix sp.*) in the state of Minas Gerais, southeast Brazil, and it was further classified as a member of genotype III, previously found only in Panama, suggesting a possible potential of dispersion of this genotype toward other populated and susceptible regions.[15] Between 2003 and 2004, two ORO fever outbreaks were described in the municipalities of Parauapebas and Porto de Moz, respectively,[3] and more recently in 2006, a large outbreak was recognized in the Bragantina region, Para state, striking the municipalities of Igarapé Açu, Maracanã, Magalhães Barata, and Viseu.[34]

The genetic analysis of the strains isolated in 2003–2004 and 2006 detected the presence of genotype II in the eastern Amazon region (Porto de Moz and municipalities of the Bragantina region, Para State). This genotype was found previously only in Peru and areas of the Brazilian western Amazon region (see Fig. 5.2).

In the last five decades, since the first isolation of the OROV in Trinidad in 1955, the virus has been responsible for outbreaks that affected thousands of people in different urban centers across tropical and subtropical areas of the Central and South Americas. Several seroepidemiological and experimental pathology studies conducted during this period contributed to the understanding of important aspects of the epidemiology and ecology of the OROV, such as transmission cycles, geographic distribution, antibody prevalence, incubation period, clinical aspects of the disease, and also the real epidemic potential of the OROV as an important human pathogen.[2,5,10,11,13,17–19,21–26,28,31,32]

Knowledge of the genetic diversity of the virus could be improved with the implementation of molecular techniques that allow genome detection and nucleotide sequencing. Currently, at least three different genotypes of the OROV have been described with active circulation in the Americas. Notably different from Trinidad, Peru, and Panama, in which only a single genotype, I, II, or III circulates, respectively,[29] in Brazil all three genotypes have been found.[3,15,29]

Regarding evolutionary aspects, the phylogenetic data currently available suggests that the OROV constitutes a monophyletic group. This is a group that shares a common ancestor in the line of time

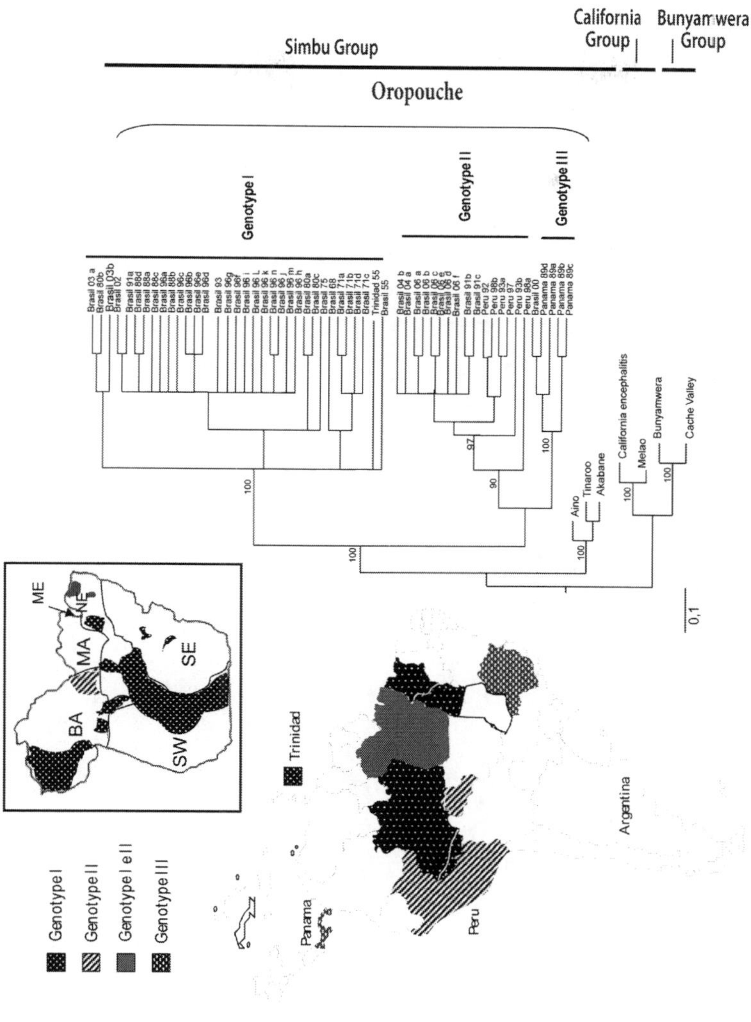

Figure 5.2 Phylogeny of the OROV and other orthobunyaviruses belonging to the Simbu, California, and Bunyamwera serogroups using the MP method. The geographic distribution of genotypes I, II, and III can be observed in the map. Bootstrap values are given above of each main node of the tree. The value under the scale bar (0,1) corresponds to 10% of nucleotide sequence divergence among the strains. Legend: BA: Baixo Amazonas region; MA: Marajó region; ME: metropolitan region of Belem municipality; NE: Northeast region of Para State; SW: Southwestern region of Para State; SE: Southeastern region of Para State. *Abbreviation*: MP, maximum parsimony.

Table 5.1 OROV strains used for phylogenetic analysis according to the host association, year of isolation, and geographic location

Virus strainl	Host association	Sample	Year of isolation	Place (city/state/country)	Legend	GenBank access number
TRVL 9760	Human	Blood	1955	Trinidad	Trinidad 55	AF164531
BeAn 19991	Bradypus trydactylus	Blood	1960	São Miguel-Pará	Brasil 60	AF164532
AR 136921	Culex (Culex) fatigans	Pool	1968	Belém-Pará	Brasil 68	NR*
BeH 271815	Human	Blood	1975	Santarém-Pará	Brasil 75	AF164533
BeAn 206119	Bradypus trydactylus	Blood	1971	Maracanã-Pará	Brasil 71a	AY993909
BeAn 208402	Bradypus trydactylus	Blood	1971	Maracanã-Pará	Brasil 71b	AY993910
BeAn 208819	Bradypus trydactylus	Blood	1971	Maracanã-Pará	Brasil 71c	AY993911
BeAn 208823	Bradypus trydactylus	Blood	1971	Maracanã-Pará	Brasil 71d	AY993912
BeH 390233	Human	Blood	1980	Manaus-Amazônas	Brasil 80c	AF164536
BeH 381114	Human	Blood	1980	Belém-Pará	Brasil 80b	AF164535
BeH 379693	Human	Blood	1980	Castanhal-Pará	Brasil 80	AF164534
BeH 472200	Human	Blood	1988	Porto Franco-Maranhão	Brasil 88a	AF164537
BeH 472204	Human	Blood	1988	Tocantinópolis-Tocantins	Brasil 88	AF164538
BeAr 473358	Culicoides paraensis	Pool	1988	Porto Franco-Maranhão	Brasil 88c	AF164539
BeH 475248	Human	Blood	1988	Tucuruí-Pará	Brasil 88d	AF164540
GLM 444477	Human	Blood	1989	Panama	Panama 89a	AF164555
GLM 444911	Human	Blood	1989	Panama	Panama 89b	AF164556
GLM 445252	Human	Blood	1989	Panama	Panama 89c	AF164557
GLM 450093	Human	Blood	1989	Panama	Panama 89d	AF164558
BeH 505514	Human	Blood	1991	Santa Isabel-Pará	Brasil 91a	AF164541
BeH 505442	Human	Blood	1991	Ouro Preto D'Oeste-Rondônia	Brasil 91b	AF164542
BeH 505663	Human	Blood	1991	Ariquemes-Rondônia	Brasil 91c	AF164543
IQT 1690	Human	Blood	1992	Peru	Peru 92	AF164549

MD 023	Human	Blood	1993	Peru	Peru 93a	AF164550
DEI 209	Human	Blood	1993	Peru	Peru 93	AF164551
BeH 521086	Human	Serum	1993	Barra do Corda-Maranhão	Brasil 93	AY704559
BeH 541863	Human	Blood	1996	Altamira-Pará	Brasil 96a	AF164544
BeH 543033	Human	Blood	1996	Oriximiná-Pará	Brasil 96b	AF164545
BeH 544552	Human	Blood	1996	Brasil Novo-Pará	Brasil 96c	AF164546
BeH 543087	Human	Blood	1996	Xapuri-Acre	Brasil 96d	AF164547
BeH 543618	Human	Blood	1996	Oriximiná-Pará	Brasil 96	AF164548
BeH 543733	Human	Serum	1996	Oriximiná-Brasil	Brasil 96f	AY704560
BeH 543745	Human	Blood	1996	Oriximiná –Pará-	Brasil 96g	NR*
BeH 543854	Human	Serum	1996	Oriximiná –Pará	Brasil 96h	NR*
BeH 543880	Human	Blood	1996	Oriximiná –Pará	Brasil 96i	NR*
BeH 543790	Human	Blood	1996	Oriximiná –Pará	Brasil 96j	NR*
BeH 543638	Human	Blood	1996	Oriximiná –Pará	Brasil 96k	NR*
BeH 543629	Human	Blood	1996	Oriximiná –Pará	Brasil 96L	NR*
BeH 543735	Human	Blood	1996	Oriximiná –Pará	Brasil 96m	NR*
BeH 543857	Human	Blood	1996	Oriximiná –Pará	Brasil 96n	NR*
IQT 4083	Human	Blood	1997	Peru	Peru 97	AF164552
01-812-98	Human	Blood	1998	Peru	eru 98a	AF164553
IQT 7085	Human	Blood	1998	Peru	Peru 98b	AF164554
BeAn 626990	Callithrix sp	Viscera	2000	Arinos- Minas Gerais	Brasil 00	AY117135
BeH 622544	Human	Blood	2002	Paranã-Tocantins	Brasil 02	EF 467368
BeH 669314	Human	Blood	2003	Parauapebas-Pará	Brasil 03a	EF467370
Be H 669315	Human	Blood	2003	Parauapebas-Pará	Brasil 03	EF467369
BeH 682426	Human	Blood	2004	Porto de Moz-Pará	Brasil 04a	EF467371

(*Contd.*)

Table 5.1 (*Contd.*)

Virus strainl	Host association	Sample	Year of isolation	Place (city/state/country)	Legend	GenBank access number
BeH 682431	Human	Blood	2004	Porto de Moz-Pará	Brasil 04b	EF 467372
H 706890	Human	Serum	2006	Igarapé-Açu-Pará	Brasil 06f	NR*
H 706893	Human	Serum	2006	Igarapé-Açu-Pará	Brasil 06d	NR*
H 707157	Human	Serum	2006	Maracanã-Pará	Brasil 06a	NR*
H 707159	Human	Serum	2006	Maracanã-Pará	Brasil 06b	NR*
H 707289	Human	Serum	2006	Magalhães Barata-Pará	Brasil 06c	NR*
H 708139	Human	Serum	2006	Magalhães Barata-Pará	Brasil 06e	NR*

*NR: not registered

when compared with other arboviruses belonging to the Simbu serogroup (group in which the OROV is included) and to other representative members of the *Orthobunyavirus* genus. It is important to note that the Brazilian strains isolated in the eastern Amazon region (Pará, Amapá, Tocantins, and also Maranhão), generally are more closely related genetically to the genotype I strains, whereas the strains isolated in the western Amazon region (Acre, Amazonas, and Rondônia) are more predominantly related to the Peruvian strains (genotype II). This finding suggests possibly different evolutionary origins for the Brazilian genotypes. Another important observation relates to the dispersion of genotypes to other geographic locations where those lineages were not detectable, previously restricted to determined geographic regions of the Americas or of the Brazilian Amazon region. As examples, genotypes II and III, previously found only in Peru and in certain states of the Brazilian western Amazon region (Acre and Rondonia states) and in Panama, were recently found in Para State (eastern Amazon) and in Minas Gerais, southeast Brazil, respectively.

These facts could be strongly related to the movement of infected individuals with a given genotype during the viremic phase from one endemic region to another or perhaps might result from the virus traffic as suggested for the yellow fever virus.[33] Another possibility for the virus dispersion could be explained by the natural movement of the elected hosts of the OROV (sloths, monkeys, and wild birds) during the viremic phase or by illegal wild animal commerce, which is a common activity in the Amazon region.

Regarding OROV evolutionary origin, the current studies provided new insights that revealed different origins for the Brazilian strains and a dynamic dispersion of the genotypes across the Amazon region. Further studies on the association of genetic and epidemiological data will provide a better understanding about the molecular epidemiology of this important human pathogen, as well as enable us to describe with more precision the evolutionary dynamics of this virus. Also, these studies will be important in assessing the possibility of virus reassortment in nature among different strains as a natural mechanism of virus diversity, a fact frequently observed among RNA viruses and, in this particular case, already demonstrated for members of the Simbu

(Jatobal virus), Bunyamwera (Ngari virus), and certain group C orthobunyaviruses.[6,16,30]

Acknowledgments

The authors are grateful to the different researchers who, in the past decade, have dedicated their time to studies involving the OROV, in particular to Dr. Francisco de Paula Pinheiro and Dr. Amelia Paes de Andrade Travassos da Rosa, as well as to the distinctive laboratory technician staff of the Department of Arbovirology and Hemorrhagic Fevers of the Evandro Chagas Institute for their valuable contribution to understanding the epidemiological and molecular aspects of the OROV.

References

1. Anderson, C., Spence, L., Downs, W. G., and Aitken, T. H. Oropouche virus: a new human disease agent from Trinidad, West Indies. 1961. *Am. J. Tropical Med. Hyg.*, **10**, 574–578.
2. Araujo, R., Pinheiro, F. P., Araujo, M. T., Peteres, D. A., and Dias, L. B. Patogenia das lesões hepáticas na infecção experimental com o vírus Oropouche (BeAn 1991): análise comparativa das curvas virêmicas e de infectividade com as alterações ultra-estruturais. 1979. *Hiléia Méd., Belém*, **1**, 7–12.
3. Azevedo, R. S. S., Nunes, M. R. T., Chiang, J. O., Bensabath, G., Vasconcelos, H. B., Pinto, A. Y. N., Martins, L. C., Monteiro, H. A. O., Rodrigues, S. G., and Vasconcelos, P. F. C. Reemergence of Oropouche fever, northern Brazil. 2007. *Emerg. Infect. Dis.*, **13**(6), 912–915.
4. Bishop, D. H. L., and Shope, R. E. Bunyaviridae. 1979. In Fraenkel-Conrat, H., and Wagner, R. R. (eds.), *Comprehensive Virology*, Vol. 14. New York: Plenum Press, 1–156.
5. Borborema, C. A., Pinheiro, F. P., Albuquerque, B. C., Travassos Da Rosa, A. P., Da Rosa, J. F., and Dourado, H. V. Primeiro registro de epidemia causada pelo vírus Oropouche no estado do Amazonas. 1982. *Revista do Instituto de Medicina Tropical. São Paulo*, **24**(3), 132–139.
6. Briese, T., Bird, B., Kapoor, V., Nichol, S. T., and Lipkin, W. I. Batai and Ngari viruses: M segment reassortment and association with severe

febrile disease outbreaks in East Africa. 2006. *J. Virol.,* **80**(11), 5627–5630.

7. Chavez, R., Colan, E., and Philips, I. Fiebre de Oropouche em Iquitos: reporte preliminar de 5 casos. 1992. *Revista de Farmacologia y Terapéutica,* **2**(1), 12–14.
8. Dixon, K. E., Travassos Da Rosa, A. P. A., Travassos Da Rosa, J. F. S., and Llewellyn, C. H. Oropouche vírus. II. Epidemiological observation during an epidemic in Santarém, Pará, Brazil, in 1975. 1981. *Am. J. Tropical Med. Hyg.,* **30**(1), 161–164.
9. Fauquet, C. M., Mayo, M. A., Maniloff, J., Desselberger, U., and Ball, A. 2005. *Virus Taxonomy: Eighth Report of the International Committee on Taxonomy of Viruses.* San Diego: Elsevier Academic Press, 1164p.
10. Freitas, R. B., Pinheiro, F. P., Santos, M. A. V., and Travassos Da Rosa, A. P. A., Travassos Da Rosa J. F. S., Freitas, E. N. Epidemia de Vírus Oropouche noleste do estado do Pará, 1979. 1980. *Revista da Fundação SESP.* Rio de Janeiro, **25**(2), 59–72.
11. Hoch, A. L., Roberts, D. R., and Pinheiro, F. P. Host-seeking behavior and sensorial abundance of Culicoides paraensis (Diptera: Ceratopogonidae) in Brazil. 1990. *J. Am. Mosquito Control Asso.,* **6**, 110–114.
12. Karabatsos, N. (ed.). 1985. *International Catalogue of Arboviruses Including Certain Other Viruses of Vertebrates.* 3rd ed., San Antônio, TX: American Society of Tropical Medicine and Hygiene. 1147p.
13. Le Duc, J. W., Hoch, A. L., Pinheiro, F. P., and Travassos Da Rosa, A. P. A. Epidemic Oropouche vírus disease in northern Brazil. 1981. *Bull. Pan Am. Health Org.,* **15**, 97–103.
14. Linley, J. R., Hoch, A. L., and Pinheiro, F. P. Biting midges (Diptera: Ceratopogonidae) and human health. 1983. *J. Med. Entomol.,* **20**(4), 347–364.
15. Nunes, M. R. T., Martins, L. C., Rodrigues, S. G., Chiang, J. O., Azevedo, R. S., Da Rosa, A. P., and Vasconcelos, P. F. C. Oropouche virus isolation, southeast Brazil. 2005a. *Emerg. Infect. Dis.,* **11**(10), 1610–1613.
16. Nunes, M. R. T., Vasconcelos, P. F. C., Rosa, A. P. A. T., Weaver, S., and Tesh, R. Molecular epidemiology of group C viruses (Bunyaviridae, Orthobunyavirus) isolated in the Americas. 2005b. *J. Virol.,* **79**, 1561–1570.
17. Pinheiro, F. P., Pinheiro, M., Bensabath, G., Causey, O. R., and Shope, R. Epidemia de vírus Oropouche em Belém. 1962. *Revista do Serviço Especial de Saúde Pública,* **12**(1), 13–23.

18. Pinheiro, F. P., Travassos Da Rosa, A. P. A., Travassos Da Rosa, J. F. S., and Bensabath, G. An outbreak of Oropouche disease in the vicinity of Santarém, Pará, Brasil. 1976. *Tropenmed. Parasitol., Germany*, **27**, 213–223.
19. Pinheiro, F. P., Hoch, A. L., Gomes, M. L. C., and Roberts, D. R. Oropouche virus. IV. Laboratory transmission by Culicoides paraensis. 1981a. *Am. J. Tropical Med. Hyg.*, **30**(1), 172–176.
20. Pinheiro, F. P., Travassos Da Rosa, A. P. A., Travassos Da Rosa, J. F. S., Ishak, R., Freitas, R. B., Gomes, M. L., Le Duc, J. W., and Olivia, O. F. Oropouche virus. I. A review of clinical, epidemiological and ecological findings. 1981b. *Am. J. Tropical Med. Hyg.*, **30**(1), 149–160.
21. Pinheiro, F. P. (ed.). 1982a. Epidemia de vírus Oropouche no leste do Estado do Pará, 1982. In *Internacional Symposium on Tropical Arvboviruses and Heamorrhagic fevers.* Rio de Janeiro: Academia Brasileira de Ciências, 419–439.
22. Pinheiro, F. P., Travassos Da Rosa, A. P. A., Gomes, M. L. C., Le Duc, J. W., and Hoch, A. L. Transmission of Oropouche virus from man to hamsters by midge *Culicoides paraensis*. 1982b. *Science*, **215**(4537), 1251–1253.
23. Pinheiro, F. P. Febre do Oropouche. 1983. *J. Bras. Med.*, **44**(4), 46–62.
24. Pinheiro, F. P., Travassos Da Rosa, A. P. A., Freitas, R. B., Travassos Da Rosa, J. F. S., and Vasconcelos, P. F. C. Arboviroses, aspectos clínico-epidemiológicos. Instituto Evandro Chagas, 50 anos de contribuição às ciências biológicas e à medicina tropical. 1986. Belém: Instituto Evandro Chagas/Fundação Serviços de Saúde Pública, 349–357.
25. Pinheiro, F. P., Travassos Da Rosa, A. P. A., and Vasconcelos, P. F. C. 1994. Arboviral zoonoses of central and South América. Part G. Oropouche fever. In Beran G. W. (ed), *Handbook of Zoonoses*. 2nd ed. Boca Raton: CRC Press, 214–217.
26. Pinheiro, F. P., Travassos Da Rosa, A. P. A., and Vasconcelos, P. F. C. 1998. An overview of Oropouche fever epidemics in Brazil and the neighbor countries. In Travassos Da Rosa, A. P. A., Vasconcelos, P. F. C., and Travassos Da Rosa, J. F. S. (eds.), *An Overview of Arbovirology in Brazil and Neighboring Countries*. Belém: Instituto Evandro Chagas, 186–192.
27. Pinheiro, F. P., Travassos Da Rosa, A. P. A., and Vasconcelos, P. F. C. 2004. Oropouche fever. In Feigin, R. D. (ed.), *Textbook of Pediatric Infectious Diseases*. 5th ed. Philadelphia: Editora Saunders, 2418–2423.
28. Roberts, D. R., Hoch, A. L., Dixon, K. E., and Llewellyn, C. H. Oropouche virus. III. Entomological observations from three epidemics in Pará, Brazil, 1975. 1981. *Am. J. Tropical Med. Hyg.*, **30**(1), 165–171.

29. Saeed, M. F., Wang, H., Nunes, M., Vasconcelos, P. F. C., Weaver, S. C., Shope, R. E., Watts, D. M., Tesh, R. B., and Barrett, A. D. T. Nucleotides sequences and phylogeny of the nucleocapsid gene of the Oropouche virus. 2000. *J. Gen. Virol.,* **81**(3), 743–748.

30. Saeed, M. F., Nunes, M., Vasconcelos, P. F. C., Travassos Da Rosa, A. P. A., Watts, D. M., Russell, K., Shope, R. E., Tesh, R. B., and Barrett, A. D. T. Diagnosis of Oropouche virus infection using a recombinant nucleocapsid protein-based enzyme immunoassay. 2001. *J. Clin. Microbiol.,* **39**(7), 2445–2452.

31. Travassos Da Rosa, S. G., Nunes, M. R. T., Magalhaes, M. T. F. Travassos Da Rosa J. F. S., and Vasconcelos P. F. C. Epidemia de febre do Oropouche em Serra Pelada, Município de Curionópolis, Pará, 1994. 1996. *Revista da Sociedade Brasileira de Medicina Tropical,* **29**, 537–541.

32. Vasconcelos, P. F. C., Travassos Da Rosa, J. F. S., Guerreiro, S. C., Degallier, N., Travassos Da Rosa, E. S., and Travassos Da Rosa, A. P. A. Primeiro registro de epidemias causadas pelo vírus Oropouche nos estados do Maranhão e Goiás, Brasil. (1989). *Revista do Instituto de Medicina Tropical de São Paulo,* **31**, 271–278.

33. Vasconcelos, P. F. C., Bryant, J. E., Travassos Da Rosa, A. P. A., Tesh, R. E., Rodrigues, S. G., and Barrett, A. D. T. Genetic divergence and dispersal of yelow fever virus, Brazil. 2004. *Emerg. Infect. Dis.,* **10**(9), 1578–1584.

34. Vasconcelos, H. B., Azevedo, R. S. S, Casseb, S. M., Nunes Neto, J. P., Chiang, J. O., Cantauria, P. C., Segura, M. N. O., Martins, L. C., Souza, M. R. S., Monteiro, H. A. O., Rodrigues, S. G., Nunes, M. R. T., and Vasconcelos, P. F. C. Oropouche fever epidemic in northern Brazil: epidemiology and molecular characterization of isolates. 2009. *J. Clin. Virol.,* **44**, 129–133.

35. Watts, D. M., Phillips, I., Callahan, J. D., Griebenow, W., Hyams, K. C., and Hayes, C. G. Oropouche virus transmission in the Amazon river basin of Peru. 1997. *Am. J. Tropical Med. Hyg.,* **56**(2), 148–152.

Chapter 6

Is Avian Influenza Subtype H5N1 a Cause for Concern? A Critical Analysis

Alan P. Zelicoff
School of Medicine, University of New Mexico, Albuquerque, NM, USA
zelicoff@slu.edu

6.1 Specter of Panzootics

Influenza A virus (IAV) infection and carrier states are well established and continuous in aquatic birds, the natural host of the virus. Influenza subtypes are also well recognized as the cause of both seasonal epidemics of disease and, much more rarely, pandemics. The latter are stochastic in their appearance and are generally associated with dramatic antigenic changes in one or both of the major virion function surface proteins, hemagglutinin (HA) and neuraminidase (NA). Over the past decade, a new subtype of influenza, H5N1, has caused an epizootic among domestic poultry (highly pathogenic avian influenza [HPAI]) and some waterfowl, including migratory birds. Several hundred documented cases of H5N1 in humans, most clearly originating as a result of direct transmission from a bird to a human host in most cases, have

Viral Diagnostics: Advances and Applications
Edited by Robert S. Marks, Leslie Lobel, and Amadou Alpha Sall

ISBN 978-981-4364-43-0 (Hardcover), 978-981-4364-44-7 (eBook)
www.panstanford.com

also occurred. The specter has been raised of a pandemic, and the expenditure of substantial resources in pandemic preparedness mostly focused on this particular virus subtype. The following are examined in this chapter: the historical course of H5N1-induced disease in humans and birds, mechanisms for human-to-human transmission, viral recombination in humans and animals, and extant immunity to disease in humans. These factors should be carefully considered in assessing the risk of a new pandemic based on some variant of the currently circulating clades of HPAI H5N1. An H5N1 pandemic appears unlikely, but systematic near real-time surveillance for severe respiratory or gastrointestinal symptoms in humans and animals is warranted, to identify disease-causing organisms that have the potential to cause pandemics or panzootics.

6.2 The Nature of Influenza A Predisposes It to Pandemics

Influenza A, B, and C virus types are members of the family *Orthomyxoviridae,* all negative-sense RNA viruses possessing a segmented genome divided into either eight or nine strands. One genus of *Orthomyxoviridae* contains influenza A and B, and a second contains influenza C. While these influenza types obviously share structural properties, their host ranges are dramatically different, with types B and C being almost exclusively human pathogens (on occasion found in pigs and seals); IAVs have a wide host range—wild birds, pigs, horses, seals and whales, and poultry. Influenza viruses routinely undergo mutational changes during host cell transcription of viral RNA. Such mutations may or may not result in a viable virus, but if they do, a virus with a changed host range or pathogenic potential may emerge.

In this chapter, we consider an IAV first recognized to cause severe disease in birds about a decade ago and that has been touted as a potential cause of another pandemic (glossary of terms), perhaps with consequences as severe as the devastating pandemic of 1918. The arguments advanced for this concern are detailed later and shall be analyzed in the context of the knowledge of influenza immunology in humans and the experimental literature on host adaptation and transmissibility of the virus. It is useful to summarize what is known about existing influenza A variants, their natural hosts, and influenza pandemics in recent human history.

Aquatic fowl of orders *Anseriformes* (including geese and ducks) and *Charadriiformes* (which includes gulls and guillemots) are widely regarded as the natural hosts of all known influenza A subtypes—defined as a certain pairing of the two major surface proteins of influenza A, known as HA and NA. Subtype designations IAVs are serologically defined and based on sufficient antigenic changes in HA and NA to affect neutralization or inhibition assays. To date, 16 HA subtypes and 9 NA subtypes have been identified, and of 144 theoretical combinations only 103 subtypes have been isolated.

HA and NA are critical to virus survival, replication, and release from host cells. HA, in any of its subtypes, binds virus particles to bird or mammalian sialic acid residues; NA releases progeny from host cell surfaces and may also be critical in helping the influenza virus navigate its way through sialic-acid-rich mucin coating respiratory or gastrointestinal epithelium, to reach the cellular surface.

Novel to IAVs is their ability to cause pandemics (though of varying clinical severity) that have appeared from time to time throughout the recorded course of human history. Pandemics are associated with large changes in viral surface proteins, especially HA, known as antigenic shifts, to which most of the host population is thought to be immunologically naive. The immune system response, including both antibody-based and cellular immunity, will be covered in more detail later.

Always on the minds of public health officials, and all influenza virologists, is that the 1918 Spanish flu—which probably originated in the American Midwest—claimed the lives of between 30 and 50 million of the then 1.6 billion people on the earth, in excess of 2% of the population. The subtype of influenza that resulted in this enormous pandemic, coincident with a world war, is denoted as H1N1—that is, possessing, by definition, the first identified HA and NA variants—though there is no question that other influenza pandemics preceded it. No other disease entity in recent memory, or in the modern historical record, can be shown to have resulted in such high mortality with near simultaneity on every populated continent across the globe. Other influenza pandemics of the past few decades, the Russian or Asian flu of 1957 and the Hong Kong flu of 1968 occurred as a result of changes to the HA protein, which were named HA2 and HA3, respectively, although the severity of these pandemics, measured as the percent mortality in the human population, was far less than that of 1918—perhaps 0.05% and 0.02%, respectively (see later).

Pandemics of the twentieth century probably arose in one of two ways: more or less direct introduction of an avian species (i.e., with all eight influenza genes derived directly from a bird-adapted strain) that then went on to acquire mutations during adaptation in some mammalian host and via the phenomenon of "reassortment" within a mammalian host. Reassortment may occur when a single host cell is co-infected with two influenza viruses at the same time. After the viral RNAs are transcribed, as the virus reassembles within the cell, genes from the two viruses may randomly mix.

As an example of a particularly fateful happenstance, human and avian influenza subtypes may co-infect a mammalian host with which humans have frequent contact, say, a pig. A novel virus may result whose genome is comprised of the HA and NA genes from an avian influenza strain, while the internal protein genes originated from the human virus. This mechanism is thought to characterize the pandemics of 1957 and 1968 (called H2N2 and H3N2 based on changes in both or one of the surface proteins), whereas the catastrophic pandemic of 1918 is thought to have resulted from an avian strain that adapted to humans via multiple mutations but without reassortment. Needless to say, with such complicated phenomena, there remains a dispute as to the mechanism by which novel influenza strains appear in the human population.[1]

Co-infection with multiple subtypes of IAVs is a common occurrence in aquatic birds,[2] so within bird species alone, virologists understandably fret over the emergence of new varieties of the virus, which may serendipitously acquire wider host range, with the possibility among new hosts of both high mortality and transmissibility. In a recent survey of migratory birds in the Netherlands, cloacal swabs of 14 of 132 mallard ducks caught in 1999 harbored simultaneously HA subtypes 1, 2, 4, 5, and 11.

Thus, during viral reproduction and virion assembly, new particles containing genomic material from multiple subtypes are probably being generated continuously.

All of the other common mechanisms underlying genomic change have been identified in avian influenza viruses (AIVs): insertion, deletion, point mutation, and crossover, although homologous crossover is distinctly rare. IAVs are subject to frequent nucleotide mutations, as RNA polymerase operates without any proofreading function as in most DNA viruses. Estimates of likelihood of mutation range from about 10^{-3} to 10^{-2} mutations per site per year.[3]

Reassortment is thought to be the dominant pathway of surface antigen shift, although recombination (or crossover) of genetic elements during viral replication is also a well-described mechanism. Insertion, deletion, and nonsynonymous nucleotide mutation or substitution are thought responsible for the phenomenon of antigenic drift, subtler changes in the antigenic (and perhaps functional) motifs of HA and/or NA to which varying degrees of preexisting immunity among humans may exist. Drift is associated with yearly, sporadic epidemics in humans where previously exposed to a closely related HA subtype vaccine or infection with a live virus provides only very limited protection against the new strains. The amount of drift sufficient to result in evasion of antibodies generated in vivo is difficult to quantitate. Drift may of course occur in influenza in birds as well, although new drift types are only rarely identified in birds. This latter fact is quite surprising, given the inherent mutability of the virus as a result of transcription errors. Some authors believe this evolutionary stasis is a manifestation of optimized adaptation to birds.[2] Modern evolutionary biology allows for the possibility of stable adaptation, depending on the host-pathogen relationship.

HA binds to sialic acid residues, but the three-dimensional relationship between sialic acid and its neighboring molecules' host-cell surfaces differs between birds and mammals. So, while it is always the case that influenza virus particles bind via surface HA to cellular receptors that contain a branched or terminal sialic acid (NeuAC) residue, in birds, sialic acid–binding sites are in an α-2,3 conformation in respiratory and gastrointestinal epithelium, whereas in humans an α-2,6 conformation of terminal sialic acid predominates in the upper respiratory tract and α-2,3 in terminal bronchioles and alveolar cells. HA H5 in highly pathogenic avian influenza has strong, binding affinity for the α-2,3 sialic acid variant, and essentially none for the α-2,6 geometry, perhaps explaining some of the host restriction of H5-containing subtypes, which avidly spread and cause oft fatal disease among tightly packed bird populations but which do so far less commonly in humans (see later).

Finally, as will be explored in some detail shortly, in humans (and perhaps other mammals), exposure to circulating virus subtypes may result in broader protection to drift strains of that particular subtype and perhaps even different subtypes of influenza A—so-called **heterologous** or **cross-subtype** immunity. Recently a wide repertoire of human immune responses to influenza has been identified that renders as far too simplistic the notion that only preexisting antibodies to HA (or HA and NA) were protective against disease.

6.3 A Brief History of the H5N1 Panzootic and Human Cases

In the spring of 1997, an IAV of subtype H5N1 was isolated from a three-year-old child in Hong Kong, who died from atypical pneumonia, Reye's syndrome, and multiorgan failure with disseminated intravascular coagulation. Before the year was out, a closely related or identical strain was found in 17 residents of Hong Kong Semiautonomous Republic (SAR), all with severe flu-like illness; 5 of these 17 victims died.

Shortly before the death of the index human case, H5N1 AIV was isolated from among 2,000 dead chickens on a farm in Hong Kong SAR. Disease among poultry (chickens) had been recognized first in March 1997,[5] though H5N1 itself had never been previously found among waterfowl or gallicenous birds in Hong Kong. Further investigation revealed H5N1 isolates in retail live bird markets, which were rapidly closed down and disinfected and importation of birds suspended. For a brief time, there was no additional clinical disease among birds, but in late December 1997 an illness among chickens consistent with highly pathogenic avian influenza (HPAI) was again recognized on a farm within Hong Kong SAR. A decision was taken to depopulate the entire bird population in markets and chicken farms, and trade in live birds was suspended for almost two months, eradicating H5N1 in Hong Kong among poultry. No subsequent cases of H5N1 disease in humans would be identified for several years, but because the small group of humans who became ill had contact with birds in live bird markets and had acquired the virus directly from birds, there was substantial worry that the H5N1 virus was adapting to humans. Given that a substantial change in the HA gene and its protein product had occurred, there was further concern that should such a virus become transmissible from human to human, another pandemic of influenza would be in the offing.

The H5N1 HPAI epizootic and human disease epidemic is believed to have originated in an influenza strain, originally known to be circulating in the Guangdong province of China,[6] which had killed a small number of geese.[7] Since many influenza viruses constantly circulate among waterfowl, as well as domesticated birds, even after the interruption of disease spread in the bird markets of Hong Kong, H5N1 probably reassorted with other influenza types. Highly pathogenic in chickens, it again reappeared in farms throughout Hong Kong in 2001. Sims et al.[33] reflected on the numerous opportunities that the live bird trade presented for H5N1 to reemerge:

> "One of the interesting features of these outbreaks has been the variable clinical picture in affected markets and farms. In some cases this may have been due to underreporting of disease. Nevertheless, we have observed cases where infection moved slowly through flocks of birds kept in cages . . . probably reflect[ing] the inefficiency of aerosol transmission of these viruses. . . . In retail outlets the pattern of disease is probably modified by the marketing methods for birds. Poultry are brought into the market and sold on a daily basis, and any birds recently infected may be sold before they have the chance to develop disease. Conversely, high mortality rates in infected markets may be related to changes in market demand. If market imbalances occur such that birds are kept longer in market stalls, the opportunity for poultry to develop disease if they acquire infection in the market would likely be increased."

Notable in Sim's description is the situation of forced crowding and high animal stress, all occurring in the setting of multiple circulating strains of influenza. Such is the environment for raising and marketing poultry throughout much of China and Southeast Asia and obviously points to the ecological conditions that might permit a random variant of influenza to take hold, in this case in domesticated birds but perhaps also in other species.

Human H5N1 disease disappeared until late in 2003 when two cases, one of them fatal, were confirmed in a Hong Kong family that had recently traveled to the Fujian province in China. A third family member also died in mainland China from a severe respiratory disease syndrome, but no laboratory analysis of samples was performed. Subsequently, H5N1 was reported among poultry for the first time in Korea in December 2003 and Vietnam in 2004. Simultaneously, two tigers and two leopards in zoological parks in Thailand died unexpectedly after being fed raw chicken meat. An H5N1 virus was isolated from the dead cats. The situation was alarming; even nonmedical but well-respected publications such as *Foreign Affairs*[8] called attention to the new virus that seemed to be able to infect a wide range of hosts. For all intents and purposes, H5N1 outbreaks in wild birds, poultry, and humans have been reported continuously and over an ever-widening geographic area on the Asian, European, and African continents. As of this writing (June 2008), 383 human cases of laboratory confirmed H5N1 have occurred in 15 countries, with a case fatality rate of approximately 63%. The vast majority of human disease has occurred in Indonesia, Vietnam, China, and Thailand. No human nor bird disease has been reported from North or South America, Western Europe, or Australia.

Most human cases are believed to have occurred as a result of direct exposure to birds—or their organs, blood, or excreta—harboring H5N1, just as with the big cats in Thailand. Some family clusters have been reported, raising the possibility of direct human-to-human exposure or a common source exposure with a particular genetic susceptibility in those families.

6.4 Review of Epidemiology of H5N1 in Humans

The World Health Organization (WHO) last reviewed the epidemiology reported cases of H5N1 in humans in June 2006,[9] which included cases reported on the WHO website from December 1, 2003, to April 30, 2006. Cases included in this summary were confirmed by polymerase chain reaction on respiratory tract specimens and/or microneutralization assays on serum specimens with confirmatory testing at WHO reference laboratories. Data on exposure to sick or dead poultry was generally reported by countries notifying WHO of human cases, but this data was incomplete.

Key findings included:

- There were a total of 203 laboratory confirmed cases.
- The median age was 20 years, with half the cases occurring among people less than 20 years of age.
- Approximately 50% of cases occurred in patients less than 20 years of age.
- Case fatality was 56%. The highest was in those aged 10–19 years; the lowest case fatality rate was 18% among those older than 50 years.

No additional summary has been published by WHO since June 2006, but individual case reports are available on the WHO website,[10] with more or less complete data on date of onset of symptoms, age, sex, and survival status of cases. Analyses of 151 cases containing data subsequent to the previous WHO review revealed:

- The overall median age was 19 years, with a median age among females as 21.3 years and among males as 19.7 years. These are not statistically different using Wilcoxon and practical equivalence tests (with a threshold practical difference of zero set at 0.25 years) and are essentially identical to the 2006 WHO summary.
- Approximately 50% of all cases occurred in patients less than 20 years of age, again consistent with the previous WHO summary.

- The overall case fatality was 63% and was highest in the age group of 10–19 years (75%). The mortality in those over 50 was also high (75%), but there were only four case reports in this group. It is difficult to know if case fatality is in fact somewhat higher or that better case findings and identification have somewhat skewed overall mortality calculation.

In summary, on the basis of the WHO 2006 review and analysis of cases reported after April 2006, case fatality appears to be highest in the 10–19-year-old age group, in contrast to the common circulating strains of influenza (H1N1 and H3N1) where fatalities are clearly the highest among the elderly.[11]

6.4.1 *The Basis of the Concern for H5N1 as a Pandemic Threat*

In view of the molecular biology and immunopathology noted before, some influenza watchers have listed the following arguments in support of the notion of avian H5N1 as a pandemic threat:

- There has been an antigenic shift in the HA, which resulted in a virus subtype that has, in fact, infected some small number of humans since 2003.
- The H5N1 subtype appears to have been introduced into humans directly from birds and is widely believed to have occurred in 1918. As with the 1918 pandemic, mortality seems to be highest in teenagers and young adults.
- Some influenza watchers believe that pandemics seem to occur at regular intervals and it has been a long time since a new pandemic occurred (1968). Put another way, we may be overdue for an influenza pandemic.
- All that remains for H5N1 to become a pandemic virus is that it adapts in such a way as to become transmissible from human to human. Since H5N1 now co-circulates with H1N1 and H3N2 in humans, and perhaps in other mammalian species, there is an increased probability of a reassortment that could lead to a virus well adapted for not only disease pathogenesis but also for human-to-human transmission.

6.4.2 *Critique of the Pandemic Hypothesis*

H5N1 represents an antigenic shift in an influenza virus that has infected some humans. Further, as may have occurred in 1918, the virus causing severe or fatal disease in humans is, for all intents and purposes, identical to the virus affecting birds, indicating direct introduction into the human population. And, as occurred during the great pandemic of 1918, young adults seem to be disproportionately affected.

Despite the passage of a decade since the first human cases of H5N1 were reported and routine contact between birds and humans—in tens of thousands of live poultry markets and countless chickens and ducks raised as poultry in tens of millions of backyards—no pandemic has appeared. This is particularly striking, given the hypermutability of H5N1 (or any) avian virus.

While it is impossible to prove the negative proposition, that is, there cannot or will not ever be a pandemic due to an H5N1 subtype of influenza, epidemiologic, clinical, and experimental data has mounted that undermines the threat to humans presented by H5N1. There are essentially five lines of evidence:

- First, statistical calculations exclude the possibility of any periodicity in pandemics, at least over the past three centuries.
- Second, on the basis of limited but growing serologic data in humans, asymptomatic H5N1 infections probably occur much more frequently than realized. Since only severe respiratory symptoms in humans result in diagnostic testing for influenza virus, mild symptoms (let alone no symptoms at all) would not warrant investigation to rule out H5N1 or any influenza virus.
- Third, there appears to be broadly distributed immunity to H5N1 based on cell-mediated immunity to internal proteins and also to highly conserved short major histocompatibility complex (MHC)-restricted epitopes in humans.
- Fourth, immunity to H5N1 can be induced in experimental animals with vaccines (delivered in a variety of contexts) that do not contain H5. This experimental evidence

recapitulates, in part, the data demonstrating immunity in humans who have lived long enough to have been exposed to circulating human influenza strains or who have received multiple vaccinations against those strains (or both).

- And finally, fifth, experimental reassortment of H5N1 with human strains fails to result in derivatives that are transmissible in the ferret model.

These considerations will be taken in turn, and because there is at least some superficial similarity between the small but potentially significant human clusters of H5N1 and the H1N1 virus of 1918, the possibility that 1918 may have been a "one off" event, characteristics of no other pandemic will be explored.

First, review the historical data on pandemics, including written descriptions from the eighteenth and nineteenth centuries, where records of multicontinent respiratory disease are generally agreed on by scholars to represent episodes of worldwide influenza.[12]

As evidenced by Potter,[12] over a period of approximately 280 years, there have been seven pandemics of influenza, but no periodicity is evident. Formal statistical calculation[13] confirms what the eye suggests: pandemics are stochastic, based on the limited number of events available for analysis. This is not to say that there is no longer-term cycle in the occurrence of pandemics, but over three centuries no shorter intrapandemic time span can be identified. Thus, the argument put forward that humanity is overdue for another pandemic is readily dismissed.

Fauci et al.[14] have recently reviewed influenza pandemics starting in the mid-1800s with a view toward assessing the periodicity of influenza (or alternatively our ability to predict the next pandemic of influenza). While they correctly conclude that influenza pandemics cannot be predicted (at least on the basis of the data we have from 1847, 1889, 1918, 1957, and 1968), they also note that:

> "Furthermore, there is no evidence that a human pandemic or even an epidemic has been caused by any previous HPAI virus reported in poultry for more than 125 years. None of the last 4 pandemics is known to have been temporally associated with a

poultry or wild bird epizootic, leaving no historical data to support the possibility that poultry are capable of serving as intermediate hosts in pandemic development."[14]

While no references are provided to substantiate this statement, given the close proximity of humans and small food animals—primarily domesticated birds in the 1700s and 1800s—it is highly unlikely that a domestic bird die-off would have been missed nor would its association (should it have occurred) with severe human respiratory and/or gastrointestinal disease been overlooked. Further, since wild fowl hunting was well established by the early 1800s, both wild water fowl and members of the order *Galliformes* were either hunted or domesticated; the disappearance or die-off of these birds would probably have also been well appreciated, as would any overlap with the three pandemics of the 1800s. In no case have prepandemic large-scale bird die-offs of any species been noted. Thus, distinct from any previous human pandemic, should H5N1 become a pandemic strain it would be the first to have caused a bird die-off before resulting in worldwide human morbidity or mortality.

It is clear that human disease is strongly associated with contact with dead or sick poultry or living in areas where neighbors or farms reported poultry die-offs, strongly supporting the hypothesis of direct spread of H5N1 from diseased fowl to humans. However, there is little evidence of H5N1 acquiring characteristics that facilitate human-to-human transmission, a key requirement for a pandemic for a new influenza entering the human population. On the basis of WHO statistics of confirmed disease, and extrapolating May 2008 data for the rest of the year, the cumulative number of cases is close to linear.

Yearly case fatality rates vary from 43% to 75%, but due to the small number of cases each year, it is difficult to show any statistical difference by year, with the possible exception of 2005, where mortality was lower than the average of approximately 65% ($p < .01$ by Fisher's exact test).

From an epidemiologic standpoint, one is left to conclude that there is no basis for any periodicity of pandemics, nor are there any clear signs of the development of human-to-human transmission on a worrisome scale. Although H5N1 clearly is transmissible from birds to humans and, in fact, causes disease in some percentage of humans so exposed, on the basis of previous observations, should avian H5N1 result in pandemic disease, it would be unique in that no previous pandemics have been reported in the face of large-scale bird die-offs in both wild and domesticated populations.

6.5 Are There Asymptomatic H5N1 Infections?

There is now substantial evidence that asymptomatic infections with avian influenzas of H5 subtypes occur in humans, including H5N1. With the discovery of HPAI H5N1, a massive four-day culling effort began in late December in both poultry markets and on farms in Hong Kong. Approximately 1.4 million chickens and hundreds of thousands of other birds raised for human consumption were slaughtered[15]—both government employees and poultry workers were utilized to cull dead or dying birds. Serological studies[16] done on poultry workers showed an approximate prevalence of 10% of anti-H5 antibodies (performed by microneutralization assay and statistical sample confirmed by Western blot). Because exposure to H5N1 among poultry workers could not be known with certainty, a nested control investigation was undertaken. This investigation revealed that more intense exposure to chickens (e.g., working in a retail poultry operation vs. working on a farm) reported mortality of more than 10% among birds that they worked with, and butchering poultry was associated with an increased probability of positive serology. The high prevalence of 10% compared to 0% among blood donors,[17] 0.7% among health care workers in general (though they may have been exposed to patients with H5N1 disease), and approximately 4% in health care workers who cared for an H5N1-infected patient.[18]

Among 293 government workers specially employed for the cull, nine (3%) were positive for anti-H5-positive serum by both microneutralization and Western blot assay. One of these nine seroconverted during the cull and reported a mild respiratory illness. No other risk factors were associated with anti-H5-positive serology.

This data is consistent with the hypothesis that prolonged or intensive exposure to birds is a risk factor for H5N1 infection in humans but not necessarily overt disease. Further, there may be transmission of sufficient H5N1 from patients with known disease due to the organism to asymptomatically infect a small percentage of health care workers.

An assessment of prevalence of flu-like illness (defined as cough and fever), over a six-month period of time in 2004 among over 45,000 people in rural Vietnam, revealed that approximately 18% reported such a symptom complex, which was in multivariate regression analysis associated with contact with sick or dead poultry,[19] after adjusting for widely varying socioeconomic status. Clustering within families of cases of flu-like illness was also more common in those households with sick or dead poultry. Further, contact with sick or dead poultry per se (independent of location) was the strongest single risk factor for flu-like illness among several settings of bird exposure.[20] Because no serology was performed, the cause or causes of flu-like illness could not be ascertained. However, HPAI was epidemic in 2004 in Vietnam among poultry in the area where humans were surveyed.

It would thus appear that asymptomatic H5N1 infections occur in humans, and even with routine, intense exposure to birds who are ill or dying in the midst of an epizootic disease, human disease is extremely rare. Further, contact between humans who are not involved in poultry processing on a large scale, but who instead purchase live birds in markets or raise them in backyards, as is common in Southeast Asia, suggests either very inefficient transmission of H5N1 from birds to humans and/or there is some measure of existing resistance to infection or progression of infectious disease in the human population.

6.6 Do Humans Have Some Immunity to H5N1?

The median age of patients suffering from H5N1 infection is much lower than that seen in yearly epidemics of H1N1 or H3N2 influenza, suggesting that there might be acquired immunity that is heterosubtypic. Several recent articles suggest that this is in fact the case.

Sandbulte et al. obtained serum samples from human volunteers to assess reactivity with AIV H5N1.[34] NA inhibition assays demonstrated reactivity with an H1N1 influenza strain (A/New Caledonia//20/99) in 31 or 38 individuals tested, with titers ranging from <20 to >320. There was measurable inhibitory activity of titers between 20 and 80 found against A/HK/213/03 H5N1 in eight individuals and in nine against the NA of A/Vietnam/1203/03. While NA antibodies are thought to be less important that HA antibodies in determining disease protection or viral shedding, in experimental settings, immunization with NA mitigates disease severity.[21]

Humans never previously exposed to H5N1 have been shown to possess CD8+ and CD4+ cells that recognize epitopes in A/HK/213/03.[22] CD4+ cells from 15 of 15 healthy subjects also naive to H5N1 had robust responses against matrix protein (M), NA, and nucleoprotein (NP) of influenza A/Vietnam/1203/2004(H5N1); productive functional responses were demonstrated by the elaboration of multiple cytokines upon incubation of CD4+ cells with the virus. Subjects having haplotypes commonly represented in the US Caucasian population (DR0101, DR0404, DR0701, DR1101, DR0301 and DR1501) were used in the latter study.

Sandbulte et al. speculate that these cell-mediated immunity phenomena are unsurprising as there are highly conserved sequences on both M and NP, while the N1 in HPAI H5N1 is closely related to the N1 in circulating strains of H1N1 among humans.[34]

Most recently, Gioia and colleagues[23] demonstrated that among 42 recipients of seasonal influenza vaccine (most were health care workers or laboratory technicians; 45% had not previously been vaccinated), antigen-specific CD4 cells for a variety of H5N1 peptides expressed in the context of common HLA-DRB1 allelles were routinely found and appeared to be directed against N1 peptides. Finally, in 13 of 38 donors, while HI titers against H5N1 were undetectable after vaccination with a trivalent vaccine, a rise in neutralization titers of more than 20 above background was found in more than a third. Gioia et al. conclude that N1 "appears to be one target for cross-type cellular immunity." Although antigens from internal proteins of the virus could not be ruled out, CD4+ T-cell responses were probably directed against NA-specific peptides.

Roti et al.[24] hypothesized that due to repeated exposure to H1N1, H3N2, and H2N2 influenza viruses (or vaccination), adult residents in the United States would possess CD4+ T-cells. These cells are reactive against a variety of conserved epitopes, present strains of the H5N1 matrix protein M1, NP, NA, perhaps even HLA-DR restricted epitopes shared by HA from H5 tetramers and those from H1, H2, or H3, and similarly for N1. In the case of NA, the NA gene of a commonly circulating H1N1 influenza virus in humans—A/New Caledonia/20/1999—and the NA gene of A/Viet Nam/1203/2004 (H5N1) have approximately an 80% sequence homology. It was expected that HA cross recognition would be least likely, given much lower sequence homology comparisons of HA between A/Vietnam/1203/H5N1 and circulating H1N1, H2N2, and H3N2 strains represented by A/New Caledonia/ 20/99 H1N1 strain (63% homology), A/Singapore/1/57 H2N2 strain (74% homology), and A/Panama/2007/99 H3N2 (41% homology).

Perhaps unsurprisingly, among 22 subjects tested with common HLA-DR haplotypes in the North American Caucasian population, CD4+ T-cells isolated from peripheral blood mononuclear cells uniformly responded to M1, NP, and NA (N1 subtype) protein epitopes expressed by A/Viet Nam/1203/2004 (H5N1). More remarkably, CD4+ T-cells evaluated for a functional response produced interferon (IFN)-γ and tumor necrosis factor (TNF)-α in the presence of H5 epitopes, as well despite much weaker conservation of gene sequences. Reflecting on similar work from Jameson et al.,[22] the authors concluded that "most healthy adults in the United States have T cells that are directed against the internal proteins of different influenza A subtypes, including the H5N1 subtype, which is completely naïve to the US population." HA-specific T-cells were identified, directed at epitopes in both HA1 and HA2 domains of H5 HA. Also observed was cross reactivity among antigenic peptides in HA with multiple amino acid mismatches. Previous exposure to H1, H2, or H3 influenza strains circulating among humans probably elicit memory T-cell responses to M1 and NP based on this study and likely to other internal proteins that are both conserved and associated with viral pathogenesis such as PA and basic polymerases.

6.7 Experimental Data: Vaccination and Challenge Experiments in Animals Using H5N1

Preexisting antibodies to NA may be protective against a challenge with HPAI of a given NA subtype. To test this hypothesis, chickens, at two weeks of age, were vaccinated with a pure recombinant N2 prepared in baculovirus encoding the Ph/MD N2 gene and again three weeks later.[25] Six weeks after initial vaccination, vaccinated and unvaccinated control chickens were inoculated with 10^6 EID50 of HPAI A/chicken/Queretaro/14588—19/95 (H5N2) and observed for mortality for two weeks. Quantitative NA inhibition (NI) assays were also performed—6/6 (100%) control chickens died; 1/8 (13%) of vaccinated chickens succumbed ($p = .0014$ by Fisher's exact test). Vaccinated chickens had an approximately 3 log increase in NA antibodies as measured by NI but suffered obvious morbidity. The authors note that inhibition of NA provides for immunity distinct from that induced by immunization with HA constructs. They speculate that in the vaccinated chickens, influenza virus may remain trapped on the surface of infected cells permitting cytotoxic T-lymphocyte killing. Such lymphocytes are known to kill readily influenza-infected cells in the respiratory tract of mice and humans.[26,27]

Thus, for HPAI of subtype H5N2, nearly full protection against mortality can be realized with immunization against the NA moiety alone.

Enhancement of cell-mediated immunity may follow from the unique effects of virus NA inhibition in vivo.

In an experiment with BALB/c mice—exquisitely sensitive to multiple wild-type strains of H5N1—immunization with a seasonal trivalent inactivated influenza vaccine prepared for the 2005–2006 season (A/NewCaledonia/20/99 (H1N1), A/NewYork/55/2004 (H3N2), and B/Shanghai/361/2002), with or without a poly(I): ploy(CU) adjuvant, was analyzed for protection against mortality from exposure to A/HK/493/97(H5N1), A/Vietnam/1194/04(H5N1), and A/Indonesia/6/2005 (H5N1). Inoculation was delivered via intranasal and subcutaneous routes. Antibody titers were also measured along with antigen-specific T-cell response.[28] Mice inoculated transnasally with adjuvant manifested cross-reactive mucosal IgA and serum IgG against H5N1 strains. Mortality against all strains was decreased at 14 days postchallenge, with 1,000 plaque-forming units of each viral strain. Viral shedding was, in general, decreased in immunized mice during the course of infection and recovery or death. Subcutaneous inoculation reduced viral shedding but did not improve survival. A systemic T-cell response to virus subtypes homologous to vaccine strains was demonstrated, but no such response was seen against heterologous H5N1 virus. The results were interpreted as consistent with a protective secretory IgA response against H5N1 to vaccination in this particular vaccine construct. Antibodies—which may have been directed at internal proteins of H5N1—could not be demonstrated despite the increased survival in immunized versus nonimmunized mice.

More recently, Lalor et al. investigated plasmid DNA vaccines in mice and ferrets challenged with A/Vietnam/1203/04 (H5N1).[35] Animals were immunized with plasmids containing various combinations of H5 and consensus sequence constructs for NP and M2. All mice were completely against mortality with the vaccine containing H5, or H5+NP+M2 pDNA. Eighty-eight percent of mice receiving NP+M2 plasmid construct vaccines survived compared with none of the controls receiving the DNA backbone alone (i.e., without gene inserts). Among ferrets, protection against a lethal challenge with the same H5N1 strain was afforded only to ferrets receiving DNA plasmid constructs containing gene sequences from H5 (save for one ferret out of six receiving HP + MP pDNA alone who survived; $p > .1$), although nasal wash viral titers were significantly reduced compared with titers in control ferrets. Be aware that the particular ferret model chosen (100 times the LD50 of the H5N1 strain, resulting in death at five days in control animals) may have accounted for differences in survival between vaccinated mice and similarly inoculated ferrets. However, the results are consistent with the hypothesis that vaccination with conserved NP and M2 pDNA constructs may protect against death and perhaps severe disease, while at the same time decreasing viral shedding before a particular pandemic subtype is isolated (including H5N1 but not necessarily restricted to this subtype) and thus before subtype-specific protection can be offered with a strain specific vaccine.

Experimental data strongly suggests that heterotypic subtype immunity can be induced in mammals by NA protein immunization, by use of the seasonal trivalent vaccines with a suitably chosen adjuvant, or by DNA plasmid vaccination. Most experiments demonstrate protection against a challenge to large doses of H5N1, despite the absence of H5 in the vaccine constructs or, at the very least, reduced virus shedding. In humans, analogous results may have already been realized by natural infection with H1N1, H2N2, or H3N2 and would be consistent with the observation that human H5N1 disease is primarily restricted to young adults who may not have had a sufficient exposure history to marshal memory T-cell responses, particularly involving cell-mediated immunity.

6.8 Transmission of Reassortment Variants of H5N1

Like humans, ferrets have a predominance of α-2,6 receptors in epithelial cells of the upper respiratory tract. Using the ferret model, Maines et al.[29] demonstrated efficient transmission of human H3N2 viruses between ferrets experimentally infected with this virus and then placed in cages adjacent to previously unexposed (and immunologically naive) ferrets. Ferrets inoculated with A/Hong Kong/465/97 (H5N1) became clinically ill and had high titers of H5N1 in nasal washings, yet they were unable to transmit the virus to nearby noninoculated animals. Similar results were obtained with genetically distinct H5N1 viruses A/Hong Kong/213/03, A/Indonesia/5/05, and A/Vietnam/HN3040805, representative of the clades of H5N1 currently circulating among birds and infecting humans. Even though most infected ferrets became clinically ill (most severely so), they were unable to transmit the H5N1 virus to ferrets in proximate cages.

To evaluate the pandemic potential of H5N1, should it undergo reassortment-based change, the same investigators created multiple variants using reverse genetics. Reassortment viruses containing either four or six human virus (H3N2) internal genes and avian H5 and N1 surface protein genes did not transmit from ferret to ferret, despite pathogenicity, as demonstrated by disease in inoculated ferrets and high replication efficiency in eggs and in Madin–Darby canine kidney (MDCK) cells.

The absence of α-2,3 sialic acid receptors in ferret and human respiratory epithelium has often been cited as a reason for lack of efficient transmission of wild-type H5N1 strains, which preferentially bind to this sialic acid residue conformation. However, the reverse genetic reassortment virus of the H5N1 subtype was found in high titers in nasal washings of experimentally infected ferrets yet did not transmit. Finally HK213 H5N1 possesses an affinity for α-2,6 receptors, which are abundantly present in ferret and human respiratory epithelium but showed no ability to transmit.

Thus, despite attempts to adapt H5N1 to ferrets by replacing the wild-type AIV internal genes with human homologues, transmissivity could not be confirmed on the H5N1 virus. While the biological requirements for transmissibility are not well understood, it appears that under a reasonable range of reassortment possibilities, H5N1 simply cannot transmit well among mammals whose respiratory epithelial cells possess HA binding similar to those found in humans.

6.9 Was the 1918 Pandemic Different from Others?

Influenza pandemics are, by definition, widespread—by wide agreement the word "pandemic" refers to human disease on all populated continents—but the actual mortality rate among influenza pandemics may differ. This certainly appears to be the case with the 1918 pandemic as compared to all others. Referring to the work of Potter, and using median estimates for world population during each of the pandemic years 1729, 1781, 1830, 1899, 1918, 1957, and 1977[30] (while utilizing simple arithmetic interpolation, as needed), with the exception of 1918, influenza pandemics have killed less than 0.1% of the world's population with each event. On the conservative end of the most recently published estimates, the 1918 pandemic resulted in the death of at least 50 million people[31] (or perhaps twice that number), or more than 2.4% of the world population at the time. The death rate from pandemics of 1968 and 1977 is apparently less than pre-1918 averages; this may have occurred because of subherotypic immunity in the case of 1968, where anti-N2 immunity (either antibody or cell mediated) was prevalent, and in 1977 (an H1N1 subtype) on the basis of circulating epidemic or endemic strains through approximately 1950.

Clearly 1918 is an outlier in its severity, although other influenza pandemics were certainly terrible enough. What might account for its extraordinary mortality?

We have already noted that the 1918 virus is believed to have been introduced into the human population directly from an avian source—that is, not necessarily involving reassortment in some intermediate multiply infected host such as a pig or a horse. The biologic or immunopathologic basis for the severity of disease brought on by infection with the 1918 H1N1 is still poorly understood. However, we may get some clue from modern evolutionary biological theories.

Organisms compete for the resources of a host, and all other things being equal, the spread of an organism from host to host depends in great measure on its ability to reproduce and then escape the initial host, with its long-term fitness in a given environment dependent on, then infecting, another host and repeating the process. Organisms with high lethality need not necessarily come to the end of their lineage if they can transmit over long distances (e.g., with vector-borne diseases, such as malaria). They can also survive for a long time on surfaces with which immunologically naive hosts are likely to have contact (e.g., variola, the causal agent of smallpox)—the so-called **sit and wait** hypothesis well demonstrated by careful studies.[32]

Similarly, highly lethal organisms, which are usually rapidly multiplying and either overwhelm host defenses or initiate an immunological cascade (sometimes called "cytokine storm"), have an advantage over less rapidly reproducing competitors if there are fresh hosts nearby. This situation is commonly realized with outbreaks of cholera in refugee camps (where many people crowd around a water source that may be used for both bathing and drinking supplies), with bacterial meningitis in college dormitories, or military boot camps, where otherwise healthy people are crowded together in small buildings.

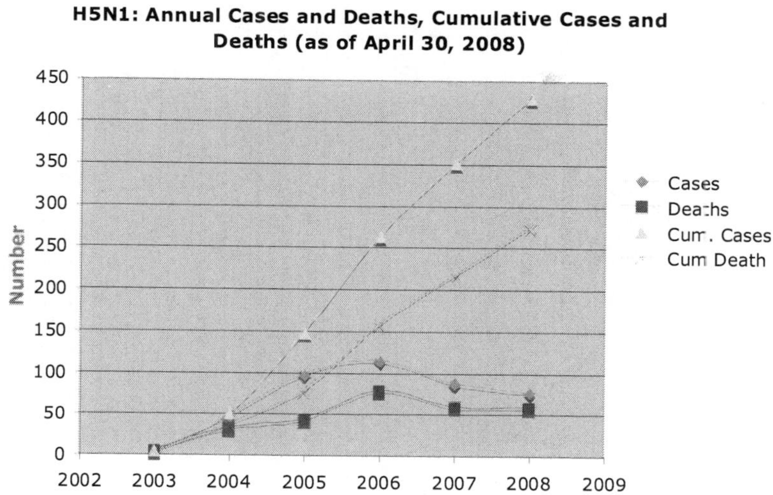

Figure 6.1 Emergency hospital during influenza epidemic, Camp Funston, Kansas, National Museum of Health and Medicine (http://nmhm.washingtondc.museum/collections/archives/agalleries/ 1918flu/Ncp1603.jpg).

Conditions of crowding were both common and extreme in 1918 in the midst of World War I. Troop ships, trenches, tenements, and hospitals were points of congregation of soldiers and civilians (see Fig. 6.1); people resided for long periods no more than a few feet apart.

Therein likely lies the secret of the virulence of 1918: a hypermutable virus—given the strange nature of its multistranded genome and reproductive fallibility due to an error-prone RNA-replicating mechanism—suddenly resulted in a highly lethal variant for humans. In usual circumstances, the virus would kill its host and perhaps the very few people who happened to be in contact with that individual. Soon the virus's opportunity to infect others would disappear, and its lineage would come to an end. However, during wartime conditions, when a given soldier (or resident of a tenement without central heating where everyone congregated around the kitchen stove and slept four to a bed) became ill, it was virtually certain that another susceptible host would be found. In this way, the continuation of this lineage of the virus was guaranteed. It may also be the case that after several passages through human victims,

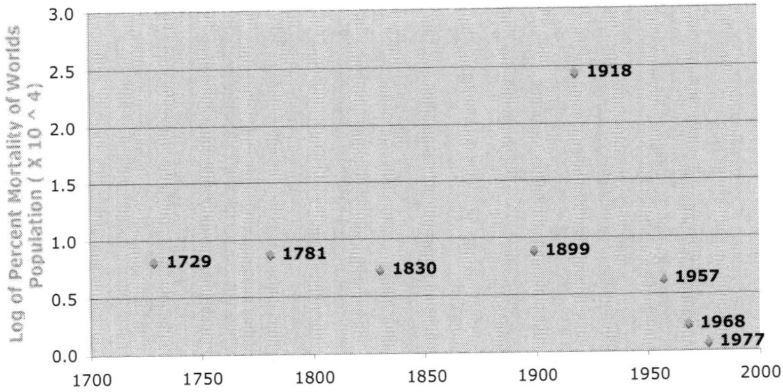

Figure 6.2 Mortality levels in known influenza epidemics.

the virus became especially well adapted to the respiratory tract of *Homo sapiens*.

The elderly were not nearly as severely affected by the 1918 flu, which is probably explained by previous exposure to either H1- or N1-containing strains, and for the most part, a two-decade hiatus in even localized outbreaks of influenza around the world before 1918. Thus, the age group most severely affected by the disease would be those who had no immunological experience with any influenza—young adults. Compared to typical influenza years, young adults were more than two orders of magnitude more likely to die from influenza in 1918 than in other years (see Figs. 6.2 and 6.3).

In short, the 1918 pandemic may have been the product of the confluence of unusual events: a world war in which millions of soldiers and hundreds of thousands of civilians supporting them were confined in close quarters; a period of perhaps two decades in which garden-variety human influenza seems to have been rare; and a random mutation in the genome of an AIV that was both transmissible and pathogenic, at least in the population that had had little or no exposure to other influenza viruses in the wild.

Is a human pandemic with avian H5N1 influenza likely?

A novel avian influenza subtype H5N1, which results in high morbidity or is fatal in birds, has also entered the human population, apparently directly from infected birds without an intermediate reassortment in other mammalian hosts. Among the approximately

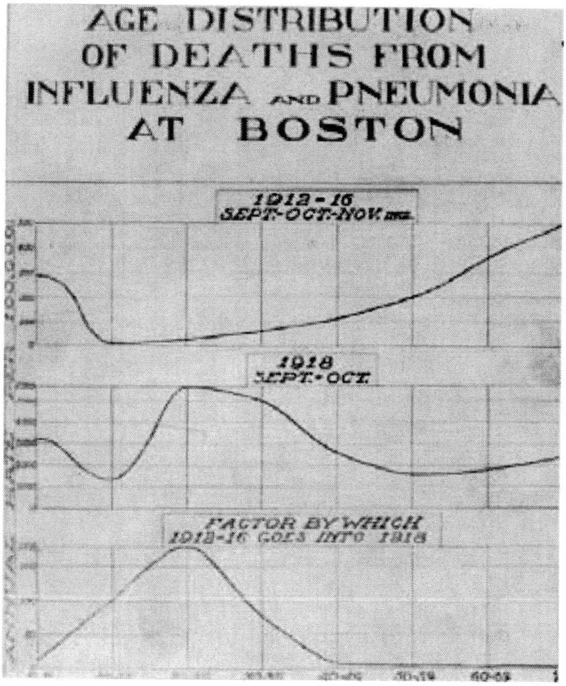

Figure 6.3 Age distribution of deaths from influenza and pneumonia at Boston, National Museum of Health and Medicine (http://nmhm.washington dc.museum/c ol le ctions/archives/agalleries/1918flu/Reeve2960.jpg).

400 symptomatic cases to date in humans, mortality has been high, in the order of 60%. The median among humans with disease is about 20 years of age. In what may be the earliest phase of human H5N1 disease, the story is reminiscent of the horrific pandemic of 1918.

There is much evidence accumulating that H5N1 is not a likely candidate for the next pandemic of influenza. Although there is only limited seroprevalence data to date, asymptomatic or minimally symptomatic infection with H5N1 occurs in humans. Despite the antigenic shift in the HA protein, existing T-cell-based-immunity-responding highly conserved epitopes in HA are common in human populations that have not been exposed to H5N1 (or to the best of our knowledge, to any H5 subtype). Cytotoxic and CD4+ T-cell responsiveness to multiple internal genes and to the NAs in the major circulating clades of H5N1 already exist, at least in humans in North America and Europe. The presence of an antibody to NA, induced by exposure to currently circulating H1N1 strains or by vaccination with trivalent seasonal vaccines in humans, may result in an immobilization of the virus on the cell surface, permitting a robust cell-mediated immune response in either an MHC-I or an MHC-II context. (Note added in proof: There is a very recent paper that demonstrates approximately 1% of villagers in Cambodia have neutralizing antibody titers of >=1:80 against H5N1.[36])

Transmission from person to person with existing H5N1 clades may have occurred but appears to be extremely rare. Reassortment events that might make such transmission possible have been created with reverse genetics, combining H5N1 surface protein genes with human influenza internal genes. No experimental isolate has been identified yet in ferret models that transmit. In some cases, such experimental reassortments also replicate less well than wild-type virus.

There can be no guarantee that a random mutation or set of mutations in an inherently hypermutable virus will not occur in such a way so as to render some variant of H5N1 both infectious and transmissible. But in the absence of severe selective pressures, as occur in enforced crowding and immobilization of hosts (as obtained in 1918), such a possibility seems remote. While it is necessary to keep a close eye on H5N1, there are a large number of influenza virus subtypes circulating in bird and mammalian populations that warrant a robust, international, near-real-time disease surveillance system.

References

1. Gibbs, M. J., and Gibbs, A. J. Molecular virology: was the 1918 pandemic caused by a bird flu? 2006. *Nature*, **440**, E8, Antonovics, J., Hood, M. E., Christi, H. B. *Nature*, **440**, E9, reply Tautenberger, J. K. et al. *Nature*, **440**, 310.
2. Munster, V., Wallensten, A., Olsen, B., Rimmelzwann, G. F., Osterhaus, A. D., and Rouchier, F. A. M., Influenza A virus surveillance in wild birds. 2005. *Avian Influenza: Prevention and Control* (Schrijver, R. S., and Koch, G. eds.) Springer.

3. Scholtissek, C., Rohde, W., Von Hoyningen, V., and Rott, R. On the origin of the human influenza virus subtypes H2N2 and H3N2. 1978. *Virology*, **87**, 13–20.
4. Ewald, P. W. Evolution of virulence. 2004. *Infect. Dis. Clin. N. Am.*, **18**, 1–15.
5. Sims, L. D., Ellis, T. M., Liu, K. K., Dyrting, K., Wong, H., Peiris, M., Guan, Y., and Shortridge, K. F. Avian influenza in Hong Kong 1997–2002. 2003. *Avian Dis.*, **47**, 832–838.
6. Xu, X., Subbaro, K., Cox, N. J., and Guo, Y. Genetic characterization of the pathogenic influenza A/Goose/Guangdong/1/96 (H5N1) virus: similarity of its hemagglutinin gene to those of the H5N1 viruses from the 1997 outbreaks in Hong Kong. 1999. *Virology*, **261**, 15–19.
7. Abolnik, Celia, Molecular epidemiology of Newcastle disease and avian influenza in South Africa. Doctoral Thesis, 2007. University of Pretoria. http://upetd.up.ac.za/thesis/available/etd-06202007-123059 (last accessed 4 June 2008).
8. Garrett, L. The next pandemic? (2005). *Foreign Aff.*, **84**(4), 3–23.
9. Epidemiology of WHO-confirmed human cases of avian influenza (H5N1) infection. 2006. *WHO Wkly. Epidemiol. Rec.*, **81**, 249–260.
10. http://www.who.int/csr/disease/avian_influenza/updates/en/index.html
11. OMS. *Avian Influenza: Assessing the Pandemic Threat, 2005.* http://www.who.int/csr/disease/influenza/WHO_CDS_2005_29/en/index.html, consulté le 22 mai 2006.
12. Potter, C. W. A history of influenza. 2001. *J. Appl. Microbiol.*, **91**, 572–579.
13. Zelicoff, A. P. *Avian Flu. Risk of a Human Pandemic from H5N1. Avian Flu Task Force, World Bank Working Paper, 2005. Appendix A.* Appendix available upon request from the author.
14. Taubenberger, J. K., Morens, D. M., and Fauci, A. S. The next influenza pandemic: can it be predicted? 2007. *JAMA*, **297**(18), 2025–2027.
15. Lee, S. Y. K., Mak, K. H., and Saw, T. A. The avian flu (H5N1): one year on. 1998. *Public Health Epidemiol. Bull.*, **8**, 1–7.
16. Bridges, C. B., Lim, W., Hu-Primmer, J., Sims, L., Fukuda, K., Mak, K. H., Rowe, T., Thompson, W. W., Conn, L., Lu, X., Cox, N. J., and Katz, J. M. Risk of influenza A (H5N1) infection among poultry works, Hong Kong, 1997–1998. 2002. *J. Infect. Dis.*, **185**, 1005–1010.
17. Centers for Disease Control and Prevention. Update: isolation of avian influenza A (H5N1) viruses from humans: Hong Kong, 1997–1998. 1998. *MMWR Morb. Mortal Wkly. Rep.*, **46**, 1245–1247.

18. Buxton Bridges, C., Katz, J. M., Seto, W. H., Chan, P. K., Tsang, D., Ho, W., Mak, K. H., Lim, W., Tam, J. S., Clarke, M., Williams, S. G., Mounts, A. W., Bresee, J. S., Conn, L. A., Rowe, T., Hu-Primmer, J., Abernathy, R. A., Lu, X., Cox, N. J., and Fukuda, K. Risk of influenza A (H5N1) infection among health care workers exposed to patients with influenza A (H5N1), Hong Kong. 2000. *J. Infect. Dis.*, **181**, 344–348.
19. Thorson, A., Petzold, M., Nguyen, T. K., and Ekdahl, K. Is exposure to sick or dead poultry associated with flulike illness?: a population-based study from a rural area in Vietnam with outbreaks of highly pathogenic avian influenza. 2006. *Arch. Intern. Med.*, **166**, 119–123.
20. Groome, P. A., and Richardson, H. Flulike illness and exposure to sick or dead poultry: reply. 2006. *Arch. Intern. Med.*, **166**(13), 1420–1421; author reply 1421.
21. Johansson, B. E., Grajower, B., and Kilbourne, E. D. Infection-permissive immunization with influenza virus neuraminidase prevents weight loss in infected mice. 1993. *Vaccine*, **11**, 1037–1039.
22. Jameson, J., Cruz, J., Terajimia, M., and Ennis, F. A. Human CD8+ and CD4+ T lymphocyte memory to influenza A viruses of swine and avian species. 1999. *J. Immunol.*, **162**, 7578–7583.
23. Gioia, C., Catillett, C., Tempestilli, M., Piacentini, P., Bordi, L., Chiappini, R., Agrati, C., Squarcione, S., Ippolito, G., Puro, V., Capobianchi, M. R., and Poccia, F. Cross-subtype immunity against avian influenza in person recently vaccinated for influenza. 2008. *Emerg. Infect. Dis.*, **14**(1), 121–128.
24. Roti, M., Yang, J., Berger, D., Huston, L., James, E. A., and Kwok, W. W. Healthy human subjects have CD4+ T cells directed against H5N1 influenza virus. 2008. *J. Immunol.*, **180**, 1758–1768.
25. Style, M. J., Hubby, B., and Suarez, D. L., Influenza neuraminidase antibodies provide partial protection for chickens against high pathogenic avian influenza infection. 2007. *Vaccine*, **25**, 3763–3772.
26. Wiley, J. A., Hogan, R. J., Woodland, D. L., and Harmsen, A. G. Antigen-specific CD8(+) T cells persist in the upper respiratory tract following influenza virus infection. 2001. *J. Immunol.*, **167**(6), 3293–3299.
27. Flynn, K. J., Riberdy, J. M., Christensen, J. P., Altman, J. D., and Doherty, P. C. In vivo proliferation of naive and memory influenza-specific CD8(+) T cells. 1999. *Proc. Natl. Acad. Sci. U S A*, **96**(15), 8597–8602. *Vaccine*, **17**(15–16), 2073–2080.
28. Ichinohe, T., Tamura, S., and Kawaguchi, A. Cross-protection against H5N1 influenza virus infection is afforded by intranasal inoculation with

seasonal trivalent inactivated influenza vaccine. 2007. *J. Infect. Dis.*, **196**, 1313–1320.

29. Maines, T. R., Chem, L. M., Matsuoka, Y., Chen, H., Rowe, T., Ortin, J., Falcón, A., Nguyen, T. H., Mai le, Q., Sedyaningsih, E. R., Harun, S., Tumpey, T. M., Donis, R. O., Cox, N. J., Subbarao, K., and Katz, J. M. Lack of transmission of H5N1 avian–human reassortant influenza viruses in a ferret model. 2006. *Proc. Natl. Acad. Sci. U S A*, **103**(32), 12121–12126.
30. *US Census Bureau: Pre-1900 Estimates*, http://www.census.gov/ipc/www/worldhis.html and post-1900 data from: http://www.census.gov/ipc/www/idb/
31. Knobler, S., Mack, A., Mahmoud, A., and Lemon, S. *The Story of Influenze, the Threat of Pandemic Influenza: Are We Ready? Workshop Summary.* 2005. Washington, DC: The National Academies Press, 60–61.
32. Walther, B. A., and Ewald, P. W. Pathogen survival in the external environment and the evolution of virulence. 2004. *Biol. Rev.*, **79**, 849–869.
33. Sims, L. D., Ellis, T. M., Liu, K. K., Dyrting, K., Wong, H., Peiris, M., Guan, Y., and Shortridge, K. F., Avian influenza in Hong Kong 1997–2002. 2003. *Avian Dis.*, **47**, 832–838.
34. Sandbulte, M. R., Jimenez, G. S., Boon, A. C., Smith, L. R., Treanor, J. J., and Webby, R. J., 2007. Cross-reactive neuraminidase antibodies afford partial protection against H5N1 in mice and are present in unexposed humans. *PLoS Med.*, **4**, e59.
35. Lalor, P. A., Webby, R. J., Morrow, J., Rusalov, D., Kaslow, D. C., Rolland, A., and Smith, L. R. Plasmid DNA-based vaccines protect mice *and* ferrets against lethal challenge with A/Vietnam/1203/04 (H5N1) influenza virus. 2008. *J. Infect. Dis.*, **197**(12), 1643–1652.
36. Vong, S., Ly, S., Van Kerkhove, M. D., Achenbach, J., Holl, D., Buchy, P., Sorn, S., Seng, H., Uyeki, T. M., Sok, T., and Katz, J. M. Risk factors associated with subclinical human infection with avian influenza A (H5N1) virus: Cambodia, 2006. 2009. *J. Infect. Dis.*, **199**(12), 1744–1752.

Chapter 7

Diagnostics of Viral Respiratory Diseases

Tamar Amir, Guy Gubi, and Leslie Lobel
Department of Virology and Developmental Genetics, Faculty of Health Sciences, Ben-Gurion University of the Negev, Beer-Sheva, Israel
tammy.amir@gmail.com, guygubi@gmail.com, llobel@bgu.ac.il

7.1 Viral Respiratory Diseases

There are various viruses, both known and unknown, that target the respiratory tract, causing infection of the upper and lower tracts in infants and young children. Such victims will experience 6–10 infections per year, and approximately one-third will develop a lower respiratory infection in the first year of their lives. The majority of these infections do not require hospitalization. Although adolescents and adults suffer from two to four respiratory viral infections per year, these viruses are an important cause of severe, lower respiratory tract disease in the elderly and immunocompromised populations. All told, they cause significant morbidity, mortality, and economic loss. In addition, since these viral pathogens cause very similar clinical symptoms, usually defined as "flu-like," it is difficult to establish a clinical diagnostic that differentiates one virus from the next. Therefore, the realization that

Viral Diagnostics: Advances and Applications
Edited by Robert S. Marks, Leslie Lobel, and Amadou Alpha Sall

ISBN 978-981-4364-43-0 (Hardcover), 978-981-4364-44-7 (eBook)
www.panstanford.com

a new, unknown pathogen is the cause of a given outbreak typically occurs after all known pathogens have been ruled out.[1–3]

The development of sensitive, reliable diagnostic tools for detection of respiratory viruses is a necessity for many reasons—the right treatment depends heavily on accurate identification of the cause of the illness,[4] for example, this helps decide accurately whether or not to give the patient antibiotics, as unnecessary antibiotic treatments lead to the emergence of resistant bacterial strains. In some cases, an accurate diagnosis is cost effective and decreases morbidity and mortality.[3] The definition of the pathogen is also necessary to reduce the potential for nosocomial transmission to susceptible individuals. For control of viral infections, a vaccine can be decided upon once the cause of the illness has been identified.[5] Since a proportion of respiratory illnesses still cannot be attributed to any known pathogen, more unknown viruses will probably be identified in the near future. The development of methods to rapidly identify infected individuals is also essential for the monitoring of the spread of virus within a population. Surveillance is vital for monitoring the success of control strategies and should focus on identifying these agents in humans, as well as in susceptible animal populations.[2]

Despite the availability of sophisticated diagnostic methods, approximately 20–30% of colds remain without a proven viral cause, probably due in part to suboptimum methods used in the collection, transportation, and assay of clinical specimens, resulting in underdetection of the viruses known to exist. However, many colds could be caused also by infectious agents yet to be identified.[6] Once it is clear that there is indeed an unknown pathogen, a case definition as wide as possible is defined. Identification then includes classification as a member of a virus family based on size, morphology, genetic makeup (RNA or DNA genome), and genome organization, which facilitates an understanding of the biological properties. This aids in the search for potential treatments and preventive strategies.[2]

7.2 Respiratory Viruses

Intensive research into the cause of respiratory illnesses during the 1950s and 1960s led to the discovery of several viruses that cause illness: adenovirus, parainfluenza virus, rhinovirus, respiratory syncytial virus, enterovirus, and coronavirus, in addition to the influenza viruses identified earlier in history. As techniques of viral detection were limited to virus isolation during that era, in only approximately 25% of patients with respiratory infections was the cause identified. Improvements in viral detection techniques during the past two decades have substantially increased the rates of viral detection in clinical specimens.[6] Indeed, in a study conducted in 1997, in 69% of patients with the common cold, a virus was found to be the infecting agent through use of virus culture, serology, and polymerase chain reaction (PCR) for viral detection.[7]

Discoveries in the last decade (resulting from the application of new and more sensitive techniques, which enabled the detection of not only unknown viruses but also those circulating undetected in the human population for years)[2] led to the identification of a number of novel human viral respiratory pathogens. For example, since 2000 five new viruses were found to be causative agents of respiratory infections: metapneumovirus (MPV), severe acute respiratory syndrome coronavirus (SARS-CoV), avian influenza virus H5N1, CoVs NL63 and HKU1, and human bocavirus.[8] Today, up to 90% of the causal agents of acute respiratory infections are thought to be viral; over 200 viral agents have been implicated in these infections.[9] It is known that respiratory syncytial viruses, influenza viruses, human parainfluenza viruses (HPIVs), human metapneumovirus (hMPV), rhinoviruses, and adenoviruses are the causes of the vast majority of identified viral respiratory infections in both children and adults.[3,10]

Respiratory syncytial virus (RSV) is a major viral pathogen that causes respiratory tract illness during infancy, especially during the first year of life and in particular during winter months in the northern hemisphere, when it is more prevalent. In addition, epidemiological studies suggest that the clinical impact of RSV in certain adult populations may approach that of nonpandemic influenza. RSV was first identified in 1956 as a chimpanzee coryza agent since it was recovered from a chimpanzee with respiratory symptoms. It was then renamed respiratory syncytial virus to reflect the giant syncytia that are formed in infected tissue cultures. Human RSV is an enveloped nonsegmented, single-strand, negative-sense RNA virus belonging to the family *Paramyxoviridae*, classified within the genus *Pneumovirus*. RSV has three transmembrane glycoproteins: G, F, and SH; antibodies directed against the F or G glycoprotein can neutralize the virus in vitro and in vivo. Human RSV isolates can be classified into two major groups, A and B, each containing several distinct subgroups. There is some evidence that heterotypic immunity exists between the RSV groups. This classification is based upon antigenic and genomic differences found in several viral proteins, but especially G. Group A and B viruses generally circulate simultaneously within geographically confined epidemics, although group A viruses are more prevalent. Some studies have found that infection in infants is more severe with group A RSV than with group B.[11,12]

Influenza viruses have evolved in avian hosts, with aquatic birds being the major reservoir of influenza A. Some subtypes have been observed to infect and even become established in certain mammals, including humans. Influenza B appears to have diverged and evolved as a uniquely human pathogen. These viruses are members of the genus *Orthomyxovirus* in the family *Orthomyxoviridae.* Their genomes are segmented and consist of seven or eight, single-stranded, negative-sense RNA segments. There are three types of human influenza viruses—A, B, and C—that are distinguished by their antigenically distinct major internal proteins: the nucleoprotein and matrix protein. Influenza A and B have eight RNA segments, while influenza C has only seven. During colder months in temperate climates, and throughout the year in tropical and subtropical countries, epidemics of influenza occur. Influenza A and B cause major outbreaks and severe disease, while influenza C is associated with common cold-like illnesses, principally in children. Influenza A and B viruses have two major surface glycoprotein antigens: the hemagglutinin (HA) and the neuraminidase (NA). Whereas separate subtypes of influenza B have not been observed, and it does not appear that influenza B is involved in pandemics, influenza A consists of 16 antigenically distinct HAs and nine NA subtypes. The HA and NA proteins undergo "antigenic drift," a gradual, progressive antigenic variation resulting from the error-prone replication of the single-strand RNA that allows the viruses to escape the host immune system and cause further outbreaks and epidemics. In addition, influenza A viruses undergo "antigenic shift" caused by their ability to produce genetic hybrids, or reassortment viruses, by the mixing of genetic segments from two viruses in a combinatorial fashion when dual infection occurs, producing a new influenza A subtype. This results in no, or at best very few, people having immunity to the influenza antigens and promotes influenza pandemics.[13]

In 2001, a new human respiratory virus, hMPV, was first isolated in nasopharyngeal specimens from children with acute respiratory infection in the Netherlands. Since then, hMPV has been recognized as a common cause of respiratory infections, ranging from upper respiratory infections to severe bronchiolitis and pneumonia, in both children and adults worldwide. It is believed to be the cause of up to 20% of respiratory infections in children, with data from analyses of archived specimens suggesting that hMPV has been circulating in the human population for at least 25 years.[2,14,15] hMPV is an RNA virus, member of the genus *Metapneumovirus* of the subfamily *Pneumovirinae* of the family *Paramyxoviridae*. There are two serotypes belonging to this family, hMPV A and hMPV B, with each serotype divided into genetic sublineages 1 and 2 and then further divided into many strains. hMPV infections occur throughout the year, although its main clinical impact is during the winter months in countries with moderate climates.

During outbreaks, co-circulation of hMPV genogroups and subgenogroups has been reported.[15] HPIVs are the second major cause of lower respiratory tract illness in infants and young children. Additionally, HPIV infection causes serious lower respiratory tract illness among immunocompromised and institutionalized elderly individuals. In healthy older children and adults, repeat infections are typically less severe. There are four recognized serotypes of HPIV: 1, 2, 3, and 4. HPIV3 is most commonly associated with serious, lower respiratory tract illness, and HPIV4 is of no real medical importance since it is rarely associated with serious illness.[16] HPIVs belong to the family *Paramyxoviridae*, classified within the genus *Paramyxoviruses* together with the mumps virus.[11]

Rhinoviruses are the most common respiratory tract viruses in older children and adults, causing about two-thirds of common colds and asthma exacerbations. Rhinoviruses belong to the family of picornaviruses. They are small, nonenveloped, icosahedral particles, consisting of a single-strand RNA genome surrounded by viral capsid proteins.[17] A birth cohort study found that rhinoviruses were the predominant viruses detected in both acute disease as well as control specimens; the latter may indicate the pervasive nature of these viruses in the community. Early childhood is a time of particular susceptibility for acute respiratory infections, and they are responsible for 50% of primary care physician consultations among preschool children.[18] More than 100 different serotypes of rhinoviruses have been identified, the relative prevalence of which seems to vary between different geographical areas and over the course of time. Although rhinoviruses can be detected throughout the year, the incidence of infections peak during autumn, with a subsequent smaller outbreak in the spring.[6]

Adenoviruses are nonenveloped viruses consisting of 10 structural proteins that build their icosahedral capsid. Adenoviruses have a single, linear, double- stranded DNA genome. There are at least 51 serotypes of the human adenoviruses, which are conveniently divided into six subgroups (subgenera A to F) based on DNA homology, hemagglutination grouping, and differential ability to transform murine primary cells.[19,20]

Although there is a winter time peak, illness occurs throughout the year.[21] Adenovirus infections can have many clinical presentations in the immunocompetent individual, and many infections are in fact subclinical. However, over half of adenovirus serotypes can cause clinical disease associated with one or multiple organs, of which the most common are the respiratory, gastrointestinal, and genitourinary tracts; the liver; and the eye. It is estimated that adenoviruses cause about 5% of acute respiratory illnesses in children under the age of five.[20]

7.3 Diagnostic Techniques

Tissue culture used to be considered the "gold standard" for diagnostics of viral respiratory viruses but has been replaced during the past decades by molecular methods. Serology in these illnesses is usually not helpful in acute diagnosis since 10–30% of patients with documented respiratory virus infections are serologically negative. Significant advancements in rapid diagnostic methods have been developed during the past 30 years, with more on the horizon. The oldest assays are antigen based and include indirect/direct immunofluorescent and enzyme immuno-, optical immuno-, and neuraminidase-activity assays. These methods are still widely used and offer the advantages of being inexpensive and simple to perform. However, they suffer from a lack of sensitivity and specificity, especially during periods of low prevalence or in special populations such as the immunocompromised, chronically ill, or elderly.[3]

New diagnostic tools incorporate PCR and microarrays.

For the isolation and detection of the causative virus, tissue culture has a long and proven track record. Typically, cells are incubated with a nasal swab or other specimen and observed for a cytopathic effect as well as with immunofluorescent assays using specific monoclonal antibodies for the different viruses.[22] The advantage of the conventional tissue culture method is that, once isolated, the stocked isolates are available at any time for further etiological and epidemiological research of acute respiratory infections in children.[9] With respect to the influenza viruses, tissue culture enables the recovery of novel or highly divergent strains missed by other tests and in addition provides an isolate for subsequent characterization of the virus to detect new strains.[23] One of the drawbacks of this technique is that not all viruses can be cultured, especially true in the case of new, unknown viruses, for which a protocol has not yet been established. The isolation of some viruses requires laborious procedures that cannot always be performed in small, underequipped laboratories. In some cases, culture and subsequent identification methods are limited in sensitivity, specificity, or both.[4] Thus when needed, the test is heavily dependent on proper isolation, transport, and storage conditions to preserve the infectivity of the virus.[24] Additionally, the biggest drawback of this technique is that it often requires the use of several cell lines and time, up to several days, for the correct identification of the viral pathogen. Consequently, the information is not provided rapidly enough to affect initial treatment.[4,22,24]

A specific tissue culture assay was developed in the mid-1980s, which consists of a 96-well tissue culture plate termed HHVM for the four cell lines it contains; the human embryonic lung fibroblast (HEF), human laryngeal carcinoma (HEp-2), African green monkey kidney (Vero), and Madin–Darby Canine Kidney (MDCK). It is useful for the isolation and identification of influenza A H3N2, influenza A H1N1, influenza B, RSV, HPIVs 1–3, mumps virus, adenovirus 1–6, poliovirus 1–3, Coxsackie virus A16 and B1–5, echovirus 3,6,7,9,11,21,22, herpes simplex virus, and cytomegalovirus. Between 1988–1998 influenza C, rhinovirus, adenovirus 7, Coxsackie virus A7,9, echovirus 15,18,25,30, and enterovirus 71 were also isolated using this system. However, it demonstrated low sensitivity to Coxsackie virus A, except for Coxsackie A16, and insufficient sensitivity to enteroviruses. As a result, in 2001 the HHVMRG plate was created by adding two cell lines, rhabdomyosarcoma (RD-18S) and green monkey kidney (GMK), to allow for more efficient isolation of enteroviruses. Subsequently, it was discovered that hMPV could be replicated in Vero E6 cells, which led to the replacement of the regular Vero cell line in 2004 with Vero E6, creating the HHVe6MRG plate for inclusion of hMPV in the screening process.[9]

R-Mix is a combination of human lung carcinoma cells (A-549) and mink lung cells (Mv1Lu) co-cultivated in a balanced single monolayer. This combination was reported to exhibit sensitivity for respiratory virus detection comparable with other viral detection methods. R-Mix was developed as a way to overcome the need to grow multiple cell lines. These cells are available on the North American continent as a fresh culture or anywhere in the world in a cryopreserved state from Diagnostic HYBRID (DHI, Athens, Ohio).[22]

The centrifugation-enhanced shell vial culture assay reduces the turnaround time and consequently is becoming one of the most frequent techniques used in clinical virology laboratories. This assay has been shown to have a good correlation with traditional tube culture for the detection of the main respiratory viruses. Navarro-Mari simplified this assay by performing a simultaneous culture of three cell lines—HEp-2, LLC-MK2, and MDCK cells—in the same vial, thus permitting growth of the main respiratory viruses. For final identification, the cells are observed for cytopathic effect and are also stained for indirect immunofluorescence with a pool of monoclonal antibodies directed against the seven respiratory viruses: RSV, influenza A and B, HPIV 1, 2, and 3, and adenovirus (Respiratory Viral Screen IFA; Light Diagnostics, Temecula, CA), followed by monoclonal antibodies against each virus (IMAGEN; Dako Diagnostic Ltd., Denmark). This new assay has some remarkable advances—it shortens the time needed to obtain results compared to the time required with traditional tissue culture and allows for diagnostics of seven common respiratory viruses in a single vial. The use of monoclonal antibody pools for the identification of viral infections greatly simplifies the assay. This technique can further be improved by using monoclonal antibody pools labeled with different fluorochromes, which would enable final viral identification in a single step.[25]

7.3.1 *Immunoassays*

Immunoassays include agglutination tests, enzyme-linked immunoassays, and direct immunofluorescent antibody assays. These assays have revolutionized the diagnosis of infectious diseases, especially since the use of monoclonal antibodies in these systems. They can also be performed in small as well as big laboratories due to their simplicity and reasonable costs, allowing the testing of more viruses than ever before within hours of receipt of the specimen. However, most laboratories use them only for quick diagnostics and still prefer tissue culture techniques as the gold standard for accurate results due to the disadvantages of immunoassays, which includes their poor sensitivity and low negative predictive value. Tissue culture methods, however, rest on accurate interpretation of the results, which are in some cases subjective and require technical skill, and the results may be affected by specimen quality (i.e., if the specimen contains infected cells). Immunoassays, on the other hand, may fail to detect emerging viruses, or viruses that have altered amino acid sequences on their outer proteins, due to a lack of monoclonal antibody recognition for those viruses.[4,5,8,24]

One particularly useful immunofluorescence is the SimulFluor Respiratory Screen (Chemicon International, Temecula, CA). It is a new respiratory virus–screening reagent that permits direct immunofluorescent assay detection of seven viruses (RSV, influenza A and B, HPIV 1, 2, and 3, and adenovirus) in one cell spot.

This system utilizes a reddish gold (rhodamine) label for RSV and an apple green (fluorescein) label for the rest of the viruses. Therefore, if cells are visually green-stained, additional slides must be stained to determine which of the last group of viruses is in the sample.[26]

7.3.2 *Molecular Techniques*

Molecular techniques have increased the speed, sensitivity, and specificity by which respiratory pathogens can be detected, and enable laboratories to identify organisms that grow slowly or do not grow at all in conventional viral cultures.[27] The published sensitivities and specificities of molecular assays are extremely high, as compared to tissue culture or antigen assays, and usually very near 100%. Nucleic acid amplification strategies, such as PCR, allow for a wide range of applications, including diagnosis of the infecting agent from a clinical specimen, strain typing, and epidemiologic tracking of the infection. These tools enable extreme specificity, high sensitivity, and sufficient rapidity of diagnostics.[4] However, one must always bear in mind that there is a constant risk of contamination when molecular amplification methods are being used.[3]

PCR has several advantages over antigen detection assays: it is more specific and sensitive,[27] and results are not dependent on infectious virus or viable cells since only the presence of viral nucleic acid is required. As a result, specimen quality is not a confounding issue as it can be with tissue culture or antigen detection assays.[24] Disadvantages of this system are that in addition to the concern of contamination, PCR requires specialized equipment and personnel and funds that are not always available to small laboratories. Paradoxically, improvement in sensitivity may pose interpretative diagnostic problems and lessen the clinical relevance of the information, for instance, in the case of viruses capable of establishing latent or persistent infection in the host.[4] In the event of a mutation in the primer-binding site, there is a strong possibility that the pathogen will go undetected.[28] PCR products can be labeled by incorporation of fluorescence-labeled primers, enabling detection and sizing (or typing) with greater sensitivity, comparable to that of nested PCR but without the added risk of DNA contamination associated with nested PCR procedures. GeneScan software analysis makes it possible to size bands with single-base-pair accuracy, minimizing the chance of detecting an incorrect but similarly sized product.[16]

Real-time PCR has also become commonplace in viral diagnostics, and it rests on a straightforward principle. Following reverse transcription (when needed) the amplification of the PCR products is monitored and reported in real time using fluorescent reporter molecules. The higher the starting copy number of the nucleic acid target, the sooner a significant increase observed in fluorescence, which corresponds with the PCR product.[1] Real-time PCR is not necessarily more sensitive than regular PCR, but it does have advantages over regular PCR. First and foremost, real-time PCR is a quantitative rather than a qualitative system and provides more information than the causing agent alone.[1]

This is especially important in some instances, as in the case of viruses that cause persistent or latent infection.[4]

Since the system uses fluorescent reporter molecules to monitor the amplification, there is no need for post-PCR processing; thus it has fewer steps and faster results with less hands on time.[1,3] Disadvantages of this system include higher startup costs, higher reagent costs, less stable commercial reagents, and a multiplex of only three to four analytes that can be processed together.[3]

PCR methods for viral detection have rapidly been multiplexed for identification of multiple viruses simultaneously. Instead of having a single reverse transcriptase polymerase chain reaction (RT-PCR) assay detecting one respiratory virus at a time, development of RT-PCR assays that detect multiple respiratory pathogens concurrently has been a recent trend. These assays can be performed in several different ways:

(1) Using primer pairs to different respiratory viruses in separate reactions that share single standard amplification conditions—in this case a few RT-PCR reactions are run at the same time.
(2) Using broadly reactive primers to highly conserved genes that encompass different virus genera (e.g., RSV and HPIV).

(3) Combining multiple primer sets into single "multiplexed" amplification reactions.[10] These multiplex PCR formats are a significant improvement over conventional PCR protocols and are primarily based on incorporation of multiple primer sets that can amplify RNA or DNA from several viruses simultaneously in a single reaction.[24]

This enables more reliable detection of the causative agent and thereby improves clinical management. These assays can also detect co-infection if such a case is presented.[27,29] The multiplex RT-PCR assay can also be coupled with detection by microplate hybridization for confirmation of results, subtype differentiation, and increased sensitivity and specificity.[5]

Multiplex PCR reactions are in general more economical as multiple pathogens can be detected in a single assay without a proportional increase in reagent costs.[27,29] In addition a low volume of specimen is required for testing in this method.[5] There are some related disadvantages: their detection sensitivities are often lower than those of equivalent monoplex PCR assays. A way to overcome this problem is to perform nested PCR, which may be associated with a higher risk of cross contamination. The presence of several pairs of primers in PCR increases the probabilities of mistaken pairing and nonspecific amplification, particularly in the formation of primer-dimers.[27] Multiplex assays to date are capable of detecting between 20 and 30 pathogens at a time, which typically includes most of the common ones. However, there are many more, less common pathogens not included in many multiplexed assays.[28]

The ligase chain reaction (LCR) has also been employed for viral diagnostics. It uses two probes that hybridize adjacent to each other on a target strand. The probes are then ligated and function as a template for further hybridizations, achieving an amplification of the original DNA. The high sensitivity and specificity of the system are achieved because a single base pair mismatch at the point of ligation inhibits the reaction. This is the assays' biggest problem, especially concerning RNA viruses that have a very high mutation rate. A further disadvantage of these assays is they are technically demanding and time consuming and therefore have not gained wide acceptance.[4]

7.3.3 *Multiplex PCR-Based Assays in Use Today*

Lam et al. have developed a multiplex assay that can simultaneously detect 21 different respiratory pathogens, including influenza A (subtypes H1, H3, and H5) and B; HPIV types 1, 2, 3, and 4; RSV; all serotypes of human adenoviruses; hMPV; human corona viruses 229E, OC43, and SARS corona virus; all serotypes of human enteroviruses; all serotypes of human rhinoviruses; and respiratory bacteria *Chlamydophila pneumoniae*, *Mycoplasma pneumoniae*, and *Legionella pneumophila*. In this assay the pathogens were arranged in five groups of four pathogens per group for multiplex nested PCR assays that included an RT step, where necessary, with the rationale for primer groupings as follows:

1. RNA and DNA pathogens were separated, that is, groups 1 to 4 for RNA and group 5 for DNA pathogens.
2. Pathogens of the same or similar family were grouped together so each member amplified within the same PCR could be easily differentiated.
3. The PCR product size was another factor affecting multiplex grouping, the size of the PCR products being amplified as suitable for visual differentiation on agarose gel.[27]

Syrmis et al. developed a combination of multiplex RT-PCR with an enzyme-linked amplicon hybridization assay (ELAHA) to detect amplification products using a colorimetric detection method. The ELAHA method can increase the sensitivity and specificity of PCR assays by detecting amplicon with a sequence specific biotinylated probe. This assay can detect simultaneously adenovirus, influenza A and B, RSV A and B, and HPIV 1, 2, and 3 and has some significant advantages such as improved sensitivity and specificity, improved result turnaround times (approximately five hours), and greater cost-effectiveness than that of the direct immunofluorescent assay or viral tissue culture.[24]

RT-PCR–enzyme hybridization assay (RT-PCR-EHA) is a commercially available multiplex RT-PCR technique (RT-PCR-EHA; Hexaplex; Prodesse Inc., Waukesha, WI) incorporating both the RT-PCR technology and microplate hybridization for confirmation of results. This combination increases the sensitivity of the assay. RT-PCR-EHA is capable of simultaneously detecting seven respiratory pathogens (HPIV 1, 2, and 3, influenza A and B, and RSV A and B) from one clinical specimen of various body fluids, including washes (nasal, throat, and tracheal), swabs (nasopharyngeal and throat), aspirates (tracheal, lung, and throat), lung biopsy specimens, and cerebrospinal fluid.[29]

A multiplex PCR assay, called the respiratory virus panel (RVP) test, was developed that can detect 20 different respiratory viruses. It employs a 96-well microtiter plate format and the Luminex 100 flow cell instrument with a multiplex PCR with 14 primer pairs, followed by a second multiplex reaction with 21 primer pairs. The RVP is capable of detecting and identifying 20 different viruses.[8] They include influenza A and B; HPIV 1, 2, 3, and 4; RSV; adenovirus; hMPV; rhinovirus; CoVs OC43 and 229E; SARS-CoV; avian influenza virus H5N1; and CoVs NL63 and HKU1. In addition to the specific detection of H1, H3, and H5 strains, RVP also detects all influenza A virus subtypes without identifying their specific types by employing a matrix gene bead. If only the matrix bead gives a signal, this indicates that a new influenza type has emerged into the human population and that there is a need for further genotyping of the isolate.

Osiowy has developed a new RT-PCR assay for the detection of three groups of respiratory viruses involved in lower respiratory tract infection and hospitalization (RSV, HPIV1 2 and 3, and adenoviruses). In a single-step multiplex RT-PCR, a five-primer set is used targeting highly conserved areas of each viral genome. The various virus types are differentiated by their unique amplicon sizes following separation of the PCR products on agarose gel. Subtype characterization of PCR products by hybridization with subtype-specific probes, restriction digestion, or sequencing is also possible.[5]

Echevarria et al. developed multiplex assays that both detect and differentiate directly from clinical specimens all three medically important HPIV serotypes. The primers used in this multiplex are designed to react with known temporally and geographically diverse HPIV isolates.[16]

Erdman et al. developed a RT-PCR assay panel based on automated fluorescent capillary electrophoresis and GeneScan software analysis (Applied Biosystems, Foster City, CA) for the detection of

RSV, HPIV 1, 2, and 3, and influenza A and B viruses from clinical specimens. This GeneScan RT-PCR assay offers a highly sensitive and specific method for detection of these six respiratory viruses in young children.[10]

DNA micro arrays offer unprecedented opportunities for multiplexing. However, due to their drawbacks that include problems with sensitivity, throughput, validation, and expense, they are not widely implemented in clinical microbiology laboratories. Sample nucleic acids are typically processed for pathogen-specific multiplex, which is more sensitive but in the case of a mismatch does not allow for amplification. Randomly primed amplification, which enables detection of a wider variety of pathogens, reduces the assay's sensitivity, however, prior to the hybridization. The majority of these respiratory arrays rely on reporter molecules directly incorporated into primers or amplification products.[28]

The GreeneChipResp system allows the sensitive detection of a wide variety of respiratory viruses and subtype identification of all influenza A viruses, HA, and NA. This assay uses random priming, enabling unbiased amplification of all templates in a sample. In addition, agent-specific primers are employed that can enrich sequences present in a low copy number or that may fail to amplify efficiently with the random primers. GreeneChipResp also uses an indirect dendrimer-labeling method whereby the signal is enhanced by the presence of 300 fluorescent reporter molecules in each probe-target hybridization.[28]

There are many assays that employ multiplex PCR with different methods for PCR product detection (e.g., enzyme hybridization, microarray, capillary electrophoresis), and more are continuously being developed. All these assays can be automated in a single machine, chip, or plastic wafer and performed in approximately one hour. They may also have open and changeable formats, making them point-of-care diagnostics.[3] If a general practitioner or emergency physician is able to quickly identify the respiratory tract infection agent, the best decision can be made and available therapy can be administered, if required. In some cases, appropriate isolation precautions can be taken in time.[30]

7.3.4 *Point-of-Care Tests*

Point-of-care tests usually produce visual results within several minutes of adding an extracted specimen, for example, on an immunochromatographic strip or membrane enzyme-linked immunoadsorbent assay. These are simple tests to perform, though, unfortunately, not available for all respiratory viruses and are often of suboptimal sensitivity. Specimen quality is a major determinant of these tests' performance. Ideally, when using rapid antigen tests, specimens for cell culture and immunofluorescence or nucleic acid testing should be submitted in parallel to improve sensitivity and to obtain viral isolates for further typing.[23,26]

Since anti-influenza medications have greater efficacy if started within 48 hours of symptom onset, rapid point-of-care diagnostics for influenza would facilitate more judicious use of these medications. With the current concern about the threat of pandemic influenza, the availability of rapid and accurate tests for influenza is particularly desirable. This could allow for the appropriate use of infection control methods, which can reduce substantially the spread of a pandemic strain. But there are some limitations on point-of-care diagnostic tools today—their sensitivity for influenza is lower than that of other tests. They also do not give information on the influenza subtype, and it is not yet clear whether many tests can adequately detect H5N1 strains. Point-of-care diagnostics generally have a higher sensitivity in children as opposed to in adults and for nasopharyngeal aspirates compared to throat swabs, which are more difficult and unpleasant to obtain. These tests have a shelf life of about 12–18 months, so cost is also an issue.[30]

Some of the more facile and rapid tests available today for influenza include the QuickVue Influenza A + B test (Quidel Corporation, San Diego, CA). This is an enzyme immunoassay that uses monoclonal antibodies, impregnated in a test strip, against various influenza A and B antigens. Following the addition of a reagent solution to the provided extraction tube, the nasal swab, fluid from nasal washings or nasopharyngeal aspirates are added to the tube. The test strip is then inserted into the tube and left for 10 minutes. Influenza virus antigens react with reagents in the test strip, with positive results signified by a pink or red line. The sensitivity and specificity depends heavily on the timing of the test relative to the stage of the illness and decreases if the test is delayed. This test is able to reliably distinguish between influenza A and B viruses.[30]

The BinaxNOW Influenza A & B test (Binax, Scarborough, ME) is an immunochromatographic test that uses monoclonal antibodies against the nucleoproteins of both influenza A and B, which are bound to a membrane. It can be used with either nasopharyngeal aspirates or a nasal swab. The clinical specimen, 100 μL, is transferred onto the white sample pad of the test card and read after 15 minutes. A blue control line at the bottom of the card window should turn pink or purple for the test to be valid. Tests that are positive for influenza A and B will have a pink or purple line in the middle or upper third of the card window. This is the simplest of the rapid tests and, like the QuickVue test, can distinguish between influenza A and B viruses.[30]

The ZstatFlu test and the ZstatFlu-II test (ZymeTx Inc, Oklahoma City, OK) are assays that detect viral neuraminidase activity from influenza A and B. A throat swab is the recommended specimen for these tests. In the ZstatFlu test a substrate for this enzyme is coupled to a chromogen, which, when cleaved by viral neuraminidase, produces the color blue. This test takes between 20 and 60 minutes. The control spot must turn blue for a valid result. A positive test is any shade of blue in the bottom of the collection device. The ZstatFlu-II test is a chemiluminescent rapid diagnostic test. When a specimen containing influenza virus is added to the device, the viral enzyme hydrolyzes a synthetic substrate, leading to release of reporter groups that can generate chemiluminescence. The test takes 30 minutes; a positive reaction releases light, which is captured on instant high speed detector film within the imaging device. This shows up as a white "plus" sign on the black background of the film. The test cannot distinguish between influenza A and B, but there is no cross reactivity with the neuraminidase of parainfluenza viruses. One small advantage of the test is that the result can be stored as a permanent record. Compared with viral culture, immunofluorescence, and PCR, the above tests have only moderate sensitivity.[30]

Point-of-care diagnostics for RSV are intended for use in neonatal or pediatric populations and hence is less accurate when used in adults. Tests available for RSV include BinaxNOW RSV (Binax, Scarborough, ME), the simplest test; SAS RSV (SA Scientific, San Antonio, TX); Clearview RSV (Unipath Limited, Bedford, UK); and QuickVue RSV (Quidel Corporation, San Diego, CA). All of these tests have moderate sensitivities compared to other diagnostic tools.[30]

A large number of patients with lower respiratory infections, probably due to a viral agent, are tested each year and still have negative test results. Reasons for this include undiscovered agents (including new viruses that emerged into the human population and old circulating viruses waiting to be discovered), a low viral copy number in the clinical sample, or, rarely, the presence of viral inhibitors in the sample. The most important cause of a false-negative test is poor specimen handling or collection.[3] Most viral diagnostic labs around the world use one or more of the techniques described before for the accurate identification of the causative agent. Usually, quick tests are accompanied with virus culture for confirmation as well as for further epidemiological studies. Point-of-care diagnostics available today and ones that are under development will make a significant improvement in patients' care, since they will aid in making the right decisions on therapy. More importantly, with the threat of new influenza pandemics upon us, and with the potential use of virulent influenza strains as bioweapons, rapid point-of-care diagnostics for influenza will serve as a formidable weapon in the arsenal to contain and control an outbreak. In addition, the development of new tools that provide greater speed, sensitivity, and specificity is essential so that highly efficient screening systems can be effectively deployed in the event of a pandemic or outbreak due to nefarious causes. Finally, all of the techniques outlined before rely on specific identification of an influenza virus strain. We also need in our arsenal tests that have heterosubtypic reactivity so that new drifted and shifted influenza A strains can be detected before they cause epidemics or pandemics.

References

1. Bustin, S. A., and Mueller, R. Real-time reverse transcription PCR (qRT-PCR) and its potential use in clinical diagnosis. 2005. *Clin Sci (Lond)*, **109**(4), 365–379.
2. Gillim-Ross, L., and Subbarao, K. Emerging respiratory viruses: challenges and vaccine strategies. 2006. *Clin Microbiol Rev*, **19**(4), 614–636.
3. Henrickson, K.J. Cost-effective use of rapid diagnostic techniques in the treatment and prevention of viral respiratory infections. 2005. *Pediatr Ann*, **34**(1), 24–31.
4. Garcia-de-Lomas, J., and Navarro, D. New directions in diagnostics. 1997. *Pediatr Infect Dis J*, **16**(3 Suppl), S43–S48.
5. Osiowy, C. Direct detection of respiratory syncytial virus, parainfluenza virus, and adenovirus in clinical respiratory specimens by a multiplex reverse transcription-PCR assay. 1998. *J Clin Microbiol*, **36**(11), 3149–3154.
6. Heikkinen, T., and Jarvinen, A. The common cold. 2003. *Lancet*, **361**(9351), 51–59.
7. Mäkelä, M. J., Puhakka, T., Ruuskanen, O., Leinonen, M., Saikku, P., Kimpimäki, M., Blomqvist, S., Hyypiä, T., and Arstila, P. Viruses and bacteria in the etiology of the common cold. 1998. *J Clin Microbiol*, **36**(2), 539–542.
8. Mahony, J., Chong, S., Merante, F., Yaghoubian, S., Sinha, T., Lisle, C., and Janeczko, R. Development of a respiratory virus panel test for detection of twenty human respiratory viruses by use of multiplex PCR and a fluid microbead-based assay. 2007. *J Clin Microbiol*, **45**(9), 2965–2970.
9. Mizuta, K., Abiko, C., Aoki, Y., Suto, A., Hoshina, H., Itagaki, T., Katsushima, N., Matsuzaki, Y., Hongo, S., Noda, M., Kimura, H., and Ootani, K. Analysis of monthly isolation of respiratory viruses from children by cell culture using a microplate method: a two-year study from 2004 to 2005 in Yamagata, Japan. 2008. *Jpn J Infect Dis*, **61**(3), 196–201.
10. Erdman, D. D., Weinberg, G. A., Edwards, K. M., Walker, F. J., Anderson, B. C., Winter, J., Gonzlez, M., and Anderson, L. J. GeneScan reverse transcription-PCR assay for detection of six common respiratory viruses in young children hospitalized with acute respiratory illness. 2003. *J Clin Microbiol*, **41**(9), 4298–4303.
11. Falsey, A. R., and Walsh, E. E. Respiratory syncytial virus infection in adults. 2000. *Clin Microbiol Rev*, **13**(3), 371–384.

12. Wright, P. F., and Cutts, F. T. Generic protocol to examine the incidence of lower respiratory infection due to respiratory syncytial virus in children less than five years of age. 2000. Field test version. World Health Organization.
13. Hampson, A. W., and Mackenzie, J. S. The influenza viruses. 2006. *Med J Aust,* **185**(10 Suppl), S39–S43.
14. Williams, J. V., Harris, P. A., Tollefson, S. J., Halburnt-Rush, L. L., Pingsterhaus, J. M., Edwards, K. M., Wright, P. F., and Crowe, J. E. Jr. Human metapneumovirus and lower respiratory tract disease in otherwise healthy infants and children. 2004. *N Engl J Med,* **350**(5), 443–450.
15. Abiko, C., Mizuta, K., Itagaki, T., Katsushima, N., Ito, S., Matsuzaki, Y., Okamoto, M., Nishimura, H., Aoki, Y., Murata, T., Hoshina, H., Hongo, S., and Ootani, K. Outbreak of human metapneumovirus detected by use of the Vero E6 cell line in isolates collected in Yamagata, Japan, in 2004 and 2005. 2007. *J Clin Microbiol,* **45**(6), 1912–1919.
16. Echevarría, J. E., Erdman, D. D., Swierkosz, E. M., Holloway, B. P., and Anderson, L. J. Simultaneous detection and identification of human parainfluenza viruses 1, 2, and 3 from clinical samples by multiplex PCR. 1998. *J Clin Microbiol,* **36**(5), 1388–1391.
17. Hyypiä, T., Puhakka, T., Ruuskanen, O., Mäkelä, M., Arola, A., and Arstila, P. Molecular diagnosis of human rhinovirus infections: comparison with virus isolation. 1998. *J Clin Microbiol,* **36**(7), 2081–2083.
18. Kusel, M. M., de Klerk, N. H., Holt, P. G., Kebadze, T., Johnston, S. L., and Sly, P. D. Role of respiratory viruses in acute upper and lower respiratory tract illness in the first year of life: a birth cohort study. 2006. *Pediatr Infect Dis J,* **25**(8), 680–686.
19. Hierholzer, J. C. Adenoviruses in the immunocompromised host. 1992. *Clin Microbiol Rev,* **5**(3), 262–274.
20. Kinchington, P. R., Romanowski, E. G., and Jerold, G. Y. Prospects for adenovirus antivirals. 2005. *J Antimicrob Chemother,* **55**(4), 424–429.
21. Rocholl, C., Gerber, K., Daly, J., Pavia, A. T., and Byington, C. L. Adenoviral infections in children: the impact of rapid diagnosis. 2004. *Pediatrics,* **113**(1 Pt 1), e51–e56.
22. Kim, J. S., Kim, S. H., Bae, S. Y., Lim, C. S., Kim, Y. K., Lee, K. N., and Lee, C. K. Enhanced detection of respiratory viruses using cryopreserved R-Mix ReadyCells. 2008. *J Clin Virol,* **42**(3), 264–267.

23. Dwyer, D. E., Smith, D. W., Catton, M. G., and Barr, I. G. Laboratory diagnosis of human seasonal and pandemic influenza virus infection. 2006. *Med J Aust,* **185**(10 Suppl), S48–S53.

24. Syrmis, M. W., Whiley, D. M., Thomas, M., Mackay, I. M., Williamson, J., Siebert, D. J., Nissen, M. D., and Sloots, T. P. A sensitive, specific, and cost-effective multiplex reverse transcriptase-PCR assay for the detection of seven common respiratory viruses in respiratory samples. 2004. *J Mol Diagn,* **6**(2), 125–131.

25. Navarro-Marí, J. M., Sanbonmatsu-Gámez, S., Pérez-Ruiz, M., and De La Rosa-Fraile, M. Rapid detection of respiratory viruses by shell vial assay using simultaneous culture of HEp-2, LLC-MK2, and MDCK cells in a single vial. 1999. *J Clin Microbiol,* **37**(7), 2346–2347.

26. Landry, M. L., and Ferguson, D. SimulFluor respiratory screen for rapid detection of multiple respiratory viruses in clinical specimens by immunofluorescence staining. 2000. *J Clin Microbiol,* **38**(2), 708–711.

27. Lam, W. Y., Yeung, A. C., Tang, J. W., Ip, M., Chan, E. W., Hui, M., and Chan, P. K. Rapid multiplex nested PCR for detection of respiratory viruses. 2007. *J Clin Microbiol,* **45**(11), 3631–3640.

28. Quan, P. L., Palacios, G., Jabado, O. J., Conlan, S., Hirschberg, D. L., Pozo, F., Jack, P. J., Cisterna, D., Renwick, N., Hui, J., Drysdale, A., Amos-Ritchie, R., Baumeister, E., Savy, V., Lager, K. M., Richt, J. A., Boyle, D. B., Garca-Sastre, A., Casas, I., Perez-Breña, P., Briese, T., and Lipkin, W. I. Detection of respiratory viruses and subtype identification of influenza A viruses by GreeneChipResp oligonucleotide microarray. 2007. *J Clin Microbiol,* **45**(8), 2359–2364.

29. Liolios, L., Jenney, A., Spelman, D., Kotsimbos, T., Catton, M., and Wesselingh, S. Comparison of a multiplex reverse transcription-PCR-enzyme hybridization assay with conventional viral culture and immunofluorescence techniques for the detection of seven viral respiratory pathogens. 2001. *J Clin Microbiol,* **39**(8), 2779–2783.

30. Charles, P. G., and Grayson, M. L. Point-of-care tests for lower respiratory infections tract. 2007. *Med J Aust,* **187**(1), 36–39.

Chapter 8

Reverse Genetics as a Tool for Detection of Negative-Stranded RNA Viruses

Pavel Naumenko,[a] Leslie Lobel,[b] and Robert S. Marks[a,c]

[a] *Department of Biotechnology Engineering, Ben-Gurion University of the Negev, Beer-Sheva, Israel*
[b] *Department of Virology, Ben-Gurion University of the Negev, Beer-Sheva, Israel*
[c] *The National Institute of Biotechnology in the Negev, Ben-Gurion University of the Negev, Beer-Sheva, Israel*
paveliko@gmail.com, llobel@bgu.ac.il, rsmarks@bgu.ac.il

8.1 Dangerous Viruses Easily Accessible

Reverse genetics (RG) was developed about 30 years ago and has been used mainly for exploring the molecular biology and pathogenesis of RNA viruses. Using RG, the need for working with infectious viral particles is obviated, thus making it possible to explore dangerous viruses in almost any laboratory. Recently, RG-based detection systems for several negative-stranded RNA (NS RNA) viruses were developed. The novel detection techniques rely on short, extremely well-conserved sequences at the ends of the viral genome, so this detection method is unaffected by the diversity of the strains within a given viral family. Due to the use of the green

Viral Diagnostics: Advances and Applications
Edited by Robert S. Marks, Leslie Lobel, and Amadou Alpha Sall

ISBN 978-981-4364-43-0 (Hardcover), 978-981-4364-44-7 (eBook)
www.panstanford.com

fluorescent protein or Luciferase reporter genes as a signal of viral replication in vitro, these detection systems can serve as a platform technology for high-throughput screening of antiviral agents and neutralizing antibodies for NS RNA viruses.

8.2 Negative-Stranded RNA Viruses

NS RNA viruses are classified into eight families.[1] Five of these (*Bornaviridae, Rhabdoviridae, Filoviridae, Paramyxoviridae*, and an unassigned family represented by Deltavirus) have single-RNA-molecule genomes, while the remaining three families (*Orthomyxoviridae, Bunyaviridae,* and *Arenaviridae*) are characterized by segmented genomes. The NS RNA viruses are a broad group of mammalian viruses that comprise several classic pathogens, including influenza, measles, mumps, and rabies, and many newly emergent and reemerging pathogens, such as Ebola, Hantaviruses, and Rift Valley fever virus (see Table 8.1). Most of these pathogens are capable of causing severe morbidity and significant mortality, being responsible for thousands of deaths every year. Some of the viruses are on the National Institute of Allergy and Infectious Diseases (NIAID) and Centers for Disease Control and Prevention (CDC) priority A list of potential bioterror pathogens.

8.2.1 *Genome Structure*

NS RNA viruses are divided into two major groups, nonsegmented negative stranded (NNS) and segmented negative stranded (SNS) (see Table 8.1). All the NNS and SNS RNA viruses have several features in common: by definition, the polarity of the genomic RNA of NS RNA viruses is opposite (negative sense) to mRNA (positive sense); all of them are enveloped viruses, and the majority replicate in the cell cytoplasm. Two exceptions to this rule are orthomyxo- and bornaviruses, which transcribe and replicate in the nucleus. The major difference between NNS and SNS RNA viruses is the structure of their genome. The genome of SNS RNA viruses consists of several segments, each segment usually encoding one protein,

Table 8.1 NS RNA viruses can be divided into two major groups. Five of the eight NS RNA virus families comprise a nonsegmented genome, while the other three families comprise a segmented genome

Genome	Family	Virus
Nonsegmented	*Bornaviridae*	Borno disease virus
	Rhabdoviridae	Rabies, vesiculovirus (VSV), bovine ephemeral
	Filoviridae	Ebola, Marburg
	Paramyxoviridae	Mumps, human parainfluenza, Newcastle disease virus, Nipah, measles, RSV
	Unclassified	Deltavirus
Segmented	*Orthomyxoviridae*	Influenza A, B, C
	Bunyaviridae	Bunyamwera, La Crosse virus, Rift Valley fever, CCHH Hantaviruses
	Arenaviridae	LCMV, Lassa virus, Junin virus (Argentine hemorrhagic fever)

Abbreviations: RSV, respiratory syncytial virus; CCHF, Crimean-Congo hemorrhagic fever; LCMV, lymphocytic choriomeningitis virus.

while NNS RNA viruses contain one segment that has several open reading frames (ORFs) encoding all of the viral proteins (see Fig. 8.1). Unlike the genomic RNA of positive-stranded RNA viruses, that is recognized by the infected cells, the genomic RNA of NS RNA viruses lacks the 5′ cap and 3′ poly(A) structures. Thus it is not recognized by the host cells and by itself is thereby not infectious.

Initiation of virus replication and transcription requires the delivery of viral RNA (vRNA)-dependent RNA polymerase (RdRp) by the infectious virus particles. The viral RdRp is a protein complex that uses the genomic vRNA as a template for synthesis of antigenomic vRNA (for further replication) and of mRNA (for viral protein expression by hosts ribosomes). During the synthesis of mRNA, the RdRp adds the cap structure at the 5′ termini of the transcript and the poly(A) at the 3′ termini as well. The genomic and antigenomic RNA of the viruses have to be assembled with the nucleocapsid protein (NP) for RdRp recognition (the complex of vRNA, NP, and RdRp is called ribonucleoprotein [RNP]). Recognition of the vRNA by the polymerase complex is accomplished by short untranslated regions (UTRs) at the 3′ and 5′ genome ends. Since

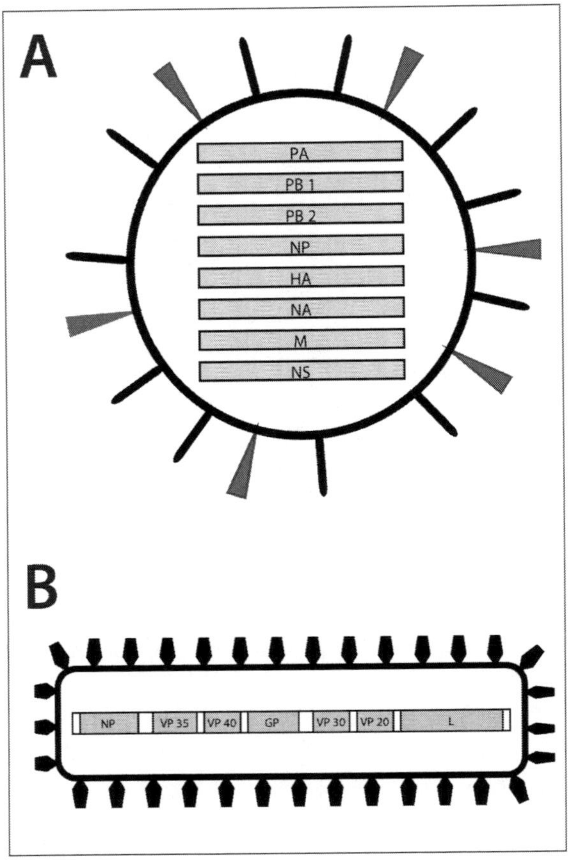

Figure 8.1 Schematic diagram of influenza (A) as an example of a segmented group and Ebola (B) as an example of a nonsegmented group of NS RNA viruses. (A) Influenza virus possesses eight genomic segments. Each one of them encodes one protein only (except the M and NS genes, which encode two proteins each). Every segment is replicated and transcribed separately. (B) On the contrary, the genome of Ebola virus contains seven genes sequentially arranged in one segment.

the essential sequences (*cis*-acting elements) that control virus replication, encapsidation, packaging, and mRNA transcription are embedded in UTRs, these regions are extremely well conserved through the evolution of influenza virus.[1,2]

8.3 Reverse Genetics System Development

By definition, RG is an approach for discovering the function of a gene. Simply put, while forward genetics seek to find the genetic basis of a phenotype, RG seeks to find the possible phenotypes that may derive from a specific genetic sequence. The simplest RG system deals with one viral gene at a time; using the full RG system makes it possible to generate infectious virus completely from cDNA. The first viruses to be generated by the use of RG methods were QB phage in 1978[3] and poliovirus in 1981.[4] As referred to previously, the genomic RNA of positive-stranded RNA viruses (poliovirus belongs to this group) is recognized by the infected cells; thus transfection of the full-length RNA of these viruses into eukaryotic cells results in viral protein expression, leading to virus replication. However, the situation for NS RNA viruses is much more complicated, since by itself, the genomic RNA is not infectious. To start replication and transcription of the vRNA, the genome has to be assembled with the nucleoprotein (NP), and the viral RdRp has to be delivered. The need for providing these additional proteins was how the full RG system for NS RNA viruses only came to be developed a decade after development of such a system for positive-stranded RNA viruses.[1,5,6] In 1994, the recombinant rabies virus was generated,[7] demonstrating for the first time the feasibility of producing an NS RNA virus entirely from cloned cDNA. In short, plasmids encoding the viral NP and the RdRp proteins, under the control of T7 RNA polymerase (T7Rp) promoter, were transfected into cells that were previously infected with recombinant vaccinia virus expressing the T7Rp (see Fig. 8.2A). An additional plasmid encoding the full-length viral cDNA under the control of a T7Rp promoter was also transfected. After synthesis of vRNA by the T7Rp, and transcription of viral RdRp and NP, the vRNA (assembled to NP) is recognized by the viral RdRp. Finally, the viral RdRp synthesizes mRNA, which is then translated to viral proteins, resulting in a generation of the infectious virus. The rescue of other NNS RNA viruses followed.[8–10]

Another breakthrough in RG development occurred in 1996, when, for the first time, the SNS RNA virus was generated completely from cDNA.[11] This improvement allowed the application of the full

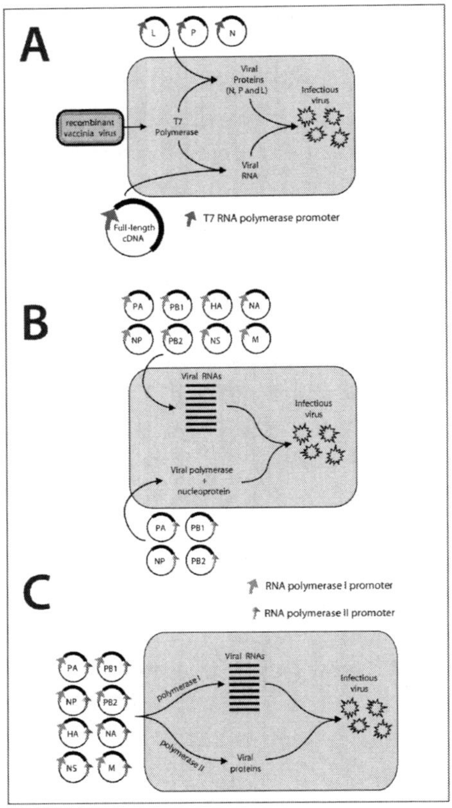

Figure 8.2 Schematic diagram of NNS (A) and SNS (B, C) RNA virus generation. (A) Cells are infected with recombinant vaccinia virus (encoding T7Rp) and cotransfected with protein expression plasmids—for the NPs and **RdRp** (L)—and with a plasmid containing full-length viral cDNA. All the plasmids are under control of the T7Rp promoter. vRNA is synthesized and assembled with nucleoprotein (N) and phosphoprotein (P); this complex is recognized by the viral polymerase (L) which initiates the virus replication cycle. (B) Cells are cotransfected with plasmids that encode all eight influenza segments under the control of RNA pol I promoter. Cellular RNA pol I synthesizes vRNAs that are assembled with the NP and then replicated and transcribed by the vRNAdependent RNA polymerase (PA, PB1, and PB2). (C) Because each plasmid contains two different promoters, both cellular Pol I and Pol II transcribe the plasmid template resulting in the synthesis of viral proteins and vRNAs, resulting in decreasing the number of plasmids to be used from 12 to only 8 (in the case of influenza). *Abbreviations*: Pol I, polymerase I; pol II, polymerase II.

RG system to many SNS RNA viruses. In 1999, the eight-genome-segmented influenza A was rescued from cDNA constructs,[12] carrying for the first time, the RNA Pol I promoter and terminator sequences to drive vRNA transcription. Briefly, each of the eight influenza segments were cloned under the Pol I recognition sequence and transfected into cells (see Fig. 8.2B). Four additional plasmids were introduced into the cells encoding PA, PB1, PB2, and NP proteins for the expression. The first three proteins assembled to give rise to the viral RdRp; the fourth one, the NP, interacted with the genomic vRNA to form a complex that is recognized and transcribed, or replicated, by the RdRp to form all viral proteins that are needed for virus assembly. A year later, a more efficient expression system was described.[13] In the new system the Pol I cassette was cloned between an RNA pol II promoter sequence and a polyadenylation signal. In this new system, the vRNA and mRNA were generated simultaneously from these constructs, thereby reducing to just 8 instead of 12, the total number of plasmids required for rescue (see Fig. 8.2C). Overall, the RG-based rescue technology has been going through continuous improvements, leading to great success in the rescue of individual virus segments from cDNAs and in the construction and rescue of full-length minigenomes.[14–18]

Nevertheless, the complexity of bringing all parts of the full RG system together (e.g., the ratio of the support plasmids is critical just as it is in the virus life cycle) resulted in only a few of the NS RNA viruses being recovered from cDNA constructs so far.

8.4 Choosing the Promoter

During the late eighties, a novel approach was successfully used to rescue individual segments of a segmented virus.[19] Each of the components necessary to reconstitute an enzymatically active RNP complex was generated inside cells by transfecting the cells with viral cDNA plasmid constructs under T7Rp promoter control and co-transfecting with a plasmid expressing T7Rp. The efficiency of the technique using constructs under the control of a wild-type T7Rp promoter was relatively low as a result of the addition of three nonviral G nucleotides at the 5′ end of the transcript. Major improvement in rescue efficiency occurred when transcripts with authentic viral ends were achieved with the use of a truncated T7Rp promoter sequence, which prevented the addition of nonviral G residues.[20] Also, instead of using T7Rp expression plasmids, the recombinant vaccinia virus was used to deliver the T7Rp, resulting in higher polymerase expression, but this complicated the method because of virus-mediated cytotoxicity. While the T7Rp-based system can be achieved in several ways, the requirement of introducing exogenous T7Rp potentially limits the system efficiency due to technical difficulties of achieving the expression of the polymerase in all cells within the culture.[2]

Several years later, the Pol I transcription system was applied.[21] It overcame a majority of the limitations of the T7Rp-driven systems. The Pol I enzyme is expressed in the nucleolus of all eukaryotic cells and therefore does not need to be provided. Moreover, it synthesizes ribosomal RNA, which like the vRNA of NS RNA viruses, does not contain a 5′ cap or 3′ poly(A) structures. Thus, transcripts generated by Pol I have precise viral ends. Depending on these properties, Pol I presents an attractive alternative to T7Rp for future minigenome system development.[5] However, one of the difficulties using a Pol I–based system is connected to the fact that a suitable cell line has to be used. Since every virus is able to infect only a narrow range of cell lines with a great degree of efficiency, and due to the fact that Pol I is species specific, different promoter and terminator sequences are needed for different cell lines, resulting in the Pol I–driven plasmid construction being much more complicated. Today, both systems (T7 and Pol I based) are used in RG studies, although the latter tends to be the first choice of method in recent years.

8.5 Applications

RG has been applied mainly to explore molecular biological questions (referring to RNA viruses), which include:

(1) the discovery of the location and boundaries of the essential sequences that control virus transcription, replication, encapsidation, and packaging;

(2) the secondary structure of promoter regions;
(3) the functional domains of the transacting proteins;
(4) cap-snatching mechanisms for transcription initiation of SNS viruses; and
(5) the role of the nonstructural genes and many others.

Even more important is the fact that using RG, the infectious particles are unnecessary for most of this research, making it possible to study many of the aspects mentioned above with regard to dangerous viruses (like Ebola) without the need for special research facilities. RG not only deals with the molecular biology of the viruses but also helps to study pathogenesis of viruses through the study of virus–host interactions. Another important application of RG was seen in the formation of novel therapeutic and disease prevention models. Since genetically engineered NS RNA viruses may induce immune responses that are longer lasting and more effective than those generated by conventional killed vaccines, and due to the ability of the recombinant viruses to stably express foreign genetic material, live weakened vaccines against NS RNA viruses can be explored and developed.[1,2,5,6,22–24] Finally, in the last few years, RG has shown potential for use as a platform technology for development of NS RNA virus detection systems.[25,26]

8.6 Detection of Negative-Stranded RNA Viruses

The detection of NS RNA viruses is, in principle, no different from detection of any other group of viruses. A wide variety of methods for viral detection exists today and can be divided into three main groups: viral culture, antigen detection, and nucleic acid detection.

Viral culture has been the most widely used method for identifying viruses from patient samples. In this technique viruses are isolated by cell culture. Unlike diagnostic microbiology, viruses require cellular machinery for replication; thus living systems must be used. One of the well-known diagnostic methods is based on the ability of some viruses to induce the cytopathic effect (CPE)—visible morphologic changes in infected cells detected microscopically. However, some viruses may produce only a minimal CPE but express on the surface of infected cells viral proteins that bind to erythrocyte membranes. These viruses, with their ability to bind red blood cells, can be detected routinely by testing the cultures in a process termed "hemadsorption." Viral culture has important advantages compared with other diagnostic methods. Culture is an amplification method that increases the amount of the virus (facilitating detection and characterization), provides an isolate of viable virus that can be characterized further when necessary, and can be stored for future studies. In contrast to immunologic or nucleic acid–based diagnostic tests (which basically detect only the specific virus to which the diagnostic reagent is directed), viral culture can discover previously unidentified viruses. However, viral culture has several significant disadvantages as a diagnostic method, including the requirement for specialized facilities, expense, and both a prolonged time of detection and a limited range of viruses that can be detected.[1, 27]

Antigen detection in clinical specimens has become an essential component of diagnostic virology. There are several methods used for the visualization of antigen detection. In fluorescent antibody (FA) staining, a fluorescent label is used. In the direct format, the antibody that recognizes the viral antigen is conjugated to that label. In the indirect format, the antiviral antibody is unlabeled and is detected by a second antibody that carries the fluorescent label, recognizing the antiviral antibody. Immunoperoxidase (IP) staining is similar to FA staining, except that instead of a fluorescent label, horseradish peroxidase is used. After the labeled antibody is incubated with the sample, a substrate for peroxidase, which changes color, is added. The advantage of this method is that the staining can be viewed by light microscopy. Finally, an enzyme immunoassay (EIA) is a widely used method that can be applied to the detection of antigens, regardless of whether they are cell associated or in fluid phase. A common assay format for antigen detection is the double antibody sandwich technique, in which a capture antibody is bound to a reaction surface and a detection antibody is added after the sample. There is no need for virus viability because whole cells are not required; this method may be advantageous when specimen transport time is prolonged. Furthermore, antigen detection methods can provide diagnostic information within a few hours, another important advantage as compared with viral culture.[1]

Nucleic acid detection techniques, which are currently revolutionizing diagnostic virology, recognize specific sequences and thus can be applied for the detection of nearly every virus. The earliest attempts at nucleic acid–based diagnosis involved direct hybridization of nucleic acid probes to viral nucleic acids presented in clinical specimens. In microarray technology, hundreds or even thousands of oligonucleotide probes can be bound to a small silicon chip, hybridization reactions carried out directly on the chip, and detection, to which sequence the probes have been bound, is performed. The use of probes representing all possible nucleotide variations within a target sequence allows very rapid determination of nucleotide sequence, but this technique suffers from low sensitivity. The development of amplification assays such as polymerase chain reaction (PCR) made it possible to overcome this sensitivity barrier. These methods are particularly attractive for viruses that are difficult or impossible to culture, for which antigen detection cannot be applied because of their diversity, or because the level of viral antigen in clinical specimens is too low to permit successful detection. Multiplex PCR refers to PCR reactions in which more than one primer set is incorporated into the reaction mix, allowing the detection of multiple targets. This has proved to be an efficient means for detecting multiple agents (within the same reaction). Recently, several new PCR assays have been developed in which the synthesis of the PCR product is detected as it occurs. These assays have been termed real-time PCR. Due to the fact that the accumulation of the PCR product is monitored in the reaction tube, no separate detection method (e.g., gel electrophoresis) is required, thus noticeably shortening the effective assay time. Because of its simplicity and broad applicability, PCR assays have been widely used in viral diagnostics.[1,28,29]

8.7 Reverse Genetics–Based Detection

Virus detection using genetically engineered cell lines expressing reporter genes (under the control of virus-derived promoters) combines the sensitivity of viral culture with the rapid detection provided by reporter genes. RG systems were applied for positive-stranded RNA virus detection in 1994[30] and for NS RNA viruses in 1998.[31] Recently, the application of RG for the detection and quantification of influenza A[25] and Ebola[26] viruses was described. The development of such systems is based on the fact that the UTRs at the 5′ and 3′ termini of viral genomic segments are conserved for the most part and have been shown to be not only necessary but also sufficient *cis*-acting elements for replication and transcription of viral[32] as well as nonviral or reporter genes.[33] First, a plasmid needs to be constructed containing a viral cDNA under Pol I (or T7Rp) promoter followed by an appropriate termination sequence. After that, the plasmid can be digested so that the region encoding for a viral protein is removed, leaving the 5′ UTR on the one end and the 3′ UTR on the other. Instead of the viral protein, a reporter gene is inserted between the viral UTRs in the negative sense (see Fig. 8.3A). The constructs are then transfected into a chosen cell line expressing the appropriate cellular Pol I (or artificially introduced, T7Rp), resulting in vRNA-like segment transcription. Since mammalian cells lack the ability to express NS vRNA, the virus must carry the RdRp with it while infecting the cells. In this manner, the vRNA-like segments (containing the reporter gene flanked by viral UTRs) are constantly expressed inside the cell by Pol I (or T7Rp) and transcribed to mRNA by viral RdRp from an infecting virus. Therefore, the reporter gene is expressed if, and only if, the cells were infected with the virus (see Fig. 8.3B). The great advantage of this system is that reporter gene activity prevents the need for antigen- or sequence-specific reagents. Instead of these reagents, the use of *cis*-acting promoter elements (viral UTRs), which are highly conserved among viral strains, provide the necessary specificity.[34–36]

To summarize the above, the RG-based detection system is grounded in well-established techniques, such as genetic engineering methods and cell culture procedures, with the result that virtually any molecular biology laboratory can develop it. Moreover, the development of such a detection system does not require the use of infectious virus particles, meaning that a detection system for lethal viruses (that require B3 or even B4 facilities) can be

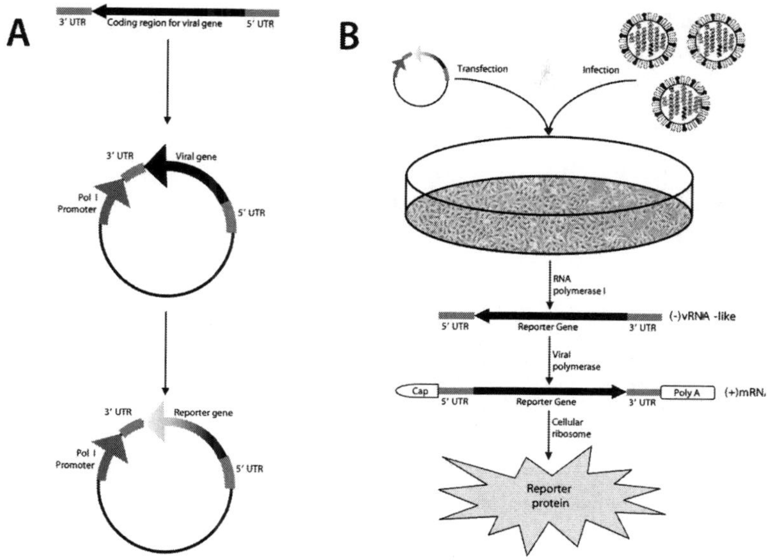

Figure 8.3 Schematic diagram of detection system development. (A) cDNA of a viral gene is inserted in a plasmid under control of RNA pol I promoter (T7Rp promoter can be used instead). A reporter gene of a choice is inserted instead of a coding region of the viral gene between the UTRs. (B) Cells are transfected with the construct, resulting in the synthesis of vRNA-like segments by cellular RNA pol I. Viral polymerase, obtained from infection, uses the vRNA-like segments to synthesize mRNA which then expressed (by cellular ribosomes) to a reporter protein.

developed in any (B2) laboratory and only the final steps of checking different parameters of the system (e.g., sensitivity, specificity) must be performed in the appropriate facility.

Except for the different promoters, the main concern about building RG-based detection system is the choice of reporter gene (see Table 8.2), which depends mainly on the nature of the assay. The first reporter gene used to detect and monitor transcriptional activity in cells was CAT. It has no endogenous activity in mammalian cells; but the assay relies on radiochemicals, causing the linear range and sensitivity of the assay to be relatively low.

Another widely used reporter gene is luciferase, which refers to a family of enzymes that catalyze the oxidation of various substrates, resulting in light emission. The most commonly used

Table 8.2 Advantages and disadvantages of most commonly used reporter genes

Reporter gene	Advantages	Disadvantages
CAT	No endogenous activity. Automated ELISA available.	Narrow linear range; use of radioisotopes; stable.
Luciferase (firefly and Renilla)	High specific activity; no endogenous activity; broad dynamic range; convenient assays.	Requires substrate.
GFP	Autofluorescent (no substrate needed); no endogenous activity; mutants with altered spectral qualities available	Requires post-translational modification; low sensitivity (no signal amplification).

Abbreviations: CAT, chloramphenicol acetyltransferase; GFP, green fluorescent protein.

luciferase for reporter gene assays is the firefly luciferase, which has become very popular because of its high sensitivity and broad linear range. The original substrate (luciferin) not only induced a flash response but was also membrane impermeable, requiring cell lysis prior to the addition of the substrate. This requirement has been removed—the development of membrane-permeable firefly luciferin and additional "glow" reagents increased the duration and stability of the flash response so that it can be detected in a scintillation counter, also making it suitable for high-throughput screening (HTS).[37] A more recently used reporter gene is the Renilla luciferase. Since it catalyzes the oxidation of coelenterazine, which is membrane permeable, it may serve as an appropriate reporter for living cell systems.[38]

Finally, GFP is autofluorescent and unique among light-emitting proteins, not requiring any cofactors or substrates in order to produce its light. Several GFP mutants are now available, exhibiting improved fluorescent properties as compared with the wild-type GFP. However, with fluorescence, it is necessary to excite the cell with an external light that is stronger than the signal. This requires special filters to read the results, making the technique less sensitive than the luciferase-based systems.[39–41]

Due to the use of living cells and the fact that luciferase provides a highly sensitive way of monitoring the expression of intracellular proteins, an RG-based system can not only detect the virus but also track the kinetics of the infection, providing an additional tool for viral pathogenesis studies. Finally, unlike traditional methods that require a new set of cells for each measurement, a recent development in Renilla luciferase substrate made it possible to track the viral infection in real time (while cell viability is maintained) using only one set of cells for all measurements performed throughout the course of an infection. This novel application enables significant improvement in the efficiency of performing kinetic studies of viral infection (see Fig. 8.4).

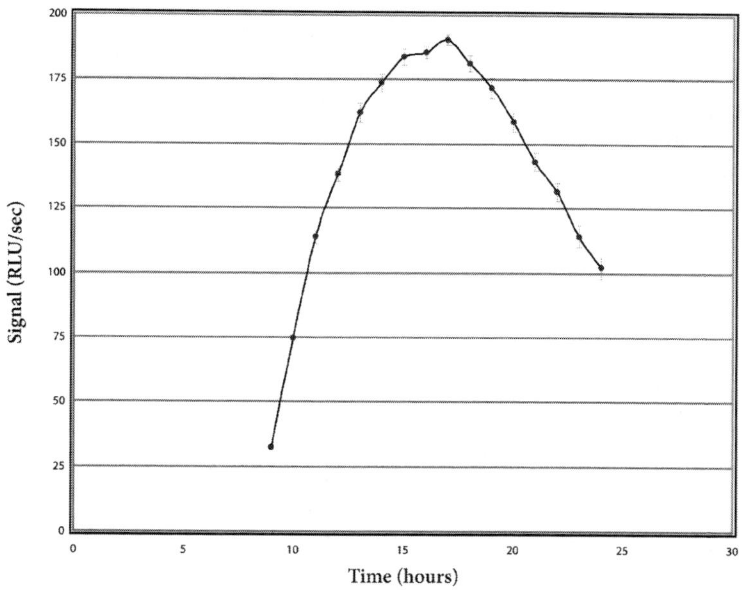

Figure 8.4 293 T-cells were transfected with a plasmid that contains the Renilla luciferase gene flanked by influenza UTRs (constructed in our laboratory). Then, 24-hour posttransfected cells were infected with A/Brisbane/59/07 influenza virus in a 96-well plate. The signal was measured every 3 h, starting 9 h after the infection, using a Luminoskan luminometer. The signal increase represents the first replication cycle of the virus. Negative control resulted in a fixed signal of less than 0.2 RLU/sec (not shown) (unpublished data, Marks laboratory). *Abbreviation*: RLU, relative light unit.

8.8 Where Do We Go from Here?

As RG was just recently applied to detection system development, there are still many parameters to be checked in this novel technology. The reliability, reproducibility, sensitivity, and adaptability of reporter gene technology to HTS have made cell-based assays an increasingly attractive alternative to in vitro biochemical assays. These cell-based assays represent model biological systems that mimic physiological conditions; thus they may also provide important information about bioavailability and cytotoxicity of possible antiviral compounds.[40–45] Moreover, the ability to maintain these cells in culture throughout the measurement procedure allows tracking the kinetics of the antiviral compounds and neutralizing human monoclonal antibodies for immunotherapy. Due to the fact that these novel detection techniques rely only on highly conserved short sequences (UTRs), they may serve in the future as a platform technology for development of biosensors that are able to detect emergent viral strains invisible to traditional detection methods. As a result, the application of such a biosensor to the bioterrorism field is entirely likely.

References

1. Knipe, D. M., and P. M. Howley. 2007. *Fields Virology* (5th ed.), Williams and Wilkins, Lipppincott.
2. Walpita, P., and R. Flick. Reverse genetics of negative-stranded RNA viruses: a global perspective. 2005. *FEMS Microbiol. Lett.*, **244**(1), 9–18.
3. Taniguchi, T., M. Palmieri, and C. Weissmann. QB DNA-containing hybrid plasmids giving rise to QB phage formation in the bacterial host. 1978. *Nature*, **274**(5668), 223–228.
4. Racaniello, V. R., and D. Baltimore. Cloned poliovirus complementary DNA is infectious in mammalian cells. 1981. *Science*, **214**(4523), 916–919.
5. Neumann, G., M. A. Whitt, and Y. Kawaoka. A decade after the generation of a negative-sense RNA virus from cloned cDNA: what have we learned? 2002. *J. Gen. Virol.*, **83**(Pt 11), 2635–2662.

6. Palese, P., H. Zheng, O. G. Engelhardt, S. Pleschka, and A. Garcia-Sastre. Negative-strand RNA viruses: genetic engineering and applications. 1996. *Proc. Natl. Acad. Sci. U S A*, **93**(21), 11354–11358.
7. Schnell, M. J., T. Mebatsion, and K. K. Conzelmann. Infectious rabies viruses from cloned cDNA. 1994. *EMBO J.*, **13**(18), 4195–4203.
8. Whelan, S. P., L. A. Ball, J. N. Barr, and G. T. Wertz. Efficient recovery of infectious vesicular stomatitis virus entirely from cDNA clones. 1995. *Proc. Natl. Acad. Sci. U S A*, **92**(18), 8388–8392.
9. Lawson, N. D., E. A. Stillman, M. A. Whitt, and J. K. Rose. Recombinant vesicular stomatitis viruses from DNA. 1995. *Proc. Natl. Acad. Sci. U S A*, **92**(10), 4477–4481.
10. Radecke, F., P. Spielhofer, H. Schneider, K. Kaelin, M. Huber, C. Dötsch, G. Christiansen, and M. A. Billeter. Rescue of measles viruses from cloned DNA. 1995. *EMBO J.*, **14**(23), 5773–5784.
11. Bridgen, A., and R. M. Elliott. Rescue of a segmented negative-strand RNA virus entirely from cloned com- plementary DNAs. 1996. *Proc. Natl. Acad. Sci. U S A*, **93**(26), 15400–15404.
12. Neumann, G., T. Watanabe, H. Ito, S. Watanabe, H. Goto, P. Gao, M. Hughes, D. Perez, R. Donis, E. Hoffmann, G. Hobom, and Y. Kawaoka. Generation of influenza A viruses entirely from cloned cDNAs. 1999. *Proc. Natl. Acad. Sci. U S A*, **96**(16), 9345–9350.
13. Hoffmann, E., G. Neumann, Y. Kawaoka, G. Hobom, and R. G. Webster. A DNA transfection system for generation of influenza A virus from eight plasmids. 2000. *Proc. Natl. Acad. Sci. U S A*, **97**(11), 6108–6113.
14. Flick, R., and R. F. Pettersson. Reverse genetics system for Uukuniemi virus (Bunyaviridae): RNA polymerase I-catalyzed expression of chimeric viral RNAs. 2001. *J. Virol.*, **75**(4), 1643–1655.
15. Flick, K., J. W. Hooper, C. S. Schmaljohn, R. F. Pettersson, H. Feldmann, and R. Flick. Rescue of Hantaan virus minigenomes. 2003. *Virology*, **306**(2), 219–224.
16. Flick, R., K. Flick, H. Feldmann, and F Elgh. Reverse genetics for crimean-congo hemorrhagic fever virus. 2003. *J. Virol.*, **77**(10), 5997–6006.
17. Lee, K. J., I. S. Novella, M. N. Teng, M. B. Oldstone, and J. C. de La Torre. NP and L proteins of lymphocytic choriomeningitis virus (LCMV) are sufficient for efficient transcription and replication of LCMV genomic RNA analogs. 2000. *J. Virol.*, **74**(8), 3470–3477.
18. Perez, M., A. Sanchez, B. Cubitt, D. Rosario, and J. C. de la Torre. A reverse genetics system for Borna disease virus. 2003. *J. Gen. Virol.*, **84**(Pt 11), 3099–3104.

19. Fuerst, T. R., E. G. Niles, F. W. Studier, and B. Moss. Eukaryotic transient-expression system based on recombinant vaccinia virus that synthesizes bacteriophage T7 RNA polymerase. 1986. *Proc. Natl. Acad. Sci. U S A*, **83**(21), 8122–8126.

20. Kato, A., Y. Sakai, T. Shioda, T. Kondo, M. Nakanishi, and Y. Nagai. Initiation of Sendai virus multiplication from transfected cDNA or RNA with negative or positive sense. 1996. *Genes Cells*, **1**(6), 569–579.

21. Zobel, A., G. Neumann, and G. Hobom. RNA polymerase I catalysed transcription of insert viral cDNA. 1993. *Nucleic Acids Res.*, **21**(16), 3607–3614.

22. Conzelmann, K. K. Nonsegmented negative-strand RNA viruses: genetics and manipulation of viral genomes. 1998. *Annu. Rev. Genet.*, **32**, 123–162.

23. Collins, P. L., S. S. Whitehead, A. Bukreyev, R. Fearns, M. N. Teng, K. Juhasz, R. M. Chanock, and B. R. Murphy. Rational design of live-attenuated recombinant vaccine virus for human respiratory syncytial virus by reverse genetics. 1999. *Adv. Virus Res.*, **54**, 423–451.

24. Garcia-Sastre, A. Negative-strand RNA viruses: applications to biotechnology. 1998. *Trends Biotechnol.*, **16**(5), 230–235.

25. Lutz, A., J. Dyall, P. D. Olivo, and A. Pekosz. Virus-inducible reporter genes as a tool for detecting and quantifying influenza A virus replication. 2005. *J. Virol. Methods*, **126**(1–2), 13–20.

26. Towner, J. S., J. Paragas, J. E. Dover, M. Gupta, C. S. Goldsmith, J. W. Huggins, and S. T. Nichol. Generation of eGFP expressing recombinant Zaire ebolavirus for analysis of early pathogenesis events and high-throughput antiviral drug screening. 2005. *Virology*, **332**(1), 20–27.

27. Ogilvie, M. Molecular techniques should not now replace cell culture in diagnostic virology laboratories. 2001. *Rev. Med. Virol.*, **11**(6), 351–354.

28. Yang, S., and R. E. Rothman. PCR-based diagnostics for infectious diseases: uses, limitations, and future applications in acute-care settings. 2004. *Lancet Infect. Dis.*, **4**(6), 337–348.

29. Kessler, N., O. Ferraris, K. Palmer, W. Marsh, and A. Steel. Use of the DNA flow-thru chip, a three-dimensional biochip, for typing and subtyping of influenza viruses. 2000. *J. Clin. Microbiol.*, **42**(5), 2173–2185.

30. Olivo, P. D., I. Frolov, and S. Schlesinger. A cell line that expresses a reporter gene in response to infection by Sindbis virus: a prototype for detection of positive strand RNA viruses. 1994. *Virology*, **198**(1), 381–384.

31. Olivo, P. D., P. L. Collins, M. E. Peeples, and S. Schlesinger. Detection and quantitation of human respiratory syncytial virus (RSV) using minigenome cDNA and a Sindbis virus replicon: a prototype assay for negative-strand RNA viruses. 1998. *Virology*, **251**(1), 198–205.

32. Fodor, E., L. Devenish, O. G. Engelhardt, P. Palese, G. G. Brownlee, and A. García-Sastre. Rescue of influenza A virus from recombinant DNA. 1999. *J. Virol.*, **73**(11), 9679–9682.

33. Flick R., G. Neumann, E. Hoffmann, E. Neumeier, and G. Hobom. Promoter elements in the influenza vRNA terminal structure. 1996. *RNA*, **2**(10), 1046–1057.

34. Huang T. S., P. Palese, and M. Krystal. Determination of influenza virus proteins required for genome replication. 1990. *J. Virol.*, **64**(11), 5669–5673.

35. Flick, R., and G. Hobom. Interaction of influenza virus polymerase with viral RNA in the "corkscrew" conformation. 1999. *J. Gen. Virol.*, **80**(Pt 10), 2565–2572.

36. Neumann G., G. G. Brownlee, E. Fodor, and Y. Kawaoka. Orthomyxovirus replication, transcription, and polyadenylation. 2004. *Curr. Top. Microbiol. Immunol.*, **283**, 121–143.

37. Williams T. M., J. E. Burlein, S. Ogden, L. J. Kricka, and J. A. Kant. Advantages of firefly luciferase as a reporter gene: application to the interleukin-2 gene promoter. 1989. *Anal. Biochem.*, **176**(1), 28–32.

38. Lorenz, W. W., M. J. Cormier, D. J. O'Kane, D. Hua, A. A. Escher, and A. A. Szalay. Expression of the Renilla reniformis luciferase gene in mammalian cells. 1996. *J. Biolumin. Chemilumin.*, **11**(1), 31–37.

39. Bronstein, I., J. Fortin, P. E. Stanley, G. S. Stewart, and L. J. Kricka. Chemiluminescent and bioluminescent reporter gene assays. 1994. *Anal. Biochem.*, **219**(2), 169–181.

40. Naylor, L. H. Reporter gene technology: the future looks bright. 1999. *Biochem. Pharmacol.*, **58**(5), 749–757.

41. Welsh, S., and S. A. Kay. Reporter gene expression for monitoring gene transfer. 1997. *Curr. Opin. Biotechnol.*, **8**(5), 617–622.

42. Noah, J. W., W. Severson, D. L. Noah, L. Rasmussen, E. L. White, and C. B. Jonsson. A cell-based luminescence assay is effective for high-throughput screening of potential influenza antivirals. 2007. *Antiviral Res.*, **73**(1), 50–59.

43. Joyeux, A., P. Balaguer, P. Germain, A. M. Boussioux, M. Pons, and J. C. Nicolas. Engineered cell lines as a tool for monitoring biological activity of hormone analogs. 1997. *Anal. Biochem.*, **249**(2), 119–130.

44. Broach, J. R., and J. Thorner. High-throughput screening for drug discovery. 1996. *Nature*, **384**(6604 Suppl), 14–16.
45. Stables, J., A. Green, F. Marshall, N. Fraser, E. Knight, M. Sautel, G. Milligan, M. Lee, and S. Rees. A bioluminescent assay for agonist activity at potentially any G-protein-coupled receptor. 1997. *Anal. Biochem.*, **252**(1), 115–126.

Chapter 9

Diagnostics of Ebola Hemorrhagic Fever Virus

Ariel Sobarzo,[a] Robert S. Marks,[b,c] and Leslie Lobel[a]

[a] *Department of Virology, Health Faculty, Ben-Gurion University of the Negev, Beer-Sheva, Israel*
[b] *Department of Biotechnology Engineering, Ben-Gurion University of the Negev, Beer-Sheva, Israel*
[c] *The National Institute of Biotechnology in the Negev, Ben-Gurion University of the Negev, Beer-Sheva, Israel*
tautau.ariel@gmail.com

9.1 Ebola Virus

The Ebola virus is a member of the family *Filoviridae*, which is the cause for Ebola hemorrhagic fever (EHF). The single-stranded, negative-sense RNA virus, which can produce high-mortality disease in humans and nonhuman primates, has caused sporadic outbreaks in Central Africa throughout the last 40 years. Public health concerns about Ebola have increased in recent years due to large natural outbreaks and its potential use as a bioweapon agent. Despite the capabilities of laboratory diagnostics, the initial diagnosis of Ebola infection is based on clinical assessment. Today, laboratories located

Viral Diagnostics: Advances and Applications
Edited by Robert S. Marks, Leslie Lobel, and Amadou Alpha Sall

ISBN 978-981-4364-43-0 (Hardcover), 978-981-4364-44-7 (eBook)
www.panstanford.com

in high-risk areas are still not equipped to diagnose Ebola infections; thus specimens must be sent to reference laboratories around the world for viral confirmation. This chapter reviews current diagnostic methods and further aspects involved in the identification of the Ebola virus. The important features of each method are discussed and compared, along with an introduction to novel advanced biosensors and future technological tools for viral detection.

9.2 Etiology and Epidemiology

Filoviruses are taxonomically classified within the order *Mononegavirales*, a large group of enveloped viruses whose genomes are composed of a nonsegmented, negative-strand (NNS) RNA molecule.[1] The unique and distinctive members of the family include the Marburg and Ebola viruses, the latter having five different subtypes: Zaire, Sudan, Reston, the Ivory Coast, and Bondibugyo.[2–9] The NNS RNA genomes of the Ebola virus show the gene arrangement 3′-NP-VP35-VP40-GP-VP30-VP24-L-5′, with a total molecular length of approximately 19 kb. The virion morphology appears either as a long, filamentous form, sometimes branched (shaped as a short U, a figure 6, or a mace) or in circular configurations.[10,5]

The first recognition of the Ebola virus happened in 1976, when two parallel epidemics occurred in Zaire near the Ebola River and in Sudan. More than 550 cases with 430 associated deaths were recorded during those outbreaks.[9,11,12] Since then, more than 17 outbreaks have occurred in periodic cycles, responsible for more than 2,000 cases, 1,400 deaths,[13–15] and a case fatality rate that ranges from 30%–90%. Despite a few isolated cases in different regions of Africa, Ebola has mainly emerged and reemerged in Central Africa in the regions of Sudan, the Democratic Republic of the Congo, the Ivory Coast, Gabon, and Uganda.[16]

9.3 Disease Transmission and Clinical Behavior

Until this day, the routes of viral transmissions of the Ebola virus are unknown, although several outbreaks over the years were clearly linked to animal carcasses.[13] Studies indicate that during outbreaks, the virus appears primarily transmitted by close and physical contact with infected persons, their clothing, bodily fluids, blood, vomit, stool, saliva, and, possibly, sweat.[13] In addition, further studies suggest that transmission may also occur through sexual and aerosol contacts.[4,17–19]

In general, the onset of infection is insidious, beginning with symptoms that resemble a cold or influenza virus infection, rapidly deteriorating into hemorrhaging from every orifice, and usually ending in death.[19] The incubation period ranges from several days to three weeks. The majority of cases become symptomatic within 5 to 12 days, and death ensues after 7 to 14 days following the onset of symptoms.[4,18–20]

Despite much research in recent years, the key elements of recovery from Ebola infection have yet to be established in either human or animal models. Studies of fatal cases of Ebola infection show high viremia and no evidence of an immune response; however, the reasons remain a mystery.

9.4 Therapy

Despite limited successes in animal models,[14,21–36] it seems that the challenges imposed by Ebola virus infection mechanisms have hampered development of an effective and safe vaccine for humans. A recently developed DNA/adenovirus prime-boost vaccine is undergoing safety testing on human volunteers.[37] Several other products have been tested or are currently in late-stage development.[34,38] Antibodies acting as prophylactics have yet to provide conclusive efficacy, demonstrating little or no effect at all.[34,38–41] Thus, no specific treatment for Ebola hemorrhagic fever is yet available, and treatment is mainly symptomatic, focused on providing adequate hydration and nutritional support.

9.5 The Fear of Ebola

Epidemics and pandemics have had a great impact on the course of human history. The emergence and reemergence of high-risk diseases to the individual and/or the community are specific interests and concerns to public health systems, be they in developed or in developing countries.[16] Ebola viruses represent a prime example of such reemerging pathogens. Although rare, Ebola outbreaks are unpredictable, with a high severity and fatality rate. These facts, together with their high infectivity and lack of vaccines and chemotherapy, elucidate why Ebola viruses have been classified as biosafety level (BSL)4 pathogens and are regarded as potential bioweapon agents by the World Health Organization (WHO) and the Centers for Disease Control (CDC).[35,42]

In comparison to other viruses affecting millions of people every year, Ebola stands out, from all other diseases, revealing an alarming gap in our immune system's ability to defend itself against it. Ebola is still a threat, although outbreaks have been relatively infrequent, largely confined to remote areas and effectively quarantined.[20] Unstable political situations worldwide and the potential threat of Ebola used as a bioweapon have dramatically changed world perspective and emphasizes why our future security against emerging pathogens such as this requires much research and further resources.

9.6 Current Methods in Ebola Diagnostics (Table 9.1)

First attempts to develop fast and efficient diagnostic tests for Ebola were unsatisfying, some demonstrating more success than others. Early methods, such as the indirect fluorescent antibody test, were used for acute diagnosis and sero-epidemiology; however, their limitations were quickly recognized.[43] Further confirmatory tests, including neutralization, also showed unsuccessful results. The important clinical finding that acutely ill patients were intensely viremic enabled the construction of an enzyme-linked immunosorbent assay (ELISA) assay, which provided a sensitive and specific method for quickly screening large numbers of samples. However, safety and specificity were still concerns, and many Ebola-infected patients still assayed negative in these tests.

Regardless of the wide range of methods and capabilities of laboratory diagnostics, the initial Ebola diagnosis is typically based first on clinical assessment and known Ebola contact. Identification

Table 9.1 Current methods of Ebola diagnostics

Method	Objective	Target protein	Strain	Reference
Culture	Virus isolation	–	All	59–61
EM	Virus	NP/VP40	All	56
ELISA	Antigen	NP/VP40/GP	Zaire	50
Capture ELISA	Antibody	VP40/NP	Zaire	48, 49, 63, 66
IFA	Antigen	–	All	75–78
IHC	Antigen	–	All	43
T-PCR		NP/GP/L	Sudan/Zaire	45, 46, 84, 87, 153
Real-time PCR	Antigen	NP/GP	Sudan/Zaire	44, 51, 52
Mass Tag PCR	VHF	All Virus	Sudan/Zaire	80, 85
ELISA	Antibody	rNP/rGP/rVP35	Zaire	57, 65, 90, 91
IFA	Antibody	rNP	Zaire	91

Abbreviations: EM, electron microscopy; IFA, immunofluorescence assay; IHC, immunohistochemistry; RT, reverse transcriptase; PCR, polymerase chain reaction; VHF, viral hemorrhagic fever; NP, nuclear protein; VP, virus protein; L, RNA-dependent RNA polymerase; GP, glycoprotein; r, recombinant.

may be difficult due to a wide variety of infectious diseases bearing similar clinical symptoms. Ebola is detected by identification of either the host's specific immune response or viral particles in an infected individual. Confirmation is typically obtained by virus isolation, electron microscopy, histological techniques, specific detection of nucleic acid, immunofluorescence, and immunoassays of both antigen and antibodies.[44–53] Although constant improvements, new reagents, and equipment have drastically reduced the detection and confirmation time of Ebola and enhanced sensitivity, they are still not used routinely and are not well established. Results can be obtained within 24 to 48 hours, depending on the method being used, once samples are received at the appropriate laboratories. Risk of transmission increases when contact has been made with a patient in the later stages of illness.[54] Early diagnosis is key for curtailing spread of the disease and providing more time for effective intervention as therapeutics are developed.

Although five different subtypes (i.e., Zaire, Sudan, Ivory Coast, Reston, and Bondibugyo) of Ebola have been defined and characterized, most diagnostic attempts have focused specifically on the Ebola Zaire strain due to its high fatality rate, which can reach 90%.[4]

However, today there is focus on new, improved assays for the detection of both Zaire and Sudan strains, which are considered the most dangerous and deadly to humans. In addition, risks from rapidly sporadic infectious outbreak locations and biothreat contaminants have encouraged researchers to develop fast, easy-to-use, sensitive, specific, and available detection systems, which could provide a useful tool for both individuals and populations at risk.

9.6.1 *Culture Virus Isolation*

Culture and other related methods remain the most sensitive and widely used for viral identification. As a result, basic culture, staining, and microscopy can be found throughout public health laboratory systems.[55,56]

Since Ebola is relatively easy to isolate and propagates well in various cell cultures,[57] virus isolation is a basic, simple, and sensitive method for diagnosis. The rate of growth in tissue culture varies, between 5 and 14 days, depending on the strain. This approach requires specimens be sent to BSL-4 laboratories, located mostly in developed countries, far from the outbreak. Confirmation and adequate response measures are therefore delayed. Further issues include the need for proper shipment, preparation of specimens for virus isolation, requirements of considerable expertise, and, often, further characterization by molecular techniques for confirmation.[58] Nonetheless, cell culture is still performed routinely during suspicious outbreaks and as part of virus identification in BSL-4 laboratories around the world.

9.6.2 *Electron Microscopy*

Transmission electron microscopy was applied for diagnosis and studies of filoviruses since the first outbreak during the late sixties[59] and is still in use today. The technique has enabled investigators to identify filoviruses during outbreaks and help understand their mechanism of infection.[60] Owning, in part, to the relatively high viremia during Ebola infection in humans, electron microscopy has been useful for diagnosis over the years.[5,59,61,62] Virus particles that are present in sera and cell culture fluids can be directly visualized by negative staining and detected easily in thin sections of infected tissues. Electron microscopy has provided rapid results, but it is usually available only in well-equipped research facilities. Thus identification or conclusive confirmation is usually achieved with other techniques.[62] Further disadvantages, such as work safety issues and the high level of expertise required for this technique, resulted in it mainly being used for research purposes and less as a tool for viral diagnostics.

9.6.3 *Serological Assays*

Serological assays are now regularly used to confirm the clinical diagnosis of Ebola.[63] They are used reliably for screening large numbers of small-volume test samples in the simplest of laboratory environments, with relatively high sensitivity and specificity, as compared to culture and gene amplification approaches. The course of Ebola virus infection is such that patients usually die before there is a significant antibody response. This suggests that serological diagnostics are suitable for infected patients likely to survive but not for those who succumb. High viremic titers are present in the blood and tissues of patients during the early stages of the illness, during a narrow window of time, indicating that detection of virus antigens is suitable for diagnosis of EHF. Assays such as these, directly intended for the detection of Ebola virus antigens, were developed and tested in clinical settings, including virus outbreaks in the Congo, 1995; Gabon, 1996; and Uganda, 2000, and were confirmed as efficacious tools for diagnosis.[46,49–51]

Most commonly used assays for antibody detection in the later stages of the disease are the direct IgG and IgM ELISAs and IgM capture ELISA.[49,50,64–67]

These systems have been developed by several groups[68–73] using the *NP*, *VP40*, and *GP* genes as target proteins directed toward the detection of anti–Ebola virus, Zaire, Sudan, and Reston antibodies.[69] The important key element in all immunological assays remains the antibodies. Their affinities and specificities are limiting factors, and those selected must be carefully evaluated. Even so, without any manipulation prior to analysis, many of these immunoassays have a sensitivity threshold of 100- to 1000-fold above the estimated titer, potentially limiting their utility.[74,75] Furthermore the necessary specificity to be useful for detection, let alone definitive identification, may be lacking. In addition to the ELISAs, Ebola diagnosis includes immunoblot, indirect immunofluorescence (IF), and immunohistochemistry (IHC), which were used for confirmation, surveillance, and screening,[44,67,76–79] each demonstrating specific advantages over the other. In the last decade, the detection of Ebola antigens or antibodies by enhanced serological assays has proven to be a useful and efficient diagnostic tool, both for the early and late stages of the disease. As such, the ELISA test, with all its varieties, is now considered the most accurate, rapid, and sensitive one for the diagnosis of EHF.

9.7 Nucleic Acid–Based Techniques

By definition, any self-replicating biological entity can be discriminated on the basis of its own unique nucleic acid sequences. Newly developed hemorrhagic fever assays[80,81] and the emergence of mobile diagnostic capabilities,[46,82] incorporating nucleic acid–based identification techniques, have become tools with enormous potential for outbreak management of Ebola infection around the world. New assays have shown a huge potential in the detection and identification of pathogenic agents like Ebola because of their specificity, sensitivity, and speed in which results can be obtained without the necessity for BSL-4 facilities.[83] The use of these systems for viral identification has increased rapidly during the last decade, under routine use in many large clinical laboratories and reference centers around the world.[84]

A majority of diagnostic PCR systems are designed to be highly specific for detection of the selected target virus. Molecular diagnostic methods for EHF,[85] including nested and RT-PCR techniques, were developed and performed well in epidemic settings during outbreaks in the Congo, 1995; Gabon, 1996; and Uganda, 2000.[46,47,86] Recent assays also include real-time quantitative RT-PCR and mass tag PCR.[45,46,52,53,81,87] Overall, these molecular diagnostic techniques for EHF detection have been proven sensitive, specific, and efficacious during Ebola outbreaks. They have demonstrated the potential for wide adoption in the near future, coming within the scope of several selected reference laboratories.[81]

Although RT-PCR assays, especially nested RT-PCR and real-time quantitative RT-PCR, have proven useful, false-positive and false-negative results must always be excluded.

Index case diagnoses of either outbreaks or imported cases should not be solely based on RT-PCR methods. The sensitivities of the RT-PCR systems in different laboratories worldwide show significant variation.[88] Despite some disadvantages of nucleic acid–based techniques, they are still considered one of the most powerful tools for detection, as described above, especially under field conditions. Current RT-PCR techniques,[46,47,53,89] along with antigen detection ELISA,[50,64,69–72,90] are considered the primary assays for the diagnosis of an acute infection of Ebola virus. While nucleic acid–based detection systems are more specific and sensitive than immunological-based detection assays, the latter are faster and more robust.

9.8 Engineered Recombinant Proteins

Technological improvements and the fear of unpredictable outbreaks have stimulated the developments of rapid advances in diagnostics. Monoclonal antibodies have shown to be more commercially reliable than polyclonal antibodies, and in the foreseeable future these will themselves be replaced by engineered recombinant antibodies. Although the construction and development of recombinant products for viral detection has focused primarily on antigen reagents, production of fully human full-sized antibodies poses promise of some benefit from a diagnostics point of view.

Considering several factors, for example, sporadic rise of Ebola outbreaks, increase in world travel, bioterrorism, and the risk of using authentic antigens from infectious viruses for diagnostics, it is clear why the development of systems using engineered proteins for Ebola detection is specifically essential in countries where BSL-4 laboratories are not available. To overcome these difficulties and to provide a suitable solution, several groups have recently developed various recombinant antigen protein-based diagnostic systems for EHF detection.[66,91,92]

These systems include the use of recombinant proteins, for example, the nucleoprotein (rNP), glycoprotein (rGP), and recombinant VP35 of Zaire Ebola virus, for the detection of IgG and IgM antibodies. These assays have demonstrated the potential to detect IgG antibodies, not only for one strain, but also for other Ebola virus species, while some showed specific strain recognition, depending on the recombinant protein used. An alternative method for detecting specific antibodies for the Ebola virus, an immunofluorescence technique, was also developed using rNP and was confirmed to be highly sensitive and specific.[93]

In contrast to the recent development of recombinant antigens for diagnostics use, efforts to produce antibodies for the same objective have been carried out to a lesser extent. Cloning and expression of whole antibodies, or antibody fragments, has so far not had a great impact on immunoassay technology for Ebola detection—but the possibilities of antibody engineering are tremendous and will have an effect on the development for immunoassays. Recently a study included the creation of single-chain antibody fragments (scFv) against the Ebola virus using a phage display method.[94] The antibodies produced demonstrated a highly specific interaction with viral NP and coat proteins VP40 and VP24 of the Ebola virus, indicating their potential use as diagnostic tools.

Recombinant proteins were confirmed to be useful as antigens for detecting specific antibodies in IgG- and IgM-ELISAs systems and demonstrated high sensitivity and specificity, not only for the detection of one virus strain, but also for the detection of several others simultaneously.[91] Today, it is clear that the ability to produce recombinant proteins for diagnostics use will ensure the safety of future detection systems with low background and high specificity, allowing for easy scale-up in the commercial manufacture of kits with high-quality assurance and accessibility,[95] certain to have a huge impact on viral detection and therapeutics.

9.9 New Trends in Ebola Diagnostics

At present, the most commonly used techniques for diagnostics of an acute Ebola infection are based mainly on RT-PCR technology and antigen capture ELISA, supplemented with antibody detection assays.[82,96] During the last decade, these assays were developed and established in national and international reference laboratories worldwide and have shown variable success with a good measure of sensitive and specific detection for all of the Ebola virus strains from previous outbreaks.[49,51,97] Although the use of these conventional virological methods for analyses has been successful and well adapted, they may not fully contend with the variety of challenges required for detection and outbreak prevention. There is still a continuous need worldwide for development of new and improved detection systems that will provide fast, accurate, direct, low-cost, and field operability for Ebola diagnostics (see Fig. 9.1).

Progress in technological devices has led to the development of several prototype biosensors enabling the production of simple, sensitive, specific, and safe detection systems. The new systems include the use of biosensor technologies, for example, surface plasmon resonance (SPR), quartz crystal microbalance (QCM) sensors, optical fiber immunosensor (OFIS), and a reverse genetics (RG) system. These new assays have shown the potential ability to provide rapid and specific disease diagnosis on-site so that clinicians can quickly determine whether treatment is needed.[98] Recently, there has been considerable development of label-free piezoelectric biosensors based on the use of QCM. QCM is an extremely sensitive mass-measuring device, which allows dynamic monitoring of hybridization events, using an oscillating crystal with a probe immobilized on its surface. The increased mass, associated with the hybridization reaction, results in a decrease of the oscillating frequency. Applications of this technology included detection and characterization of viruses, including Ebola.[81,99,100] These systems not only eliminate the need for labels, but also offer the potential advantage of rapid, real-time monitoring of specific hybridization, as well as high sensitivity and specificity. They have previously shown high promise in sensor research and gained momentum in detection

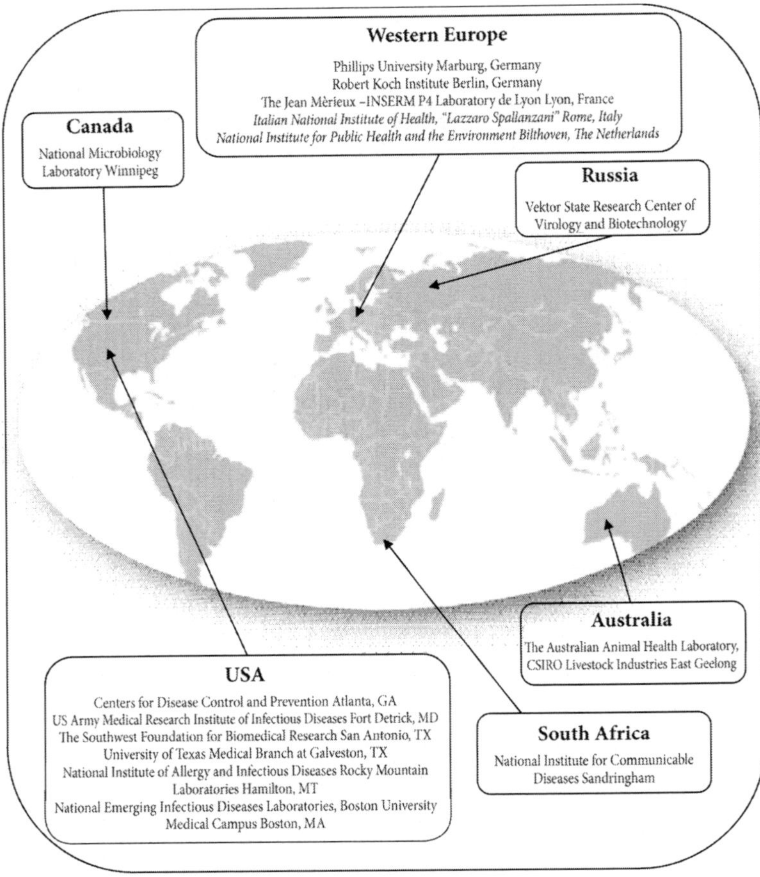

Figure 9.1 BSL-4 facilities around the world. Note: The map is not intended to serve as a definitive source on the status of BSL-4 facilities around the world. The list of facilities was taken from a recent review done by Feldmann et al. (2007). Underlined are facilities that were expected to open in the year 2010.

of viral samples, with a limitation around the subnanogram region. The development of such a sensor for Ebola antigen detection[98] has produced binding results and sensitivity comparable to ELISA. Its significant promise is a result of its rapid detection, ease of use, low cost of operation, and deployment ability to outbreak areas.[98,99,101]

In a recent study,[102] a newly developed OFIS for the detection of antibodies to Ebola virus strains Zaire and Sudan was presented.

This new device showed high sensitivity and specificity compared to a standard-luminescence ELISA. The results of the study suggested that the detection of Ebola virus antibodies using this technique would contribute significantly to serological and epidemiological studies. With some modification, an easy-to-use procedure for Ebola virus antibody or antigen detection will be available in the future as a field operable clinical tool.[102]

Although not defined as a biosensor or detection device per se, RG technology has shown the potential to be used as such, combined with an appropriate detection system. Use of RG has been rapidly increasing, and systems using this technology have been developed for many of the negative-stranded RNA viruses,[103,104] including the Ebola virus.[105–107]

This technique has enabled the generation of artificial replication systems and recombinant mutant viruses that were used mainly to study different aspects of virus biology but could also have an important role in viral detection. Recently two groups[106,108] have developed RG systems that demonstrate a new EboZ-eGFP virus that was useful for detecting horizontal and vertical transmission following experimental infections of candidate reservoir species, with a fluorescence signal that was more robust than that observed by standard IHC.

Despite the search of new breakthrough devices for future Ebola diagnosis, improvements of existing methodologies are continually being developed. New protocols and methods in nucleic acid detection and sera assays are being introduced and tested under field conditions in order to provide a simple and rapid solution. A recent example of an assay such as this has been seen in a swift and somewhat less sophisticated method developed by Lunchet.[90] The assay based on a new immuno-filtration technique for the detection of Ebola virus antigen has provided a novel tool for a future field response to outbreaks of EHF. Being less sensitive than other tests but with a rapidity far less prone to technical complications,[90] the assay demonstrated for the first time the applicability of the immuno-filtration system to detect viral antigen and its usefulness under field conditions with a sensitivity similar to that of the widely used antigen detection ELISA.[50,51]

9.10 Future Diagnostics

There has been increased interest in developing rapid and reliable methods of detection for microorganisms, which naturally pose a very high risk for human and animal populations and have the potential to be used as bioterrorism agents. Application of the new diversity of assays for viral detection includes a variety of techniques and devices only recently developed especially for diagnostic properties. Although not intended specifically for Ebola detection diagnostics, these devices may well be adapted in the near future. It should be pointed out that this review does not intend to reflect the full scope of methods and systems that have been or are undergoing development but rather presents a glimpse of future diagnostics, as summarized in Table 9.2.

New advances in technology, biosensor devices, and molecular biological systems are constantly influencing the field of viral detection. Label-free optical biosensors are among the most desirable for bioassays because they utilize light as their detection mechanism.[109,110] Assays using this method are fast, in real time, and with high sensitivity to surface modifications. A study done by Huang et al.[111] demonstrated a compact, label-free, optoelectronic biosensor system utilizing a photonic crystal as the sensor surface and a tunable VCSEL-based detection platform of human antidengue antibodies from serum samples. The sensor proved highly sensitive to surface modifications, with the ability to detect biomolecular interactions of human antibodies and potentially a versatile tool for clinical diagnostics in resource-poor environments, where infectious disease monitoring is most critical. The results obtained indicated comparable detection sensitivity as the predominant technologies, such as ELISA, with sufficient signal differentiation for diagnosis but with shorter and simpler assay preparation.[111]

One of the most highlighted and fast-moving interfaces of nanotechnology is the application of QDs in biology. The unparalleled advantages of the size-tunable fluorescent emission and the simultaneous excitation at a single wavelength make QDs a great possibility for use in optical encoding detection. A new development of QD-based biosensors has consisted of mixtures of green- and

Table 9.2 Future technological tools for viral diagnostics

Technology	Target	Developer	Category	Reference
SPR & QCM	Ebola	Yu et al.	Biosensor	97
OFIS	Ebola	Petrosova et al.	Biosensor	100
VCSEL	Dengue	Huang et al.	Biosensor	109
QDs	Influenza	Deng et al.	Biosensor	110
Electrochemical using conductive polypyrole polymers	HIV, HBV	Bouchet et al.	Biosensor	112
CNTs	FMDV	Jithesh et al.	Biosensor/ nanoarray	114
SAW	Sin Nombre virus	Bisoffi et al.	Biosensor	118
Electrochemical	HBV	Ding et al.	Biosensor	119
Phage display on gold surface	–	Gervais et al.	Biosensor	120
Phage display on optical fiber	WNV	Herrman et al.	Biosensor	121
DPN	Influenza	Liu et al.	Biosensor/ nanoarray	122
PDA microarray gold nanoparticles	Papilloma virus	Baek et al.	Biochip	141
ICS	Influenza	Oh et al.	Biosensor	123
Functional polymers (PDA-polydiacetylene)	Influenza	Amano et al.	Biosensor	96
Nanowire	Influenza	Patolsky et al.	Biosensor	132
Fluorescent microbeads	FMDV	Perkins et al.	Microarray	154
Nanometer-sized gold particles using microfluidic immunoassay	HBV	Liang et al.	Lab on chip	155
Magnetic bead with metal oxide semiconductor	Dengue virus	Aytur et al.	Lab on chip	135
DNA hybridization using paramagnetic nanoparticles	HCV	Fuentes et al.	Lab on chip	156
Bioluminescence reporter gene	Anthrax	Rider et al.	Biosensor	142

Abbreviations: VCSEL vertical-cavity surface emitting laser; DPN, dip-pen nanolithography; PDA, photodiode array; QD, quantum dot; CNT, carbon nanotube; SAW, surface acoustic wave; ICS, ion channel switch; HIV, human immunodeficiency virus; HBV, hepatitis B virus; FMDV, foot-and-mouth disease virus; HCV, hepatitis C virus.

orange-QD-labeled chromatophores for the detection of different kinds of viruses. This sensor could be used as the convenient, cheap, reversible, and effective fluorescent probe for dual, simultaneous, and independent detection of viruses based on antibody-antigen reactions.[112]

Electrochemical detection has emerged as a challenging and well-adapted technique for the realization of cheap, single-use, and portable diagnoses systems.[113]

A promising strategy based on the use of conducting polymers has presented a new, real-time, multidetection biosensor for biological samples using the electrochemical properties of cylinder shaped conducting polypyrrole, grown on miniaturized electrodes. The new developed technique could easily be generalized for the elaboration of peptide or protein microarrays and thus be adapted for miniaturization and integration for lab-on-chip systems, including viral detection.[114]

There has been significant interest in biological applications of novel inorganic nanomaterials, such as nanotubes, with the motivation to create new types of analytical tools for life sciences and biotechnology. The electrical conductance of nanotube components is highly sensitive to its environment and varies significantly with changes in electrostatic charges and surface adsorption.[115] Consequently, nanotube-based biosensors were developed with the capability for selective detection of proteins, such as antibodies in solution. Recent work has demonstrated the ability of CNTs to act as a platform for immunosensor devices.[116] Using label and nonlabel antibodies that are attached to the CNTs, the immunosensor can then be visualized by microscopy techniques, for example, atomic force, electron scanning (scanning electron microscopy [SEM]), and confocal. CNT technology provides a miniaturized, multiplexed, immunosensor assay for point-of-care testing that could be of use for viral detection.

A new biosensor based on a shear, horizontal surface acoustic wave (SH-SAW) device was developed and recently presented. This sensor enabled label-free, sensitive, and cost-effective real-time detection of bacterial and viral DNA.[117–119] Bisoffi et al.[120] have shown the ability to use SAW technology based on a protein interaction device to detect category A viral agents, such as the Sin Nombre virus. The assay demonstrated high sensitivity and selectivity of the detector that was not compromised by the presence of other viruses, indicating its usefulness for specific detection in complex solutions. The platform allowed detection of viral bioagents without preprocessing of the analyte and with the potential to be applied in portable field conditions. Further developments of diagnostics devices have a label-free biosensor for the detection of hepatitis B virus,[121] a sensor based on the phage display technique designed for highly sensitive and specific platforms for detection and identification of a pathogenic agent,[122,123] and a nanopattern the F0F1-ATPase biosensor for viral detection.[124]

With a somewhat different approach, a new rapid point-of-care test using a novel ion channel switch biosensor (ICSB) for viral detection of influenza A was recently presented.[125] This sensor uses a simple electrical reader for objective measurement, without the required time-consuming steps of specimen extraction. The assay demonstrated the ability to achieve high sensitivity and specificity for the detection of a virus such as influenza as compared to the traditional method, such as ELISA, with the possibility for multiple analyses in a single test.[125]

The use of colorimetric sensors employing functional polymers has allowed direct analysis of target analytes through a color change without requiring additional instruments.[98] Theses sensors based on smart materials with physical, optical, or electrical properties have shown an ability to respond to specific environmental stimulation. Photodiode array (PDA)-conjugated polymer sensors, which can undergo significant optical property changes in response to fluctuations in environmental conditions,[126–128] have been used for viral detection with promising results.[129–131]

In a recent publication,[132] several immunosensing assays for viral detection were reviewed. The study describes current development and future potential technologies, including the use of nanowires, microbeads, microarrays, and a whole-cell-based immunoassay-sensing system. Nanowire technology, based on a metallic multistriped nanowire platform, enabled rapid fabrication of addressable barcodes. This multistriped system could be coated with anti-Ebola specific antibodies, allowing simultaneously an efficient and accurate multiplex detection assay for several biowarfare agents.[133] Further improvements have expanded the use of this technique for viral detection[134–136] using nanowires configured as field effect transistors that can detect slight variations on the nanowire surface that cause a change in conductivity. Micrometer-sized beads have proven their utility in bio- and viral sensing applications over the years.[10,137] The great advantages of these devices are their ability to be engineered by diverse surface chemistry, and they are easily manipulated by either magnetic forces or electrical fields. Combined with a flow cytometry device, the system would be able to detect multiplexed antigen binding and be used for efficient pathogen detection. Further work presented an improved biobarcode system, which included reporter beads, decorated with specific coding DNA sequences. Although DNA-based microarray chip technology is used in diagnostics, new developments have included antibody[138] and aptamer[139–141] microarrays, which when combined with technologies such as SPR[142] and PDA biochip using gold nanoparticles could enable better sensitivities.[143]

Finally, a promising and unusual pathogen detection system, using a whole-cell immunoassay sensor (CANARY) was recently presented.[144] Engineered B lymphocyte cells with pathogen-sensing membrane-bound antibodies and an associated light-emitting reporting system were conveniently expressed in vivo. The system expressed a calcium-sensitive bioluminescent protein named Aquaria. When exposed to targeted biowarfare pathogenic compounds, an increase in photons was observed within the B lymphocyte cells that later could easily be detected using an inexpensive optical system. In addition, several other groups extended the potential of the methodology by further developing whole-virus particles or a bacteriophage-sensing system for diagnostics use.[145–147] Overall, technological progress has made a significant impact on traditional and new developments of pathogen and viral diagnostics during the past few years. The ability to miniaturize and adapt current assays and protocols to fully automated devices will be certain to hold tremendous promise for new multiplex, efficient, cost-effective, and accurate viral detection systems for pathogens such as Ebola.

9.11 The Effort Continues

Historically, the Ebola virus has affected populations primarily throughout Central Africa. Although it has only affected a limited geographic region, its effects have reverberated worldwide due to its extremely high mortality rate and the lack of reservoir identification. Today it is clear that Ebola has evoked much fear in populations living outside Central Africa and has become a biodefense concern.[17,34,88,148] The early identification and diagnosis of emerging diseases, such as Ebola, are of great importance with respect to treatment, containment, and public health control. The importance of early detection lies not only with clinicians and public health experts but also with veterinarians, animal scientists, and wildlife ecologists[149,150] who require effective communication and collaboration for the etiological identification and epidemiological assessment of the virus.

Since its discovery 40 years ago, several methodologies and protocols were developed and adapted for diagnostics of the Ebola virus. Today, detection is done traditionally by laboratory diagnosis, which is defined by common methods used to confirm the clinical observations of a physician by evaluation of standard clinical specimens, for example, blood, serum, exudates, saliva, stool, and tissues.[34] Despite the success of methods such as RT-PCR and ELISA for viral detection, standardization and evaluation of diagnostic procedures are still difficult because of the restricted availability of virologic and clinical material. In addition, outbreaks of the virus usually occur in remote areas where sophisticated medical support systems are limited and timely diagnostic services are extremely difficult to provide, constituting major problems. When considering the various issues during a virus outbreak and its huge impact on both an individual and his or her surroundings, it is clear why further research for development of diagnostic tools must continue.

The production of recombinant engineered reagents, along with the development of innovative devices and instruments, enables investigators to use the Ebola virus outside BSL-4 lab facilities in developing countries and under field conditions. The rapid growth of improvements in existing methods, along with the development of revolutionary technologies, will be certain to have a great impact on Ebola detection. Fast and portable tools for Ebola diagnostics will eventually lead to a more rapid control of the epidemic. In addition, early detection of the virus using portable and sensitive assays will allow scientists to quickly identify the index case(s) and as a result to detect the potential reservoir. The early detection and diagnostics of Ebola is an essential step if intervention is to occur at a point where the prognosis can still be influenced and selection of optimum therapeutic strategies can still be relevant.

The threat of the Ebola virus as a potential bioterrorism agent has been previously established and discussed.[35,42,88] Although new technologies mainly for detecting the release of specific biothreat agents have been widely introduced, further improvements and development are needed. The recent outbreaks of the virus in Central Africa, including the one in 2007,[18,151–154] have clearly emphasized the virus's importance not only as a biothreat agent but also as a natural cause of death to humans. Due to its ability to develop new lethal strains, such as the recent one, 2007–2008,[152] or to distinguish engineered strains from other pathogens that may produce similar symptoms, continuous efforts toward developing new, advanced assays, which will be quick and reliable with minimum sample handling and laboratory skill requirement, are a necessity.[98] Today, traditional detection systems, such as virus culture, electron microscopy, and IF, are still considered key elements in the confirmation and detection of Ebola and will continue to be so despite their risk to lab workers, the requirement for special facilities, and the high cost in time and labor intensity. It is clear that future diagnostics of viruses, such as Ebola, will evolve depending on new developments of cutting-edge technologies, as well as a greater understanding of Ebola immunology[155,156] which will produce detection systems able to cope with and overcome difficulties existing today and in the future.

References

1. Feldmann, H., et al. Ebola virus ecology: a continuing mystery. 2004. *Trends Microbiol.*, **12**(10), 433–437.
2. Pringle, C. R. The Bunyaviridae and their genetics: an overview. 1991. *Curr. Top. Microbiol. Immunol.*, **169**, 1–25.
3. Feldmann, H., and H. D. Klenk. Marburg and Ebola viruses. 1996. *Adv. Virus Res.*, **47**, 1–52.
4. Colebunders, R., and M. Borchert. Ebola haemorrhagic fever: a review. 2000. *J. Infect.*, **401**, 16–20.
5. Murphy, F. A., G., van der Groen, S. G. Whitfield, and J. V. Lange. 1978. Ebola and Marburg virus morphology and taxonomy, in *Ebola Virus Haemorrhagic Fever* (Pattyn, S. R., ed), Elsevier, Amsterdam, pp. 61–84.

6. Kiley, M. P., et al. Filoviridae: a taxonomic home for Marburg and Ebola viruses? 1982. *Intervirology*, **181**(2), 24–32.
7. Cox, N. J., et al. Evidence for two subtypes of Ebola virus based on oligonucleotide mapping of RNA. 1983. *J. Infect. Dis.*, **147**(2), 272–275.
8. Buchmeier, M. J., et al. Comparative analysis of the structural polypeptides of Ebola viruses from Sudan and Zaire. 1983. *J. Infect. Dis.*, **147**(2), 276–281.
9. Bowen, E. T., et al. A comparative study of strains of Ebola virus isolated from southern Sudan and northern Zaire in 1976. 1980. *J. Med. Virol.*, **6**(2), 129–138.
10. Aytur, T., et al. A novel magnetic bead bioassay platform using a microchip-based sensor for infectious disease diagnosis. 2006. *J. Immunol. Methods.*, **3141**(2), 21–29.
11. Johnson, K. M. Ebola virus and hemorrhagic fever: Andromeda strain or localized pathogen? 1979. *Annu. Intern. Med.*, **91**(1), 117–119.
12. Heymann, D. L., et al. Ebola hemorrhagic fever: Tandala, Zaire, 1977–1978. 1980. *J. Infect. Dis.*, **142**(3), 372–376.
13. Pourrut, X., et al. The natural history of Ebola virus in Africa. 2005. *Microbes Infect.*, **77**(8), 1005–1014.
14. Feldmann, H., et al. Effective post-exposure treatment of Ebola infection. 2007. *PLoS Pathog.*, **3**(1), e2.
15. Feldmann, H., T. Geisbert, and Y. Kawaoka. Filoviruses: recent advances and future challenges. 2007. *J. Infect. Dis.*, **196**(Suppl 2), S129–S130.
16. Feldmann, H., et al. Emerging and re-emerging infectious diseases. 2002. *Med. Microbiol. Immunol.*, **191**(2), 63–74.
17. Peters, C. J., and J. W. LeDuc. An introduction to Ebola: the virus and the disease. 1999. *J. Infect. Dis.*, **179**(Suppl 1), ix–xvi.
18. Okware, S. I., et al. An outbreak of Ebola in Uganda. 2002. *Trop. Med. Int. Health*, **71**(2), 1068–1075.
19. Bruce, J., and P. Brysiewicz. Ebola fever: the African emergency. 2002. *Int. J. Trauma Nurs.*, **8**(2), 36–41.
20. Geisbert, T. W., and P. B. Jahrling. Exotic emerging viral diseases: progress and challenges. 2004. *Nat. Med.*, **10**(12), S110–S121.
21. Xu, L., et al. Immunization for Ebola virus infection. 1998. *Nat. Med.*, **4**(1), 37–42.
22. Wilson, J. A., et al. Vaccine potential of Ebola virus VP24, VP30, VP35, and VP40 proteins. 2001. *Virology*, **286**(2), 384–390.

23. Wilson, J. A., C. M. Bosio, and M. K. Hart. Ebola virus: the search for vaccines and treatments. 2001. *Cell Mol. Life Sci.*, **5812**(13), 1826–1841.
24. Wang, D., et al. Development of a cAdVax-based bivalent ebola virus vaccine that induces immune responses against both the Sudan and Zaire species of Ebola virus. 2006. *J. Virol.*, **80**(6), 2738–2746.
25. Vanderzanden, L., et al. DNA vaccines expressing either the GP or NP genes of Ebola virus protect mice from lethal challenge. 1998. *Virology*, **246**(1), 134–144.
26. Sullivan, N. J., et al. Development of a preventive vaccine for Ebola virus infection in primates. 2000. *Nature*, **408**(6812), 605–609.
27. Sullivan, N., Z. Y. Yang, and G. J. Nabel. Ebola virus pathogenesis: implications for vaccines and therapies. 2003. *J. Virol.*, **77**(18), 9733–9737.
28. Reed, D. S., and M. Mohamadzadeh. Status and challenges of filovirus vaccines. 2007. *Vaccine*, **25**(11), 1923–1934.
29. Pushko, P., et al. Recombinant RNA replicons derived from attenuated Venezuelan equine encephalitis virus protect guinea pigs and mice from Ebola hemorrhagic fever virus. 2000. *Vaccine*, **19**(1), 142–153.
30. Leroy, E. M., et al. Sequence analysis of the GP, NP, VP40 and VP24 genes of Ebola virus isolated from deceased, surviving and asymptomatically infected individuals during the 1996 outbreak in Gabon: comparative studies and phylogenetic characterization. 2002. *J. Gen. Virol.*, **83**(Pt 1), 67–73.
31. Jones, S. M., et al. Assessment of a vesicular stomatitis virus-based vaccine by use of the mouse model of Ebola virus hemorrhagic fever. 2007. *J. Infect. Dis.*, **196**(Suppl 2), S404–S412.
32. Geisbert, T. W., et al. Evaluation in nonhuman primates of vaccines against Ebola virus. 2002. *Emerg. Infect. Dis.*, **8**(5), 503–507.
33. Feldmann, H., et al. Ebola virus: from discovery to vaccine. 2003. *Nat. Rev. Immunol.*, **3**(8), 677–685.
34. Burnett, J. C., et al. The evolving field of biodefence: therapeutic developments and diagnostics. 2005. *Nat. Rev. Drug Discov.*, **4**(4), 281–297.
35. Borio, L., et al. Hemorrhagic fever viruses as biological weapons: medical and public health management. 2002. *JAMA*, **287**(18), 2391–2405.

36. Riemenschneider, J., et al. Comparison of individual and combination DNA vaccines for B. anthracis, Ebola virus, Marburg virus and Venezuelan equine encephalitis virus. 2003. *Vaccine*, **2125**(26), 4071–4080.

37. Martin, J. E., et al. A DNA vaccine for Ebola virus is safe and immunogenic in a phase I clinical trial. 2006. *Clin. Vaccine Immunol.*, **13**(11), 1267–1277.

38. Fox, J. L. Antivirals become a broader enterprise. 2007. *Nat. Biotechnol.*, **25**(12), 1395–1402.

39. Oswald, W. B., et al. Neutralizing antibody fails to impact the course of Ebola virus infection in monkeys. 2007. *PLoS Pathog.*, **3**(1), e9.

40. Mupapa, K., et al. Treatment of Ebola hemorrhagic fever with blood transfusions from convalescent patients. International Scientific and Technical Committee. 1999. *J. Infect. Dis.*, **179**(Suppl 1), S18–S23.

41. Jahrling, P. B., et al. Ebola hemorrhagic fever: evaluation of passive immunotherapy in nonhuman primates. 2007. *J. Infect. Dis.*, **196**(Suppl 2), S400–S403.

42. Niiler, E. Bioterrorism: biotechnology to the rescue? 2002. *Nat. Biotechnol.*, **20**, 21–25.

43. Van der Groen, G., and S. R. Pattyn. Measurement of antibodies to Ebola virus in human sera from N. W. Zaire. 1979. *Annu. Soc. Belg. Med. Trop.*, **59**(1), 87–92.

44. Zaki, S. R., et al. A novel immunohistochemical assay for the detection of Ebola virus in skin: implications for diagnosis, spread, and surveillance of Ebola hemorrhagic fever. Commission de Lutte contre les Epidemies a Kikwit. 1999. *J. Infect. Dis.*, **179**(Suppl 1), S36–S47.

45. Weidmann, M., E. Muhlberger, and F. T. Hufert. Rapid detection protocol for filoviruses. 2004. *J. Clin. Virol.*, **30**(1), 94–99.

46. Towner, J. S., et al. Rapid diagnosis of Ebola hemorrhagic fever by reverse transcription-PCR in an outbreak setting and assessment of patient viral load as a predictor of outcome. 2004. *J. Virol.*, **78**(8), 4330–4341.

47. Sanchez, A., et al. Detection and molecular characterization of Ebola viruses causing disease in human and nonhuman primates. 1999. *J. Infect. Dis.*, **179**(Suppl 1), S164–S169.

48. Leroy, E. M., et al. Human asymptomatic Ebola infection and strong inflammatory response. 2000. *Lancet*, **355**(9222), 2210–2215.

49. Ksiazek, T. G., et al. ELISA for the detection of antibodies to Ebola viruses. 1999. *J. Infect. Dis.*, **179**(Suppl 1), S192–S198.

50. Ksiazek, T. G., et al. Clinical virology of Ebola hemorrhagic fever (EHF): virus, virus antigen, and IgG and IgM antibody findings among EHF patients in Kikwit, Democratic Republic of the Congo, 1995. 1999. *J. Infect. Dis.*, **179**(Suppl 1), S177–S187.

51. Ksiazek, T. G., et al. Enzyme immunosorbent assay for Ebola virus antigens in tissues of infected primates. 1992. *J. Clin. Microbiol.*, **30**(4), 947–950.

52. Gibb, T. R., et al. Development and evaluation of a fluorogenic 5' nuclease assay to detect and differentiate between Ebola virus subtypes Zaire and Sudan. 2001. *J. Clin. Microbiol.*, **39**(11), 4125–4130.

53. Drosten, C., et al. Rapid detection and quantification of RNA of Ebola and Marburg viruses, Lassa virus, Crimean-Congo hemorrhagic fever virus, Rift Valley fever virus, dengue virus, and yellow fever virus by real-time reverse transcription-PCR. 2002. *J. Clin. Microbiol.*, **40**(7), 2323–2330.

54. Dowell, S. F., et al. Transmission of Ebola hemorrhagic fever: a study of risk factors in family members, Kikwit, Democratic Republic of the Congo, 1995. Commission de Lutte contre les Epidemies a Kikwit. 1999. *J. Infect. Dis.*, **179**(Suppl 1), S87–S91.

55. Henchal, E. A., J. D. Teska, and J. W. Ezzell. Responding to biological terrorism: a role for the clinical laboratory. 2000. *Clin. Lab. News*, **26**, 14–18.

56. Gilchrist, M. J. A national laboratory network for bioterrorism: evolution from a prototype network of laboratories performing routine surveillance. 2000. *Mil. Med.*, **1657**(Suppl 2), 28–31.

57. Guimard, Y., et al. Organization of patient care during the Ebola hemorrhagic fever epidemic in Kikwit, Democratic Republic of the Congo, 1995. 1999. *J. Infect. Dis.*, **179**(Suppl 1), S268–S273.

58. Saijo, M., et al. Laboratory diagnostic systems for Ebola and Marburg hemorrhagic fevers developed with recombinant proteins. 2006. *Clin. Vaccine Immunol.*, **13**(4), 444–451.

59. Siegert, R., et al. On the etiology of an unknown human infection originating from monkeys. 1967. *Dtsch. Med. Wochenschr.*, **92**(51), 2341–2343.

60. Geisbert, T. W., and P. B. Jahrling. Use of immunoelectron microscopy to show Ebola virus during the 1989 United States epizootic. 1990. *J. Clin. Pathol.*, **43**(10), 813–816.

61. Geisbert, T. W., J. B. Rhoderick, and P. B. Jahrling. Rapid identification of Ebola virus and related filoviruses in fluid specimens using indirect immunoelectron microscopy. 1991. *J. Clin. Pathol.*, **44**(6), 521–522.

62. Geisbert, T. W., and P. B. Jahrling. Differentiation of filoviruses by electron microscopy. 1995. *Virus Res.*, **392**(3), 129–150.

63. Murray, P. R. 2003. *Manual of Clinical Microbiology*, ASM Press, Washington, DC.

64. Ksiazek, T. G. Laboratory diagnosis of filovirus infections in nonhuman primates. 1991. *Lab. Anim.*, **20**, 34–46.

65. Ikegami, T., et al. Immunoglobulin G enzyme-linked immunosorbent assay using truncated nucleoproteins of Reston Ebola virus. 2003. *Epidemiol. Infect.*, **130**(3), 533–539.

66. Groen, J., et al. Serological reactivity of baculovirus-expressed Ebola virus VP35 and nucleoproteins. 2003. *Microbes Infect.*, **5**(5), 379–385.

67. Becker, S., et al. Evidence for occurrence of filovirus antibodies in humans and imported monkeys: do sub-clinical filovirus infections occur worldwide? 1992. *Med Microbiol. Immunol.*, **181**(1), 43–55.

68. Saijo, M., et al. Characterization of monoclonal antibodies to Marburg virus nucleoprotein (NP) that can be used for NP-capture enzyme-linked immunosorbent assay. 2005. *J. Med. Virol.*, **76**(1), 111–118.

69. Niikura, M., et al. Detection of Ebola viral antigen by enzyme-linked immunosorbent assay using a novel monoclonal antibody to nucleoprotein. 2001. *J. Clin. Microbiol.*, **39**(9), 3267–3671.

70. Lucht, A., et al. Production of monoclonal antibodies and development of an antigen capture ELISA directed against the envelope glycoprotein GP of Ebola virus. 2004. *Med. Microbiol. Immunol.*, **193**(4), 181–187.

71. Lucht, A., et al. Development, characterization and use of monoclonal VP40-antibodies for the detection of Ebola virus. 2003. *J. Virol. Methods*, **111**(1), 21–28.

72. Ikegami, T., et al. Antigen capture enzyme-linked immunosorbent assay for specific detection of Reston Ebola virus nucleoprotein. 2003. *Clin. Diagn. Lab. Immunol.*, **10**(4), 552–557.

73. Shahhosseini, S., et al. Production and characterization of monoclonal antibodies against different epitopes of Ebola virus antigens. 2007. *J. Virol. Methods*, **143**(1), 29–37.

74. Franz, D. R., et al. Clinical recognition and management of patients exposed to biological warfare agents. 2001. *Clin. Lab. Med.*, **21**(3), 435–473.

75. Franz, D. R., et al. Clinical recognition and management of patients exposed to biological warfare agents. 1997. *JAMA*, **278**(5), 399–411.
76. Zaki, S. R., and C. D. Goldsmith. Pathologic features of filovirus infections in humans. 1999. *Curr. Top. Microbiol. Immunol.*, **235**, 97–116.
77. Lloyd, E. S., et al. Long-term disease surveillance in Bandundu region, Democratic Republic of the Congo: a model for early detection and prevention of Ebola hemorrhagic fever. 1999. *J. Infect. Dis.*, **179**(Suppl 1), S274–S280.
78. Elliott, L. H., M. P. Kiley, and J. B. McCormick. Descriptive analysis of Ebola virus proteins. 1985. *Virology*, **147**(1), 169–176.
79. Elliott, L. H., et al. Improved specificity of testing methods for filovirus antibodies. 1993. *J. Virol. Methods*, **43**(1), 85–89.
80. Zhai, J., et al. Rapid molecular strategy for filovirus detection and characterization. 2007. *J. Clin. Microbiol.*, **45**(1), 224–226.
81. Palacios, G., et al. MassTag polymerase chain reaction for differential diagnosis of viral hemorrhagic fever. 2006. *Emerg. Infect. Dis.*, **12**(4), 692–695.
82. Grolla, A., et al. Laboratory diagnosis of Ebola and Marburg hemorrhagic fever. 2005. *Bull. Soc. Pathol. Exot.*, **98**(3), 205–209.
83. Garcia-de-Lomas, J., and D. Navarro. New directions in diagnostics. *Pediatr.* 1997. *Infect. Dis. J.*, **163**(Suppl 3), S43–S48.
84. Whelen, A. C., and D. H. Persing. The role of nucleic acid amplification and detection in the clinical microbiology laboratory. 1996. *Annu. Rev. Microbiol.*, **50**, 349–373.
85. Sobarzo, A., E. Perelman, A. Groseth, O. Dolnik, S. Becker, J. J. Lutwama, J. Dye, V. Yavelsky, L. Lobel, and R. S. Marks. Profiling the native specific human humoral immune response to Sudan ebolavirus (Gulu) by chemiluminescense ELISA. 2013. *Clin. Vaccine Immunol.*, **19**(11), 1844–1852.
86. Leroy, E. M., et al. Diagnosis of Ebola haemorrhagic fever by RT-PCR in an epidemic setting. 2000. *J. Med. Virol.*, **60**(4), 463–467.
87. Panning, M., et al. Diagnostic reverse-transcription polymerase chain reaction kit for filoviruses based on the strain collections of all European biosafety level 4 laboratories. 2007. *J. Infect. Dis.*, **196**(Suppl 2), S199–S204.
88. Niedrig, M., et al. First international quality assurance study on the rapid detection of viral agents of bioterrorism. 2004. *J. Clin. Microbiol.*, **42**(4), 1753–1755.

89. Sanchez, A. F. H. 1995. Detection of Marburg and Ebola virus infections by polymerase chain reaction assays, in *Protocols for Diagnosis of Human and Animal Virus Diseases*, pp. 411–419.
90. Lucht, A., et al. Development of an immunofiltration-based antigen-detection assay for rapid diagnosis of Ebola virus infection. 2007. *J. Infect. Dis.*, **196**(Suppl 2), S184–S192.
91. Saijo, M., et al. Enzyme-linked immunosorbent assays for detection of antibodies to Ebola and Marburg viruses using recombinant nucleoproteins. 2001. *J. Clin. Microbiol.*, **39**(1), 1–7.
92. Prehaud, C., et al. Recombinant Ebola virus nucleoprotein and glycoprotein (Gabon 94 strain) provide new tools for the detection of human infections. 1998. *J. Gen. Virol.*, **79**(Pt 11), 2565–2572.
93. Saijo, M., et al. Immunofluorescence method for detection of Ebola virus immunoglobulin g, using HeLa cells which express recombinant nucleoprotein. 2001. *J. Clin. Microbiol.*, **39**(2), 776–778.
94. Nissim, A., et al. Antibody fragments from a "single pot" phage display library as immunochemical reagents. 1994. *EMBO J.*, **13**(3), 692–698.
95. Chen, H., et al. Evaluation of a nucleoprotein-based enzyme-linked immunosorbent assay for the detection of antibodies against infectious bronchitis virus. 2003. *Avian Pathol.*, **32**(5), 519–526.
96. Strong J. E., P. B. Jahrling, and H. Feldmann. 2006. Filo-viruses and arenaviruses, in *Manual of Molecular and Clinical Laboratory Immunology*, pp. 774–790.
97. Rowe, A. K., et al. Clinical, virologic, and immunologic follow-up of convalescent Ebola hemorrhagic fever patients and their household contacts, Kikwit, Democratic Republic of the Congo. Commission de Lutte contre les Epidemies a Kikwit. 1999. *J. Infect. Dis.*, **179**(Suppl 1), S28–S35.
98. Amano, Y., and Q. Cheng. Detection of influenza virus: traditional approaches and development of biosensors. 2005. *Anal. Bioanal. Chem.*, **381**(1), 156–164.
99. Yu, J. S., et al. Detection of Ebola virus envelope using monoclonal and polyclonal antibodies in ELISA, surface plasmon resonance and a quartz crystal microbalance immunosensor. 2006. *J. Virol. Methods*, **137**(2), 219–228.
100. Dell'Atti, D., et al. Development of combined DNA-based piezoelectric biosensors for the simultaneous detection and genotyping of high risk Human Papilloma Virus strains. 2007. *Clin. Chim. Acta*, **383**(1–2), 140–146.

101. Gerdon, A. E., D. W. Wright, and D. E. Cliffel. Quartz crystal microbalance detection of glutathione-protected nanoclusters using antibody recognition. 2005. *Anal. Chem.*, **77**(1), 304–310.

102. Petrosova, A., T. Konry, S. Cosnier, I. Trakht, J. Lutwama, E. Rwaguma, A. Chepurnov, E. Mühlberger, L. Lobel, and R. S. Marks. Development of a highly sensitive, field operable biosensor for serological studies of Ebola virus in central Africa. 2007. *Sens. Actuators, B,* **122**, 578–586.

103. Ebihara, H., A. Groseth, G. Neumann, Y. Kawaoka, and H. Feldmann. The role of reverse genetics systems in studying viral hemorrhagic fevers. 2005. *Thromb. Haemost.*, **94**(2), 240–253.

104. Neumann, G., and Y. Kawaoka. Reverse genetics systems for the generation of segmented negative-sense RNA viruses entirely from cloned cDNA. 2004. *Curr. Top. Microbiol. Immunol.*, **283**, 43–60.

105. Volchkov, V. E., et al. Recovery of infectious Ebola virus from complementary DNA: RNA editing of the GP gene and viral cytotoxicity. 2001. *Science,* **291**(5510), 1965–1969.

106. Towner, J. S., et al. Generation of eGFP expressing recombinant Zaire Ebola virus for analysis of early pathogenesis events and high-throughput antiviral drug screening. 2005. *Virology*, **332**(1), 20–27.

107. Neumann, G., et al. Reverse genetics demonstrates that proteolytic processing of the Ebola virus glycoprotein is not essential for replication in cell culture. 2002. *J. Virol.*, **76**(1), 406–410.

108. Ebihara, H., et al. In vitro and in vivo characterization of recombinant Ebola viruses expressing enhanced green fluorescent protein. 2007. *J. Infect. Dis.*, **196**(Suppl 2), S313–S322.

109. Ramsden, J. J. Optical biosensors. 1997. *J Mol. Recognit.*, **10**(3), 109–120.

110. Cooper, M. A. Optical biosensors in drug discovery. 2002. *Nat. Rev. Drug Discov.*, **1**(7), 515–528.

111. Huang, M. C. Y., C. F. R. M. Jonathan, E. Foley, R. Beatty, B. T. Cunningham, and C. J. Chang-Hasnain. VCSEL optoelectronic biosensor for detection of infectious diseases. 2008. *IEEE Photonics Technol. Lett.*, **20**, 443–445.

112. Deng, Z., et al. Green and orange CdTe quantum dots as effective pH-sensitive fluorescent probes for dual simultaneous and independent detection of viruses. 2007. *J. Phys. Chem. B.*, **111**(41), 12024–12031.

113. Drummond, T. G., M. G. Hill, and J. K. Barton. Electrochemical DNA sensors. 2003. *Nat. Biotechnol.*, **21**(10), 1192–1199.

114. Bouchet, A., et al. Cylinder-shaped conducting polypyrrole for labelless electrochemical multidetection of DNA. 2007. *Biosens. Bioelectron.*, **23**(5), 735–740.

115. Chen, R. J., et al. Noncovalent functionalization of carbon nanotubes for highly specific electronic biosensors. 2003. *Proc. Natl. Acad. Sci. U S A*, **100**(9), 4984–4989.

116. Veetil, J. V., and K. Ye. Development of immunosensors using carbon nanotubes. 2007. *Am. Chem. Soc. Am. Inst. Chem. Eng.*, **23**, 517–531.

117. Berkenpas, E., P. Millard, and M. Pereira da Cunha. Detection of Escherichia coli O157:H7 with langasite pure shear horizontal surface acoustic wave sensors. 2006. *Biosens. Bioelectron.*, **21**(12), 2255–2262.

118. Moll, N., et al. A. Love wave immunosensor for whole E. coli bacteria detection using an innovative two-step immobilisation approach. 2007. *Biosens. Bioelectron.*, **229**(10), 2145–2150.

119. Branch, D. W., and S. M. Brozik. Low-level detection of a bacillus anthracis simulant using Love-wave biosensors on 36 degrees YX LiTaO3. 2004. *Biosens. Bioelectron.*, **19**(8), 849–859.

120. Bisoffi, M., et al. Detection of viral bioagents using a shear horizontal surface acoustic wave biosensor. 2008. *Biosens. Bioelectron.*, **23**(9), 1397–1403.

121. Ding, C., et al. Hybridization biosensor using 2,9-dimethyl-1,10-phenantroline cobalt as electrochemical indicator for detection of hepatitis B virus DNA. 2008. *Bioelectrochemistry*, **72**(1), 28–33.

122. Gervais, L., M. Gel, B. Allain, M. Tolba, L. Brovko, M. Zourob, R. Mandeville, M. Griffiths, S. Evoy. Immobilization of biotinylated bacteriophage on biosensor surfaces. 2007. *Sens. Actuators B*, **125**, 615–621.

123. Herrmann, S., et al. T7 phage display of Ep15 peptide for the detection of WNV IgG. 2007. *J. Virol. Methods*, **141**(2), 133–140.

124. Liu, X., J. Yue, and Z. Zhang. WITHDRAWN: Generation of F0F1-ATPase nanoarray by Dip-Pen Nanolithography and its application as biosensors. 2008. *Arch. Biochem. Biophys.* in press.

125. Oh, S. Y., et al. Rapid detection of influenza A virus in clinical samples using an ion channel switch biosensor. 2008. *Biosens. Bioelectron.*, **23**(7), 1161–1165.

126. Leclerc, M. Optical and electrochemical transducers based on functionalized conjugated polymers. 1999. *Adv. Mater.* **11**, 1491–1498.

127. Jung, Y. K., H. G. Park, and J. M. Kim. Polydiacetylene (PDA)-based colorimetric detection of biotinstreptavidin interactions. 2006. *Biosens. Bioelectron.*, **21**(8), 1536–1544.

128. Charych, D., et al. A "litmus test" for molecular recognition using artificial membranes. 1996. *Chem. Biol.*, **3**(2), 113–120.

129. Dore, K., et al. Fluorescent polymeric transducer for the rapid, simple, and specific detection of nucleic acids at the zeptomole level. 2004. *J. Am. Chem. Soc.*, **126**(13), 4240–4244.

130. Charych, D. H., et al. Direct colorimetric detection of a receptor-ligand interaction by a polymerized bilayer assembly. 1993. *Science*, **261**(5121), 585–588.

131. Reichert, A., J. O. Nagy, W. Spevak, and D. Charych. Polydiacetylene liposomes functionalized with sialic acid bind and colorimetrically detect influenza virus. 1995. *J. Am. Chem. Soc.*, **117**, 829–830.

132. Fischer, N. O., T. M. Tarasow, and J. B. Tok. Heightened sense for sensing: recent advances in pathogen immunoassay sensing platforms. 2007. *Analyst*, **132**(3), 187–191.

133. Tok, J. B., et al. Metallic striped nanowires as multiplexed immunoassay platforms for pathogen detection. 2006. *Angew. Chem., Int. Ed. Engl.*, **45**(41), 6900–6904.

134. Wang, W. U., et al. Label-free detection of small-molecule-protein interactions by using nanowire nanosensors. 2005. *Proc. Natl. Acad. Sci. U S A*, **102**(9), 3208–3212.

135. Patolsky, F., G. Zheng, and C. M. Lieber. Nanowire sensors for medicine and the life sciences. 2006. *Nanomedicine*, **1**(1), 51–65.

136. Patolsky, F., et al. Electrical detection of single viruses. 2004. *Proc. Natl. Acad. Sci. U S A*, **101**(39), 14017–14022.

137. Wilson, R., A. R. Cossins, and D. G. Spiller. Encoded microcarriers for high-throughput multiplexed detection. 2006. *Angew. Chem., Int. Ed. Engl.*, **45**(37), 6104–6117.

138. Gehring, A. G., et al. Antibody microarray detection of Escherichia coli O157:H7: quantification, assay limitations, and capture efficiency. 2006. *Anal. Chem.*, **78**(18), 6601–6607.

139. Lubin, A. A., et al. Sequence-specific, electronic detection of oligonucleotides in blood, soil, and foodstuffs with the reagentless, reusable E-DNA sensor. 2006. *Anal. Chem.*, **78**(16), 5671–5677.

140. Baker, B. R., et al. An electronic, aptamer-based small-molecule sensor for the rapid, label-free detection of cocaine in adulterated samples and biological fluids. 2006. *J. Am. Chem. Soc.*, **128**(10), 3138–3139.

141. Lee, S., et al. Chip-based detection of hepatitis C virus using RNA aptamers that specifically bind to HCV core antigen. 2007. *Biochem. Biophys. Res. Commun.*, **358**(1), 47–52.

142. Endo, T., et al. Multiple label-free detection of antigen-antibody reaction using localized surface plasmon resonance-based core-shell structured nanoparticle layer nanochip. 2006. *Anal. Chem.*, **78**(18), 6465–6475.

143. Baek, T. J., et al. Development of a photodiode array biochip using a bipolar semiconductor and its application to detection of human papilloma virus. 2008. *Anal. Bioanal. Chem.*, **390**(5), 1373–1378.

144. Rider, T. H., et al. A B cell-based sensor for rapid identification of pathogens. 2003. *Science*, **301**(5630), 213–215.

145. Williams, D. D., O. Benedek, and C. L. Turnbough, Jr. Species-specific peptide ligands for the detection of Bacillus anthracis spores. 2003. *Appl. Environ. Microbiol.*, **69**(10), 6288–6293.

146. Sapsford, K. E., et al. A cowpea mosaic virus nanoscaffold for multiplexed antibody conjugation: application as an immunoassay tracer. 2006. *Biosens. Bioelectron.*, **21**(8), 1668–1673.

147. Martin, B. D., et al. An engineered virus as a bright fluorescent tag and scaffold for cargo proteins: capture and transport by gliding microtubules. 2006. *J. Nanosci. Nanotechnol.*, **6**(8), 2451–2460.

148. Cunha, B. A. Anthrax, tularemia, plague, ebola or smallpox as agents of bioterrorism: recognition in the emergency room. 2002. *Clin. Microbiol. Infect.*, **8**(8), 489–503.

149. Murphy, F. A. Emerging zoonoses. 1998. *Emerg. Infect. Dis.*, **4**, 429–435.

150. Chomel, B. B. Control and prevention of emerging zoonoses. 2003. *J. Vet. Med. Educ.*, **30**(2), 145–147.

151. Onyango, C. O., et al. Laboratory diagnosis of Ebola hemorrhagic fever during an outbreak in Yambio, Sudan, 2004. 2007. *J. Infect. Dis.*, **196**(Suppl 2), S193–S198.

152. Outbreak news. Ebola haemorrhagic fever, Uganda: end of the outbreak. 2008. *Wkly. Epidemiol. Rec.*, **83**(10), 89–90.

153. Outbreak news. Ebola virus haemorrhagic fever, Democratic Republic of the Congo. 2007. *Wkly. Epidemiol. Rec.*, **82**(38), 329.

154. Outbreak of Ebola haemorrhagic fever, Uganda, August 2000–January 2001. 2001. *Wkly. Epidemiol. Rec.*, **76**(6), 41–46.

155. Sobarzo, A., D. E. Ochayon, J.J. Lutwama, S. Balinandi, O. Guttmn, R.S., Marks, A. I. Kuehne, J. M. Dye, V. Yavelsky, E. C. Lewis, and L. Lobel.

Persistent immune responses after Ebola virus infection. 2013. *N. Engl. J. Med.*, **369**(5), 492–493.

156. Sobarzo, A., A. Groseth, O. Dolnik, S. Becker, J. J. Lutwama, E. Perelman, V. Yavelsky, M. Muhammad, A. I. Kuene, R. S. Marks, J. M. Dye, and L. Lobel. Profile and persistence of the virus-specific neutralizing humoral immune response in human survivors of Sudan ebolavirus (Gulu). 2013. *J. Infect. Dis.*, doi: 10.1093/infdis/jit162.

Chapter 10

Pathogen Detection Using Spatially Focused Microwaves and Metal-Enhanced Fluorescence

Kadir Aslan and Chris D. Geddes
Institute of Fluorescence, Laboratory for Advanced Medical Plasmonics and Laboratory for Advanced Fluorescence Spectroscopy, Medical Biotechnology Center, University of Maryland Biotechnology Institute, MD 21202, USA
aslan@umbi.umd.edu, geddes@umbi.umd.edu

10.1 Ultrafast and Sensitive Detection of Anthrax with Focused Microwave and Metal-Enhanced Fluorescence

This chapter summarizes recent work on ultrafast and sensitive bioassays based on the combined use of focused microwaves, low-power microwave heating, and metal-enhanced fluorescence (MEF) for the extraction and detection of biological targets of interest from pathogens and cells. When exposed to microwaves, metallic structures focus the fields to their immediate surroundings, resulting in increased electric fields at the peripheries. Subsequently, substantially increased electric field gradients increases the temperature of

Viral Diagnostics: Advances and Applications
Edited by Robert S. Marks, Leslie Lobel, and Amadou Alpha Sall

ISBN 978-981-4364-43-0 (Hardcover), 978-981-4364-44-7 (eBook)
www.panstanford.com

materials placed near these structures. The temperature increases can be controlled within a range of 30°C to 130°C, affording for DNA hybridization events to occur in several seconds at low temperatures and the extraction of biological materials from pathogens and cells at high temperatures. Subsequently, ultrafast and sensitive detection of proteins or DNA extracted from pathogens and cells can be carried out readily using MEF and low-power microwave heating. The real-life application of the combined use of these technologies was demonstrated for *Bacillus anthracis* (*B. anthracis*) spores and vegetative cells, the causative pathogen for anthrax.

10.2 Metal-Enhanced Fluorescence

Since our platform technology utilizes the benefits of the MEF phenomenon, it is pertinent to present a brief description of MEF for the sake of completeness for this book chapter. Early descriptions of MEF can be traced back to the observations made with the interactions of fluorescent species with planar metallic surfaces. When placed near a planar metal surface, the spontaneous emission of a fluorescent species was thought to follow radiative and/or nonradiative decay paths, which are mainly dependent on the distance (R) between the emitter (dipole) and the metal, as well as the orientation of the dipole of the fluorescent species with respect to the metal surface.[1] Two effects have been demonstrated as a result of the distance dependence of the emissive rate:[2] i) the emission rate oscillates as the distance is increased, as the phase of the surface-reflected field changes with distance and ii) the amplitude of the oscillation decreases. The significance of the dipole orientation can be seen when we consider that the metal surface produces an image dipole on the surface. For a very small distance between the emitter and the metal surface, a dipole that is parallel to the surface is cancelled out by its image and a perpendicular dipole is enhanced. In this regard, the distance-dependent spontaneous emission rate can be predicted, assuming that the reflecting surface is perfect and the dipole moment of the emitter rotates rapidly within the excited state lifetime.

In addition to the processes described above derived for planar metal geometries, one can find studies in the literature offering a description and applications of the spontaneous emission rate near metallic nanoparticles similar to the description for planar metal geometries.[3–11] The major difference between the planar and particulate systems is the inclusion of localized modes occurring in particles. The localized modes in particulate systems results in the omission of the oscillations of the decay rate observed for the planar systems. The frequency of the localized modes depends on the both the size and the shape of the metallic nanoparticles. For a single emitter placed near metal nanoparticles (using a dipole-dipole model) the nonradiative decay rate is shown to follow an R^{-3} dependence; the radiative decay rate follows an R^{-6} dependence.[6] As shown in Fig. 10.1A, the energy is partially transferred (via nonradiative coupling, a distance-dependent phenomenon) from the excited state of the fluorescent species to induced surface plasmons of the metallic nanoparticles and is then subsequently radiated by the nanoparticles themselves.[12] The extent of the radiation of the coupled energy by the metallic nanoparticles is also thought to be related to the scattering efficiency of nanoparticles, a component of their complex extinction spectrum.[13,14]

In addition to plasmon coupling, fluorescent species placed near-to metallic nanoparticles experience an increase in their absorption of light (a modification in their absorption cross section), which is due to increased electric fields around and in between the nanoparticles themselves.[11] An example for the enhanced absorption phenomenon is given in Figs. 10.1B–D. Figure 10.1B shows a calculated electric field distribution using an FDTD numerical approach for a two-silver nanoparticle array (diameter = 100 nm, gap = 2 nm) illuminated at 400 nm. The maximum electric field intensity reaches up to 1000-fold that of the free-space value, in between the silver nanoparticles. It is well known that the interaction of light with metallic nanoparticles is via the free electrons of the metal and is wavelength dependent.[15,16] Figure 10.1C shows the calculated the maximum electric field intensity versus wavelength of incident light for 2- and 12-silver nanoparticle arrays. The maximum electric field intensity for 2- and 12-nanoparticle arrays occurs at 400 nm and 450 nm, respectively.

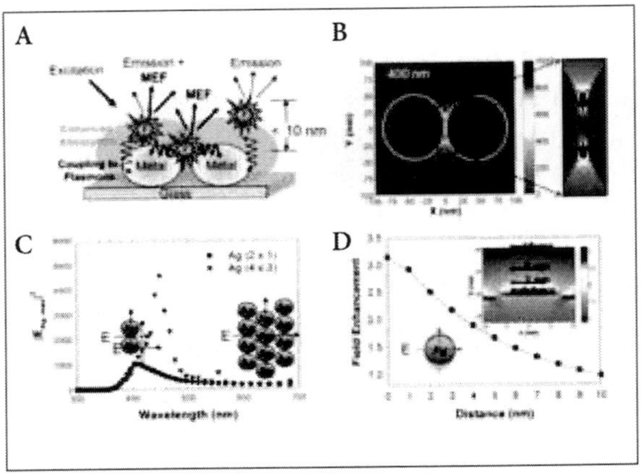

Figure 10.1 (A) Schematic representation of the MEF phenomenon. (B) FDTD calculations for two silver nanoparticle arrays (d = 100 nm). (C) Wavelength-dependent calculated $|E|^2$ maximum intensity for silver nanoparticle arrays (d = 100 nm). Geometries and incident field polarization (p polarized) and propagation direction are shown in the insets. The gap between the nanoparticles was assumed to be 2 nm in the calculations. (D) Calculated field enhancement as a function of distance for a single silver nanoparticle (d = 100 nm). The inset shows these results as an FDTD $|E|^2$ image above the nanoparticle. *Abbreviation*: FDTD, finite-difference time domain.

For the 12-nanoparticle silver nanoparticle array (4 × 3 staggered), the maximum electric field intensity is calculated to be 5000-fold the free-space value due to the interaction of multiple silver nanoparticles with the incident light. FDTD calculations also reveal that the electric field enhancement around the surface of silver nanoparticles is at a maximum at short distances (<1 nm) and completely diminishes at 10 nm from the metal surface, as shown in Fig. 10.1D and Fig. 10.1D (inset).

Over the last two decades, many surfaces have been developed for MEF on the basis of different metallic nanoparticles, such as those comprised of silver nanoparticles[17,18] and even gold,[19] copper,[20] aluminum,[21] and zinc particulate films.[22,23] Several modes of silver deposition have also been developed, such as by wet chemistry,[24] a layer-by-layer deposition technique,[25] deposition by light,[26] electrochemically,[27] on glass,[12] plastic substrates,[28] and even on indium tin oxide.[29]

Silver island films (SIFs) have been the substrate of choice in many publications to date.[12,30–35] In a typical SIF preparation on a glass support, a solution of silver nitrate (0.5 g in 60 mL of deionized water) in a clean 100 mL glass beaker, equipped with a Teflon-coated stir bar, is prepared and placed on a stirring/hot plate. While stirring at the quickest speed, 200 μL of freshly prepared 5% (w/v) sodium hydroxide solution is added. This results in the formation of dark brown precipitates of silver particles. Approximately 2 mL of ammonium hydroxide is then added, drop by drop, to redissolve the precipitates. The clear solution is cooled to 5°C by placing the beaker in an ice bath, followed by soaking the amino silane–coated glass slides in the solution. While keeping the slides at 5°C, a fresh solution of *D*-glucose (0.72 g in 15 mL of water) is added. Subsequently, the temperature of the mixture is then warmed to 30°C. As the color of the mixture turns from yellow green to yellow brown, and the color of the slides become green, the slides are removed from the mixture, washed with water, and sonicated for one minute at room temperature. SIF-deposited slides are then rinsed with deionized water several times and dried under a stream of dry nitrogen gas.

A typical photograph for SIFs deposited onto a glass microscope slide is given in Fig. 10.2A (top). One can employ the above procedure to any planar surface that contains amine functional groups. In this regard, Fig. 10.2A (bottom) shows a photograph of SIFs deposited onto a plastic surface.[28,36] The characterization of SIFs on a surface is typically done by employing optical absorption spectroscopy and microscopy techniques, such as AFM and scanning electron microscopy. Figure 10.2B shows the absorbance spectra for SIFs used in MEF applications. The surface plasmon resonance (SPR) peaks for silver occur in the blue spectral region. Figure 10.2C,D shows the AFM images of SIFs on glass and plastic supports, respectively. While the distribution and size of the silver nanoparticles on the glass are fairly homogeneous, a wide range of sizes of silver nanoparticles are deposited onto the plastic support. This is due to the fact that the chemistries of the surface of glass are well established unlike plastic surfaces, and therefore the covalent immobilization of silver nanoparticles onto glass is less arduous and is subsequently more homogeneous.

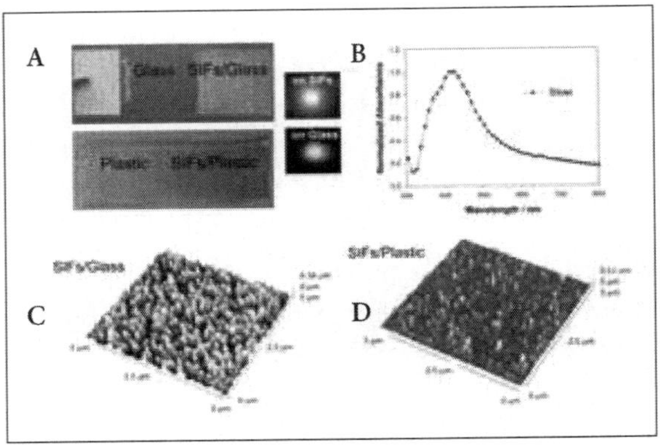

Figure 10.2 (A) Photographs of SIFs deposited onto glass and plastic supports. (B) Normalized absorbance of silver nanostructured particles on a glass support. AFM images of SIFs on (C) glass and (D) plastic supports. (Adapted from Refs. 37 [glass] and 28 [plastic]). *Abbreviation*: AFM, atomic force microscopy.

10.3 Microwave-Accelerated Metal-Enhanced Fluorescence

In the past few years our research laboratory has been working on the applications of a new technique, called microwave-accelerated metal-enhanced fluorescence (MAMEF), as applied to fluorescence-based bioassays. The MAMEF technique couples the benefits of low-power microwave heating with MEF, to address the two major shortcomings of fluorescence-based bioassays currently in use today, that is, bioassay sensitivity and rapidity. In MAMEF,[37] the MEF phenomenon increases the sensitivity of the assays, while the use of low-power microwave heating kinetically accelerates assays to completion within only a few seconds. There are three major components of the MAMEF technique: 1) plasmonic nanoparticles (e.g., silver, gold, copper, nickel, aluminum, and zinc), 2) microwaves, and 3) an aqueous-based assay medium. The plasmonic nanoparticles serve as (i) a platform for the attachment of one of the biorecognition partners (anchor probes), (ii) as an enhancer of the close-proximity fluorescence signatures via surface plasmons (i.e., the MEF effect),[12] and (iii) a material not heated by microwaves for the selective heating of the aqueous media by microwave energy.

Since the MAMEF technique utilizes the benefits of low-power microwave heating, it is also important to discuss the interactions of microwaves and metallic nanoparticles. In microwave heating, the electromagnetic energy interacts with the materials at the molecular level, where the electromagnetic energy is transferred and converted to heat through the frustrated motion of the molecules. The result is rapid and uniform heating of materials throughout their volume (also referred to as volumetric heating), especially when the size of the materials is smaller than the wavelength of the microwaves. In addition, microwave heating can also be utilized for selective heating of materials within a mixture or a composite. This arises from the differences in dielectric properties of the materials; in a mixture or a composite, microwaves selectively couple to the material with a higher dielectric loss factor (ε''). This selective heating is indeed the crux of the MAMEF technique, where the assay medium is selectively heated to a higher temperature than the metallic nanoparticles: the thermal gradient rapidly driving the biorecognition events to completion, where plasmon enhancement additionally amplifies any localized fluorescence labels. The reader is referred to the review article by Thostenson[38] for the description of dielectric properties and for a summary of electromagnetic theory.

10.3.1 *Proof-of-Principle Demonstration of the MAMEF Technique*

The proof-of-principle of the MAMEF technique, which couples the benefits of MEF with low-power microwave heating to kinetically accelerate the bioassays, was first demonstrated with a model protein-fluorophore system,[37] with biotinylated bovine serum albumin (BSA) and fluorophore-labeled streptavidin, as shown in Fig. 10.3A. The biotin-streptavidin binding step typically takes up to 30 minutes to complete at room temperature on the assay surface, while low-power microwave heating reduces the assays' run time to a staggering 20 seconds. The assay was constructed on a glass support (other materials such as paper,[39] plastics,[36] and high-throughput screening (HTS) wells[40] can also be used) coated

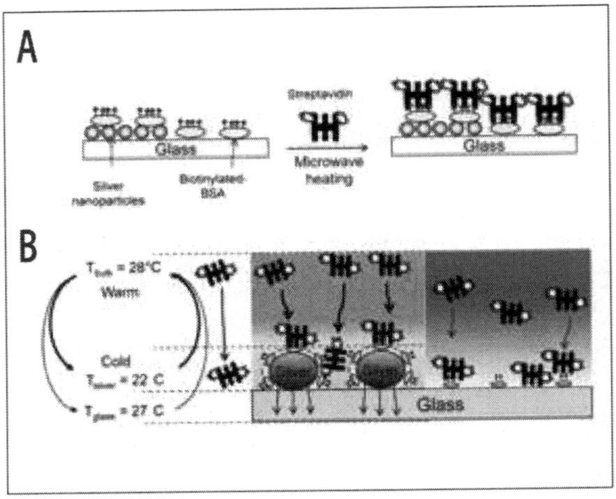

Figure 10.3 (A) A model MAMEF-based protein detection assay and (B) a schematic representation of the effect of microwave heating on the protein detection surface assay.

with silver nanoparticles. In this surface configuration, the use of silver nanoparticles results in 1) enhancement of fluorescence as compared to a blank glass surface (increased emission intensity) and 2) creation of a temperature gradient between the bulk and the silver nanoparticles themselves.

In the MAMEF technique, while water and glass are selectively heated with microwaves, the silver nanoparticles remain virtually at the same temperature as compared to before the microwave heating was initiated. As shown in Fig. 10.3B, the selective heating of these assay components creates a temperature gradient between the water, the glass, and the silver nanoparticles, which results in the rapid transfer of streptavidin molecules from the warmer bulk to the colder surface due to temperature-driven mass transfer. Since the thermal conductivity of silver (429 W/mK) is much larger than that of glass (1.05 W/mK), the transfer of streptavidin happens more efficiently toward silver nanoparticles than glass. For the duration of the microwave heating (20 sec), as heat is transferred from the bulk to the surface, and the temperature on the assay surface reaches an equilibrium temperature lower than the bulk, water molecules are continuously circulated between the warmer and colder regions of the bulk medium. The temperature of the assay components during the microwave heating process was determined to be ≈28°C using a thermal imaging camera; hence no protein denaturation occurs at these temperatures.

In a model MAMEF-based protein assay, the protein-fluorophore system was coated equally on one half of a silvered glass substrate, the other half of the glass substrate left intentionally blank to compare the benefits of the MEF phenomenon (see Fig. 10.3A). Biotinylated BSA is attached to the surface and is allowed to bind to its binding partner (fluorescein-labeled streptavidin) by incubation at room temperature or through low-power microwave heating. Figure 10.4A shows that the fluorescein emission intensity from the silvered substrate is approximately sixfold greater than that from the glass substrate for the assay run at room temperature for 30 minutes.

Figure 10.4B shows the combined effect of both low-power micro wave heating (20 sec) and the MEF effect for the identical

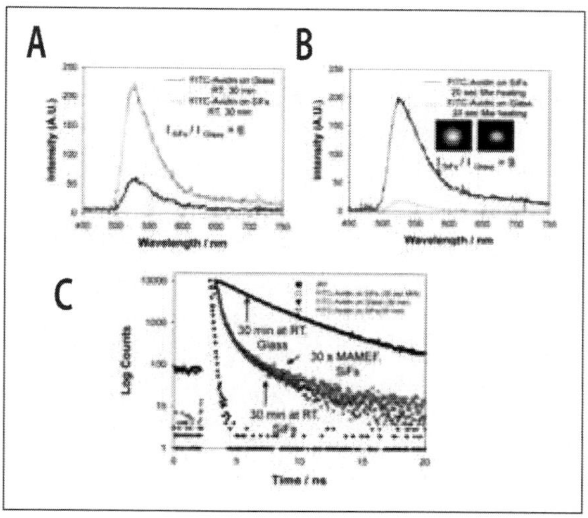

Figure 10.4 Emission spectra of FITC for a model protein assay run (A) at room temperature for 30 min and (B) with microwave heating for 20 sec. (C) Intensity decays for FITC-Avidin on both glass and SIFs, both before and after exposure to low-power microwave heating. The intensity decays on the SIFs are almost identical. (Adapted from Ref. 37.) *Abbreviations*: IRF, instrument response function; FITC, fluorescein isothiocyanate.

assay run at room temperature. The microwave-accelerated assay yielded similar fluorescence intensity after just 20 seconds as compared to the assay run at room temperature for 30 minutes. The emission intensity from the silvered substrate is approximately ninefold greater than that of the glass substrate for the microwave-accelerated assay. In addition, control experiments were also run to confirm the efficacy of the MAMEF assays. In these control experiments, where biotinylated BSA is omitted from the assay that was run at room temperature for 30 minutes, no fluorescence emission intensity was detected (data not shown here).[37] When the identical control assay was run with microwave heating, once again there was little/no fluorescence emission.

These control experiments demonstrated that the nonspecific binding of the fluorescein-labeled streptavidin to the surface was minimal.

The fluorescence lifetime of a fluorophore is indicative of its environment and was proven a useful tool to show the benefits of MEF.[2] It was previously shown that when placed near-to metallic nanostructures, fluorophores have shorter lifetimes as compared to free-space solutions or on the glass substrates. In a MAMEF-based assay, the lifetime information can be useful to assess the extent of completion of the assay: if the assay run at room temperature and with microwave heating goes to more than 95% completion (30 min and 20 sec, respectively), then the lifetime of the fluorophores would be expected to be very similar. Figure 10.5C shows that the fluorescence intensity decay curves for fluorescein after 30-minute incubation and 30-second microwave heating were almost identical and significantly reduced as compared to the glass control. These results strongly indicate that the assays are virtually identical after 30-minute incubation at room temperature and after 30-second microwave heating.

10.3.2 *Application of the MAMEF Technique to Pathogen Detection Based on DNA Hybridization Assays*

Fluorescence-based DNA hybridization assays are routinely used in many diagnostics applications,[41] on gene chips,[42] and fluorescence in situ hybridization.[43] In all of these applications, usually fluo-

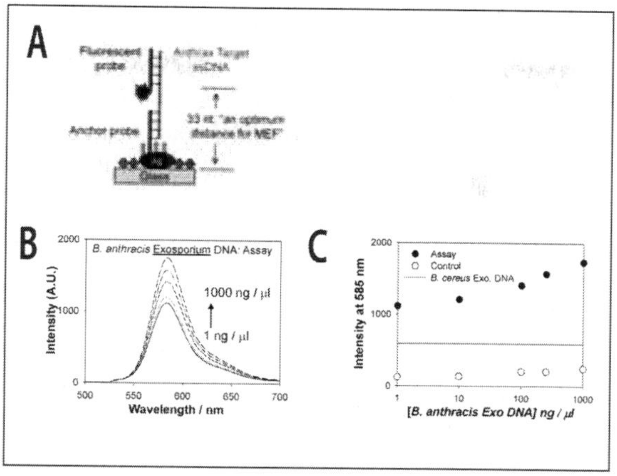

Figure 10.5 (A) Experimental design, depicting the organization of the DNA oligomers on SIFs used for the detection of *B. anthracis*. (B) Emission spectra of TAMRA-Oligo as a function of *B. anthracis* exosporium concentration after 30 sec low-power microwave heating. (C) Plot of the fluorescence emission intensity at 585 nm for TAMRA-Oligo as a function of target concentration. Data for *B. cereus* (a noncausative strain) is also shown for comparison. (Adapted from Ref. 47.) *Abbreviation*: TAMRA, carboxytetramethylrhodamine.

rophores with high-quantum yields are employed to increase the sensitivity of the hybridization assays, raising several issues such as high background emission and photostability of the fluorophores. Ideally, to maximize the efficiency and sensitivity of the MEF DNA hybridization assays, it would be beneficial to employ low-quantum-yield fluorophores, which can withstand the long exposure to excitation light.[17]

The application of MEF in DNA hybridization assays has been shown to offer improvements in the sensitivity of the DNA hybridization assays as well as in the photostability of the fluorophores.[17,24,44] However, one typically has little or no control over the rapidity of the DNA hybridization assays. In this regard, the applicability of the MAMEF technique for rapid and sensitive DNA hybridization assays was recently demonstrated,[45,46] where two complementary oligonucleotides (one labeled with a fluorophore) were hybridized on silver nanoparticles within 20 seconds, after low-power microwave heating with the hybridization believed to be over in less than 5 seconds.[45,46]

Recently, the MAMEF technique was applied to the detection of target DNA encoding a region of the *PA* gene essential for biological activity of *B. anthracis*.[47] Figure 10.5A shows the experimental details of the target *B. anthracis* DNA assay constructed on silver-deposited glass microscope slides. In this DNA detection scheme, the anchor probe is attached to silver nanoparticles through sulfhydryl groups. To minimize the nonspecific binding of DNA to the assay platform, silver nanoparticles and glass are both modified with additional surface-protective chemicals.[47] The subsequent hybridization of the fluorescent probe and the target B. anthracis DNA contained in the exosporium (isolated from *B. anthracis* by sonication and subsequent centrifugation/filtration steps), a loose-fitting balloon-like layer surrounding the spore, was carried out in a single microwave heating step, followed by an orange emission at 585 nm through an emission filter when excited with a green laser (532 nm). The intensity of the emission at 585 nm was directly related to the concentration of the target *B. anthracis* DNA[47] (Fig. 10.5B).

Control experiments, where the anchor probe was omitted from the assay surface, showed that the nonspecific binding of target DNA was significantly less than the lowest concentration of exosporium sample detected in the actual assay (c.f. Fig. 10.5C).[47] The ability of the MAMEF technology to distinguish between *Bacillus cereus* (*B. cereus*), a close relative of *B. anthracis* was also demonstrated.[47] Fluorescence emission intensity at 585 nm from this control assay shows constant emission intensity over a wide range of concentrations, indicating that the MAMEF assay platform clearly can distinguish between the two closely genetically related strains.

In addition to the experiments performed for both strains separately, as described before, to demonstrate the specificity of the assay for a sample that contains a mixture of DNA from both strains at once, a mixture of (same initial concentrations, 50% v/v) exosporium from *B. anthracis* and *B. cereus* was tested for successful detection of target DNA (data not shown here). Concentration-dependent fluorescence emission data for the assay showed a significant increase in fluorescence emission as the concentration of the exosporium (and fluorescent probe) is increased, while the control assay shows only a slight increase in intensity. A plot of concentration-dependent emission intensity at 585 nm showed the range of detection for *B. anthracis* DNA in a mixture of exosporium DNA from two *Bacillus* strains varied between 1 and 500 μg/mL.

10.4 Spatially Focused Microwaves and Metal-Enhanced Fluorescence for Pathogen and Virus Detection

The extraction of proteins and DNA from pathogens of interest is time consuming and involves arduous biological separation processes carried out in a laboratory setting. In some cases, for example, during a biological terror attack when almost instantaneous detection and treatment is required, isolation of specific proteins and DNA using conventional methods is inadequate. Thus, there is an ever-increasing need for faster techniques for separation and detection of relevant proteins and DNA in a point-of-care setting where the pathogens and viruses are suspected to be encountered. In this regard, our research laboratory has recently introduced a technique,[48] where the combined use of microwaves and metallic mini-antennae afforded for the extraction of proteins and DNA from anthrax spores and vegetative cells within 30 seconds. Subsequently, the target protein and DNA were detected within an additional 30 seconds using the MAMEF technique.[48]

It is important to briefly discuss the design of the metallic miniantennae and their interactions with microwaves to better describe the extraction of proteins and DNA using these antennae. It was previously shown that metallic structures, especially aluminum thin films with sharp corners (see Fig. 5.6A), focuses the incident electric fields to the corners of the structures, where an increase in the electric fields is predicted.[49] Figure 10.6B shows that the normalized electric field (calculated using FDTD simulations) is focused in between and at the corners of the aluminum "bow-tie" structures. The intensity of electric fields in between the aluminum structures was predicted to vary with the gap size and was calculated to be up to 50,000-fold larger as compared to a location inside the aluminum structures (see Fig. 10.6C). As a

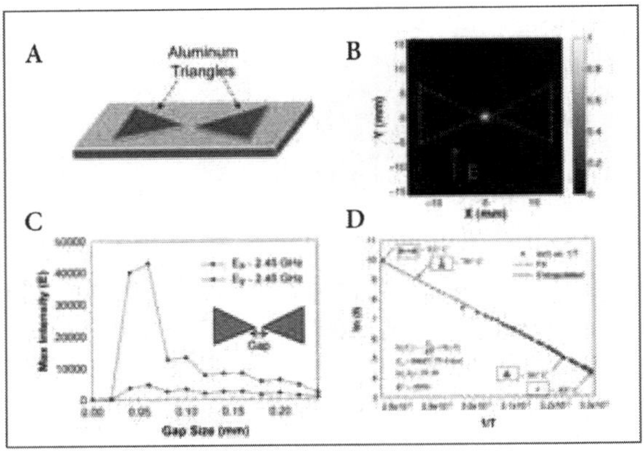

Figure 10.6 (A) A schematic representation of the aluminum "bow-tie" structures used to extract biological material form anthrax spores and vegetative cells, (B) normalized FDTD simulation of field intensity distributions ($Ix + Iy$) for equilateral aluminum triangles in a simulated 2.45 GHz TE-polarized TFSF that propagates across the geometries from bottom to top, (C) maximum intensity calculated for aluminum "bow-tie" structures for a variety of gap sizes, and (D) actual temperature of a solution experimentally measured near aluminum structures and plotted in an Arrhenius calibration curve. *Abbreviations*: TE, transverse electric; TFSF, total-field scattered-field.

result of increased electric fields, the temperature of a solution placed in between the aluminum structures is expected to rise significantly when a microwave source is used. In this regard, the actual temperature of a solution was experimentally measured and plotted in an Arrhenius calibration curve, as shown in Fig. 10.6D. For a five-second microwave heating process, the temperature of an aqueous solution varied between 36°C inside a single aluminum triangle and 93°C inside the gap of "bow-tie" structures. The heating rates for an aqueous solution placed at the tip of a single aluminum triangle and in between aluminum "bow-tie" structures were calculated to be ≈15.1°C/s and 34.6°C/s, respectively. It was predicted that the temperature of an aqueous solution can reach up to 130°C in between aluminum "bow-tie" structures.[49] This implies that pathogens or viruses can be exposed to high temperature and increased electric fields in between aluminum "bow-tie" structures and relevant proteins and DNA extracted within only a few seconds of low-power microwave heating (see Fig. 10.6).

The real-time temperature changes of an aqueous solution between the aluminum "bow-tie" structures can be monitored using thermal imaging techniques. Figure 10.7A,B shows the thermal images of aluminum "bow-tie" structures before and after two seconds of microwave heating. An increase of 30°C is recorded between the aluminum "bow-tie" structures after only two seconds of microwave heating. With the focused microwave energy and subsequent rapid heating of the solutions in the gaps of the aluminum "bow-tie" structures, this technology was employed as an alternative to the complex and long incubation periods typically required for releasing DNA from *B. anthracis* spores. In this regard, 50 μL of spores (1.0×10^5 spores/mL) were placed in the gap of the aluminum triangles (see Fig. 10.6A) and was irradiated for 20 seconds in a microwave cavity in the presence and absence of the aluminum "bow-tie" structures. Using transmission electron microscopy (TEM), the morphological changes of the spores after 20 seconds of microwave heating on a blank glass microscope slide and on aluminum triangles were compared (see Fig. 10.7B). *B. anthracis* form elongated rod-shaped spores approximately 1,000 nm in length, which are surrounded by a structure called the exosporium (transparent membrane around the spores (see Fig. 10.7B). During microwave heating of the spores in the absence of the aluminum "bow-tie" structures, no obvious structural changes in the spore and the exosporium were observed (see Fig. 10.7C). However, in the presence of the aluminum "bow-tie" structures, a significant morphological change to the exosporium was seen, while the changes in the cortex of the spores were more subtle (see Fig. 10.7B). Similarly, exposure of the vegetative cells to microwave heating resulted in cellular disruption (see Fig. 10.8A,B).

The TEM images provide the visual evidence for the physical disruption of the spore and the vegetative cells and the release of DNA using focused microwave heating. From the TEM images, the DNA release was initially thought to be from the core and the exosporium of the vegetative cells and the spores, respectively. To determine if this is indeed the case, a quantitative evaluation of the release of DNA following focused microwave heating was

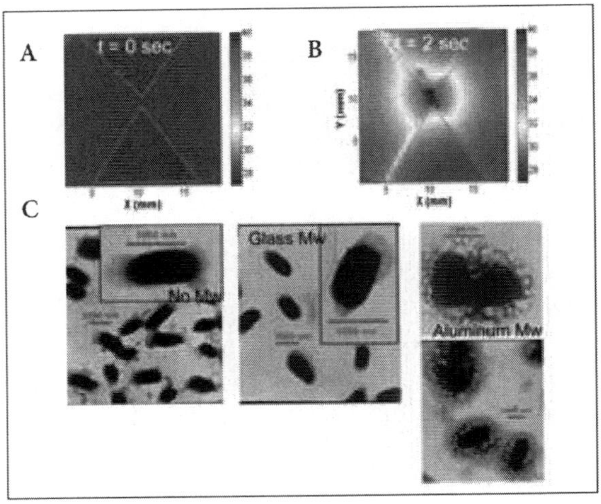

Figure 10.7 Thermal images of aluminum "bow-tie" structures (A) before and (B) after 2 sec of microwave heating. (C) TEM images of *B. anthracis* spores (left) before, (middle) on glass, and (right) on aluminum triangles after 20 sec of low-power microwave heating. While microwave heating on glass slides does not appear to cause any structural change in the spores, microwave heating of spores from between the aluminum triangles results in the rapid deformation of the spore and the release of DNA. *Abbreviations*: No Mw, no microwave heating; Mw, microwave heating.

undertaken using a gel electrophoresis technique (c.f. Fig. 10.8C; see Figs. 10.7 and 10.8).

The gel results showed that the DNA extraction protocol performed on those vegetative cells exposed to focused microwaves generated DNA extract yields similar to those from vegetative cells not exposed to focused microwaves based on the intensity of the bands in lane 3 and lane 5, respectively. In addition, the DNA extraction procedure performed on microwaved spores also yielded DNA (data not shown here).[48] It was also reported that DNA recovered from the *B. anthracis* spores originates from the cortex of the spore as the exosporium is disrupted by the microwave heating.[48]

In contrast, the DNA recovered from the vegetative organisms was reportedly from the cells themselves as these organisms lack the exosporium. The extraction of DNA using spatially focused mi-

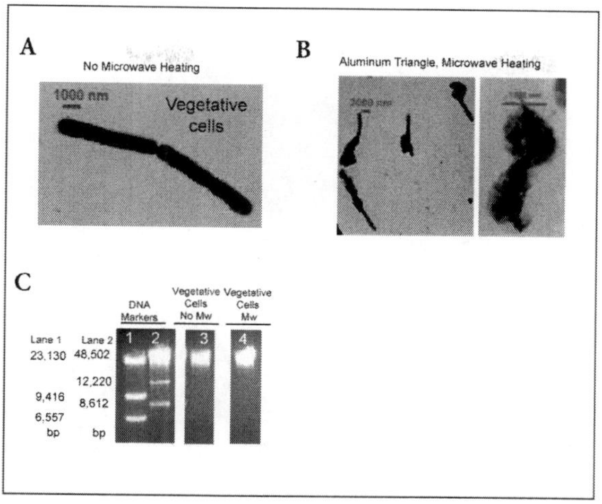

Figure 10.8 TEM images of vegetative organisms both (A) before and (B) after a 20 sec exposure to microwave heating from between the aluminum triangles, and (C) gel results for vegetative cells exposed to microwave heating. *Abbreviations*: No Mw no microwave heating; Mw, microwave heating.

crowaves was attributed to the dramatic differences in the dielectric properties of spores and the vegetative cells; it was thought that the exosporium of the spores absorbs microwave irradiation more effectively as compared to the core of the spores. From the TEM images (see Figs. 10.7A,B), it is evident that the exosporium is locally heated, creating cavitations. Since the water-based exosporium has a large dielectric constant compared to the dehydrated high carbon content of the cell wall of the organisms, it was thought that localized heating of the exosporium concentrates the heating of the spore's cortex, which facilitates the subsequent release of the DNA from the core. The gel results showed that the total DNA released from the spores and vegetative cells was mostly larger than 48,000 base pairs. Although the presence of DNA with a smaller size was not evident from the gel results due to a very intense band for larger DNA, the sample was thought to contain smaller DNA. In the MAMEF technique used for target anthrax DNA detection,[47] the anchor probe and the fluorescent probe hybridize with different regions of the target DNA in close proximity to the silver surface,

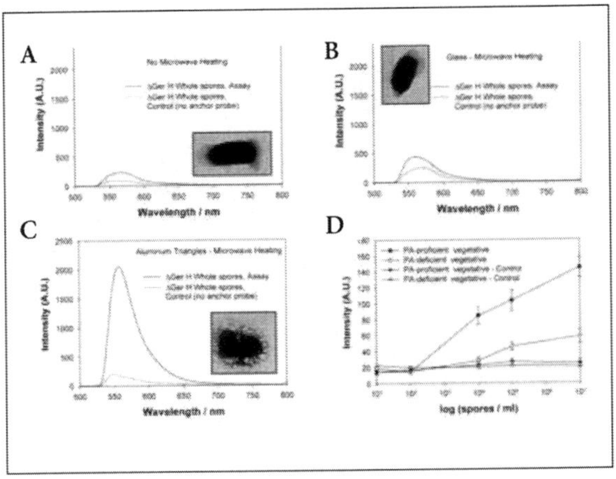

Figure 10.9 (A) Room-temperature assay and (B, C) MAMEF assays for *B. anthracis* spores. *B. anthracis* spores (10^5 spores/mL) were microwave-heated for 20 sec on the MAMEF assay, and the same number of spores were also incubated on the assay platform for 2 hr without microwave heating in the roomtemperature assay. The anchor probe is omitted in the control assays. (D) MAMEF assays for vegetative organisms. *Abbreviation*: A.U., arbitrary fluorescence units.

where the fluorescence emission is plasmon enhanced, ultimately yielding an increased assay sensitivity. In addition, the MAMEF-based three-piece DNA detection scheme[47] was adapted to detect DNA released from *B. anthracis* following exposure to spatially focused microwave irradiation. The effectiveness of this approach was determined by assaying the following samples: (A) *B. anthracis* spores without microwave heating (room-temperature assay), (B) *B. anthracis* spores exposed to microwave irradiation on a blank glass microscope slide, and (C) *B. anthracis* spores exposed to focused microwave irradiation between aluminum triangles. Control experiments were also performed, whereby the anchor probe was omitted from the metal surface (c.f. Fig. 10.9). These control experiments demonstrated the extent of nonspecific binding of target DNA to the assay surface.

Incubation of *B. anthracis* spores at room temperature in the presence of a fluorescent probe specific for the protective antigen (PA) gene of *B. anthracis* located on the MAMEF assay platform, in the absence of microwave heating, resulted in a fluorescence emission signature similar to that of a control assay (where the anchor probe was omitted), as shown in Fig. 10.9A.

Exposure of spores to microwave irradiation for 20 seconds at room temperature on glass resulted in a slightly higher fluorescence emission when run on the assay, which is indistinguishable from its corresponding control sample assay (no anchor probe) (see Fig. 10.9B). In the MAMEF assay for *B. anthracis* spores, which was exposed to focused microwave irradiation between aluminum triangles a significantly larger fluorescence emission signature was observed (see Fig. 10.9C). This implies that the sample containing anthrax DNA after the exposure to focused microwave heating is detected by the MAMEF assay. To ensure the specificity of the assay, spores of the genetic close relative *B. cereus* were exposed to microwave heating also between aluminum triangles. Following irradiation, the treated spores were mixed with the fluorescent probe and irradiated for a further 20 seconds (data not shown here). Fourfold-lower fluorescence emission intensity was measured from the MAMEF assay for *B. cereus* as compared to the MAMEF assay for *B. anthracis*, which is due to the fact that the MAMEF assay components are specifically designed and are specific for *B. anthracis*.

The ability of the MAMEF technique to detect the vegetative organisms themselves was investigated (see Fig. 10.9D). As a further confirmation of the specificity of the assay, an isogenic variant of *B. anthracis* which lacked the PA gene (labeled as PA-deficient vegetative in Fig. 10.9D) was included. A concentration-dependent fluorescence emission from the vegetative *B. anthracis* cultures (labeled as PA-proficient vegetative in Fig. 10.9D) was observed to be significantly larger than the fluorescence emission from the PA-deficient vegetative *B. anthracis* variant when the irradiated vegetative cultures are mixed with the fluorescent probe and are incubated on the MAMEF assay platform for 20 seconds. The control experiments, where the anchor DNA probe is omitted from the surface of the MAMEF assay platform, showed a minimal background fluorescence emission (signal-to-noise ratio of 7). Using this approach, the detection limit of much less than 1,000 vegetative cells per mL was achieved.

10.5 Summary and Future Outlook

This chapter summarizes our previous work on the bioassay platform technology for the ultrafast and sensitive detection of proteins and DNA of interest on the basis of the amalgamation of low-power microwave heating and MEF. The MEF component affords for improvements in the assay sensitivity and photostability of fluorophores. In MEF, a fluorophore is placed near-to plasmonic nanostructures and two significant observations are made: significantly increased fluorescence emission and a reduced lifetime of the fluorophore. The reduction in the fluorescence lifetime results in fluorophores becoming more photostable near-to the plasmonic nanoparticles. More specifically, the excited fluorophores partially transfer energy to the plasmonic nanoparticles, where the energy is in essence amplified, as the emission from the fluorophore-nanoparticle "system" radiates more efficiently than the emission from the fluorophores alone. In addition, the use of low-power microwave heating kinetically accelerates assays to completion within only a few seconds. The crux of the our bioassay platform technology is the selective heating of water, which occurs as the assay medium is heated by microwaves to a higher temperature than the plasmonic nanoparticles, creating a temperature gradient between the warmer aqueous media and the colder nanoparticles, rapidly driving the biorecognition events to completion. The major criteria for the choice of metal to be employed in our bioassay platform technology are that the metals support plasmon resonance and demonstrate MEF.

Microwaves can also be focused into a specific location on a bioassay surface using metallic antennae. In our previous work, we have demonstrated that aluminum "bow-tie" structures are predicted to focus microwaves up to 50,000-fold to the gap in between the structures. The implications of focused microwaves were also demonstrated for triggered chemiluminescence reactions, where 500-fold increase in chemiluminescence intensity as compared to the absence of microwave heating was observed. The temperature of a solution placed in between the aluminum triangles were also shown to be controlled by changing the gap size, the duration of microwave heating, etc. The change in temperature ranged from 30°C to 130°C, affording DNA hybridization to occur in several seconds at low temperatures and the extraction of biological materials from pathogens and cells at high temperatures. Subsequently, ultrafast and sensitive detection of proteins or DNA extracted from pathogens and cells were carried out using MEF and low-power microwave heating. In this regard, as low as 100 copies of *B. anthracis* spores and vegetative cells were detected in less than one minute (total time) using the spatially focused microwaves and MEF in a combined format. We note that our platform technology is generic and applicable to many other pathogens, viruses, and cells. A collaborative program is underway to test the efficacy of this method with viral pathogens and will be reported in due course.

Acknowledgments

This work was supported by the Middle Atlantic Regional Center of Excellence for Biodefense and Emerging Infectious Diseases Research (NIH NIAID—U54AI057168). Salary support to authors from UMBI, MBC, and the IoF is also acknowledged.

References

1. Amos, R. M., and W. L. Barnes. Modification of the spontaneous emission rate of Eu^{3+} ions close to a thin metal mirror. 1997. *Phys. Rev. B*, **55**, 7249–7254.
2. Geddes, C. D., and J. R. Lakowicz. Metal-enhanced fluorescence. 2002. *J. Ffluoresc.*, **12**, 121–129.
3. Gersten, J., and A. Nitzan. Spectroscopic properties of molecules interacting with small dielectric particles. 1981. *J. Chem. Phys.*, **75**, 1139–1152.
4. Das, P. C., and A. Puri. Energy flow and fluorescence near a small metal particle. 2002. *Phys. Rev. B*, **65**, 155416.
5. Dulkeith, E., M. Ringler, T. A. Klar, J. Feldmann, A. M. Javier, and W. J. Parak. Gold nanoparticles quench fluorescence by phase induced radiative rate suppression. 2005. *Nano Lett.*, **5**, 585–589.

6. Carminati, R., J. J. Greffet, C. Henkel, and J. M. Vigoureux. Radiative and non-radiative decay of a single molecule close to a metallic nanoparticle. 2006. *Opt. Commun.*, **261**, 368–375.
7. Zhang, Y. X., K. Aslan, M. J. R. Previte, and C. D. Geddes. Metal-enhanced S-2 fluorescence from azulene. 2006. *Chem. Phys. Lett.*, **432**, 528–532.
8. Chowdhury, M. H., S. K. Gray, J. Pond, C. D. Geddes, K. Aslan, and J. R. Lakowicz. Computational study of fluorescence scattering by silver nanoparticles. 2007. *J. Opt. Soc. Am. B: Opt. Phys.*, **24**, 2259–2267.
9. Hatling, T., P. Reichenbach, and L. M. Eng. Near-field coupling of a single fluorescent molecule and a spherical gold nanoparticle. 2007. *Opt. Exp.*, **15**, 12806–12817.
10. Aslan, K., and V. H. Perez-Luna. Nonradiative Interactions between biotin functionalized gold nanoparticles and fluorophore-labeled antibiotin. 2006. *Plasmonics*, **1**, 111–119.
11. Hayakawa, T., S. T. Selvan, and M. Nogami. Field enhancement effect of small Ag particles on the fluorescence from Eu3+-doped SiO2 glass. 1999. *Appl. Phys. Lett.*, **74**, 1513–1515.
12. Aslan, K., Z. Leonenko, J. R. Lakowicz, and C. D. Geddes. Annealed silver-island films for applications in metal-enhanced fluorescence: interpretation in terms of radiating plasmons. 2005. *J. Fluoresc.*, **15**, 643–654.
13. Zhang, Y., K. Aslan, M. J. Previte, and C. D. Geddes. Metal-enhanced S2 fluorescence from azulene. 2006. *Chem. Phys. Lett.*, **432**, 528–532.
14. Zhang, Y., K. Aslan, M. J. Previte, and C. D. Geddes. Metal-enhanced fluorescence: surface plasmons can radiate a fluorophores structured emission. 2007. *Appl. Phys. Lett.*, **90**, 053107.
15. Yguerabide, J., and E. E. Yguerabide. Light-scattering submicroscopic particles as highly fluorescent analogs and their use as tracer labels in clinical and biological applications: I. Theory. 1998. *Anal. Biochem.*, **262**, 137–156.
16. Yguerabide, J., and E. E. Yguerabide. Light-scattering submicroscopic particles as highly fluorescent analogs and their use as tracer labels in clinical and biological applications: II. Experimental characterization. 1998. *Anal. Biochem.*, **262**, 157–176.
17. Aslan, K., I. Gryczynski, J. Malicka, E. Matveeva, J. R. Lakowicz, and C. D. Geddes. Metal-enhanced fluorescence: an emerging tool in biotechnology. 2005. *Curr. Opin. Biotechnol.*, **16**, 55–62.
18. Chen, Y., K. Munechika, and D. S. Ginger. Dependence of fluorescence intensity on the spectral overlap between fluorophores and plasmon resonant single silver nanoparticles. 2007. *Nano Lett.*, **7**, 690–696.

19. Aslan, K., S. N. Malyn, and C. D. Geddes. Metal-enhanced fluorescence from gold surfaces: angular dependent emission. 2007. *J. Fluoresc.*, **17**, 7–13.

20. Zhang, Y., K. Aslan, M. J. R. Previte, and C. D. Geddes, 2007. *Metal-Enhanced Fluorescence from Copper Substrates* (vol. 90), AIP, p. 173116.

21. Ray, K., M. H. Chowdhury, and J. R. Lakowicz. Aluminum nanostructured films as substrates for enhanced fluorescence in the ultraviolet-blue spectral region. 2007. *Anal. Chem.*, **79**, 6480–6487.

22. Dorfman, A., N. Kumar, and J. Hahm. Nanoscale ZnO-enhanced fluorescence detection of protein interactions. 2006. *Adv. Mater.*, **18**, 2685–2690.

23. Dorfman, A., N. Kumar, and J. I. Hahm. Highly sensitive biomolecular fluorescence detection using nanoscale ZnO platforms. 2006. *Langmuir*, **22**, 4890–4895.

24. Aslan, K., J. Huang, G. M. Wilson, and C. D. Geddes. Metal-enhanced fluorescence-based RNA sensing. 2006. *J. Am. Chem. Soc.*, **128**, 4206–4207.

25. dos Santos, D. S., and R. F. Aroca. Selective surface-enhanced fluorescence and dye aggregation with layer-by-layer film substrates. 2007. *Analyst*, **132**, 450–454.

26. Geddes, C. D., A. Parfenov, and J. R. Lakowicz. Photodeposition of silver can result in metal-enhanced fluorescence. 2003. *Appl. Spectrosc.*, **57**, 526–531.

27. Geddes, C. D., A. Parfenov, D. Roll, J. Y. Fang, and J. R. Lakowicz. Electrochemical and laser deposition of silver for use in metal-enhanced fluorescence. 2003. *Langmuir*, **19**, 6236–6241.

28. Aslan, K., P. Holley, and C. D. Geddes. Metal-enhanced fluorescence from silver nanoparticle-deposited polycarbonate substrates. 2006. *J. Mater. Chem.*, **16**, 2846–2852.

29. Park, H. J., D. Vak, Y. Y. Noh, B. Lim, and D. Y. Kim. Surface plasmon enhanced photoluminescence of conjugated polymers. 2007. *Appl. Phys. Lett.*, **90**, 161107.

30. Aslan, K., J. R. Lakowicz, H. Szmacinski, and C. D. Geddes. Enhanced ratiometric pH sensing using SNA- FL-2 on silver island films: metal-enhanced fluorescence sensing. 2005. *J. Fluoresc.*, **15**, 37–40.

31. Aroca, R., G. J. Kovacs, C. A. Jennings, R. O. Loutfy, and P. S. Vincett. Fluorescence Enhancement from Langmuir-Blodgett monolayers on silver island films. 1988. *Langmuir*, **4**, 518–521.

32. Garoff, S., D. A. Weitz, C. D. Hanson, T. J. Gramila, and J. I. Gersten. Deexcitation channels of excited molecules on silver island films. 1981. *J. Opt. Soc. Am.*, **71**, 1552–1552.
33. Kummerlen, J., A. Leitner, H. Brunner, F. R. Aussenegg, and A. Wokaun. Enhanced dye fluorescence over silver island films: analysis of the distance dependence. 1993. *Mol. Phys.*, **80**, 1031–1046.
34. Hayakawa, T., S. T. Selvan, and M. Nogami. Enhanced fluorescence from Eu3+ owing to surface plasma oscillation of silver particles in glass. 1999. *J. Non-Cryst. Solids*, **259**, 16–22.
35. Shang, L., H. Chen, and S. Dong. Electrochemical preparation of silver nanostructure on the planar surface for application in metal-enhanced fluorescence. 2007. *J. Phys. Chem. C*, **111**, 10780–10784.
36. Aslan, K., R. Badugu, J. R. Lakowicz, and C. D. Geddes. Metal-enhanced fluorescence from plastic substrates. 2005. *J. Fluoresc.*, **15**, 99–104.
37. Aslan, K., and C. D. Geddes. Microwave-accelerated metal-enhanced fluorescence: Platform technology for ultrafast and ultrabright assays. 2005. *Anal. Chem.*, **77**, 8057–8067.
38. Thostenson, E. T., and T. W. Chou. Microwave processing: fundamentals and applications. 1999. *Compos. A: Appl. Sci. Manuf.*, **30**, 1055–1071.
39. Zhang, Y., K. Aslan, M. J. R. Previte, and C. D. Geddes. Metal-enhanced fluorescence from paper substrates: modified spectral properties of dyes for potential high-throughput surface analysis and assays and as an anti-counterfeiting technology. 2008. *Dyes Pigm.*, **77**, 545–549.
40. Aslan, K., P. Holley, and C. D. Geddes. Microwave- accelerated metal-enhanced fluorescence (MAMEF) with silver colloids in 96-well plates: application to ultra fast and sensitive immunoassays, high throughput screening and drug discovery. 2006. *J. Immunol. Methods*, **312**, 137–147.
41. Morrison, L. 2003. Fluorescence in nucleic acid hybridization assays, in *Topics in Fluorescence Spectroscopy* (vol. 7, Lakowicz, J. R., ed), Kluwer Academic Publishers/Plenum Press, New York, pp. 69–103.
42. Brown, P. O., and D. Botstein. Exploring the new world of the genome with DNA microarrays. 1999. *Nat. Genetics*, **21**, 33–37.
43. Difilippantonio, M., and T. Ried, 2003. Technicolor genome analysis, in *Topics in Fluorescence Spectroscopy* (vol. 7, Lakowicz, J. R., ed), Kluwer Academic Publishers/Plenum Press, New York, pp. 291–316.
44. Malicka, J., I. Gryczynski, and J. R. Lakowicz. DNA hybridization assays using metal-enhanced fluorescence. 2003. *Biochem. Biophys. Res. Commun.*, **306**, 213–218.

45. Aslan, K., S. N. Malyn, G. Bector, and C. D. Geddes. Microwave-accelerated metal-enhanced fluorescence: an ultra-fast and sensitive DNA sensing platform. 2007. *Analyst*, **132**, 1122–1129.

46. Aslan, K., S. N. Malyn, and C. D. Geddes. Fast and sensitive DNA hybridization assays using microwave-accelerated metal-enhanced fluorescence. 2006. *Biochem. Biophys. Res. Commun.*, **348**, 612–617.

47. Aslan, K., Y. Zhang, S. Hibbs, L. Baillie, M. J. Previte, and C. D. Geddes. Microwave-accelerated metal-enhanced fluorescence: application to detection of genomic and exosporium anthrax DNA in <30 seconds. 2007. *Analyst*, **132**, 1130–1138.

48. Aslan, K., M. J. Previte, Y. Zhang, T. Gallagher, L. Baillie, and C. D. Geddes. Extraction and detection of DNA from Bacillus anthracis spores and the vegetative cells within 1 min. 2008. *Anal. Chem.*, **80**, 4125–4132.

49. Previte, M. J., K. Aslan, and C. D. Geddes. Spatial and temporal control of microwave triggered chemiluminescence: a protein detection platform. 2007. *Anal. Chem.*, **79**, 7042–7052.

Chapter 11

Lyssavirus Surveillance and Diagnostics: Focus on Africa

Wanda Markotter and Louis H. Nel

Department of Microbiology and Plant Pathology, Faculty of Natural and Agricultural Sciences, University of Pretoria, Pretoria, South Africa
wanda.markotter@up.ac.za

11.1 Introduction

Rabies is caused by all members of the *Lyssavirus* genus in the family *Rhabdoviridae*, a group of bullet-shaped, single-stranded, negative-sense ribonucleic acid (RNA) viruses. Since the initial discovery of the *Lyssavirus* genus, there have been multiple virus species added, and currently this genus consists of 12 species and 2 putative members. Classical rabies viruses, including the rabies vaccine strains, are grouped under rabies virus (RABV), and this group occurs almost worldwide, whereas the remaining species are also known as rabies-related lyssaviruses. Apart from RABV, Lagos bat virus (LBV), Mokola virus (MOKV), Duvenhage virus (DUVV), Shimoni bat virus (SHIBV), and Ikoma virus (IKOV) have been reported exclusively from the African continent. The fluorescent antibody test (FAT) is considered the gold standard for lyssavirus

Viral Diagnostics: Advances and Applications
Edited by Robert S. Marks, Leslie Lobel, and Amadou Alpha Sall

ISBN 978-981-4364-43-0 (Hardcover), 978-981-4364-44-7 (eBook)
www.panstanford.com

diagnostics, recommended by both the World Organization for Animal Health and the World Health Organization. Other diagnostic methods such as histological examination, immunohistochemistry, reverse transcriptase polymerase chain reaction (RT-PCR), and isolation of the virus in cell culture or suckling mouse brain, as well as rapid immunodiagnostic tests have also been developed. For most countries in Africa no epidemiological data for lyssaviruses are known; however, an estimated 55,000 human deaths still occur annually due to rabies and 40% of these are presumed to occur in Africa. Lyssavirus diagnostics and surveillance in the developing world are seriously hampered due to the lack of diagnostic facilities, logistical support, and infrastructure. These difficulties in obtaining a diagnostic result from field specimens have led to serious underreporting of the disease and ultimately resulted in a lack of commitment to control the disease. The levels of commitment for the control of rabies in Africa are generally very low and the surveillance for lyssaviruses inadequate, to say the least. Several reasonably simple diagnostic methods are available to identify lyssaviruses and to improve surveillance. These methods will be discussed in more detail in the following sections—in the context of their utility in the developing world.

11.2 The Etiological Agent

Previous known lyssavirus genotypes were recently reclassified, and the *Lyssavirus* genus currently consists of 12 lyssavirus species and 2 putative species.[89] RABV (gt 1) is the prototype lyssavirus, and the remaining species are known as rabies-related lyssaviruses (see Table 11.1).

All lyssaviruses cause the severe neurotropic disease known as rabies. Apart from rabies virus, five other species have been isolated exclusively from the African continent: LBV, MOKV, DUVV, SHIBV, and IKOV. A lyssavirus is a bullet-shaped, single-stranded, negative-sense, nonsegment RNA virus within the order *Mononegavirales* and family *Rhabdoviridae*.[89] The RNA genome, together with the nucleoprotein (N), phosphoprotein (P), and RNA-dependent RNA polymerase (L), is condensed into a helical nucleocapsid. A

Table 11.1 Members of the *Lyssavirus* genus

Viral species	Geographical distribution	Species affected
RABV	Almost worldwide	It is infectious to warm-blooded animals, including humans, and has established itself in a vast variety of carnivores throughout the world. In Africa it also occurs in several wildlife species, including mongooses.
LBV	Africa	It infects frugivorous bats. There is one report from an insectivorous bat. Spillover infections have occurred in cats and dogs, as well as a mongoose.
MOKV	Africa	The reservoir species is unknown. It has been isolated from dogs, cats, a rodent, and shrews. One fatal human infection has been reported.
DUVV	Africa	It infects insectivorous bats. Fatal human infections have been reported.
EBLV-1	Europe	It infects insectivorous bats. Spillover infections have been reported in sheep and a stone marten. Human infections have been reported.
EBLV-2	Europe	It infects insectivorous bats. Human infections have been reported.
ABLV	Australia	It infects frugivorous bats and various insectivorous bats. Human infections have been reported.
Aravan virus	Central Asia	It infects insectivorous bats. There are no human cases.
Khujand virus	Central Asia	It infects insectivorous bats. There are no human cases.
Irkut virus	East Siberia	It infects insectivorous bats. One human infection has been reported.
WCBV	Caucasian region	It infects insectivorous bats. There are no human cases.
SHIBV	Africa (Kenya)	It infects insectivorous bats. There are no human cases.
BBLV	Europe (Germany)	It infects insectivorous bats. There are no human cases.
IKOV	Africa (Tanzania)	It infects the African civet. There are no human cases.

Abbreviations: EBLV, European bat lyssavirus; ABLV, Australian bat lyssavirus; WCBV, West-Caucasian bat virus; BBLV, Bokeloh bat lyssavirus.

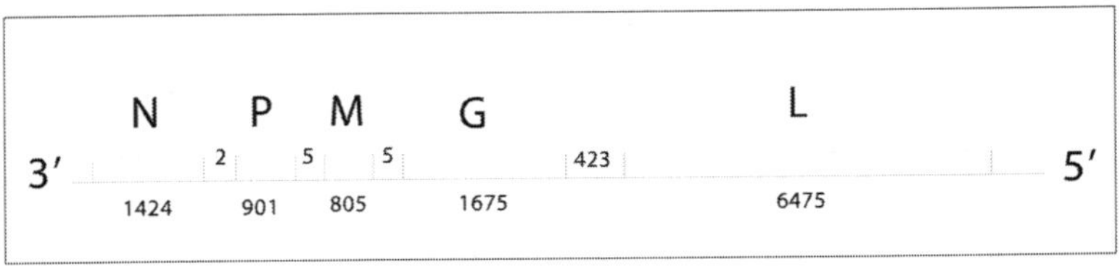

Figure 11.1 Organization of the RABV genome (PV strain). The lengths of the genes and intergenic regions are indicated. *Abbreviation*: PV, Pasteur virus.

lipoprotein envelope, derived from the host's cell membrane during viral budding, surrounds the nucleocapsid. Surface projections, consisting of glycoprotein monomers, extend from the envelope and the matrix protein (M) form a layer between the envelope and the nucleocapsid. The 3′ end of the RNA genome encodes a short leader RNA, followed by the nucleo (N), phospho (P), matrix (M), glyco (G), and RNA polymerase (L) gene (see Fig. 11.1), each encoding a protein. Due to the conserved nucleotide sequence of the N gene, it is also the lyssavirus antigen targeted in diagnostic methods used to identify lyssaviruses[80] or to differentiate between different lyssavirus species and serotypes.[28] RABV was believed to be antigenically unique until the identification of the first rabies-related lyssavirus from Africa in 1956.[10,82] After the recognition of the rabies-related lyssaviruses, the *Lyssavirus* genus was first divided into four serotypes (RABV, LBV, MOKV, and DUVV)[80] on the basis of sero-neutralization and monoclonal antibody reactions.[23] EBLVs were first believed to belong to serotype 4 (DUVV)[79] but were subsequently shown to represent a separate serotype—serotype.[5,23] With the development of molecular biology techniques, the *Lyssavirus* genus was divided into six gts (gt 1–6) based on the amino acid identity of the nucleoprotein,[12] and in 1996 a new lyssavirus was isolated in Australia, constituting gt 7 (ABLV).[32] Since then a number of new lyssavirus isolates were identified from Eurasia: Irkut,[9] Aravan,[1,50] Khujand,[50] and WCBV,[9] as well as from Europe (BBLV)[104] and Africa (IKOV and SHIBV).[105,106] Lyssaviruses are now classified as species and not according to genotypes and 12 unique lyssavirus species exist, with IKOV and BBLV currently listed as tentative species in the *Lyssavirus* genus.[89] Current rabies vaccines are based on RABV and do not protect against diverse lyssavirus species, such as MOKV, WCBV, and LBV, on the basis of experimental studies performed in murine models.[33,66]

11.3 Lyssaviruses in Africa

For RABV, the domestic dog (canine) is the major reservoir and vector and responsible for the majority of animal and human rabies cases in the developing world. In humans, rabies only occurs as spillover dead-end infections. Feline species can also be effective vectors but are not effective reservoirs, given their particular social behavior and low population densities. Rabies in wildlife has been reported in the Ethiopian wolf (*Canis simensis*)[83] and African wild dogs (*Lycaon pictus*),[34,37,38] as well as black-backed jackals (*Canis mesomelas*),[5,6] bat-eared foxes (*Otocyon megalotis*),[87] and the kudu antelope (*Tragelaphus strepsiceros*).[4,40,55] A unique RABV variant, which is distinct from the canine virus in Africa, is well adapted to various herpestid or mongoose species in southern Africa.[68] RABV infections on the African continent are severely underreported in humans and animals. A study in Malawi indicated that in 11% of 133 children diagnosed with cerebral malaria, rabies infection was diagnosed in retrospect.[54]

Apart from RABV—as mentioned earlier—LBV, MOKV, DUVV, SHIBV, and IKOV have been reported exclusively from the African continent.[86] The incidence of RABV is underreported from the African continent, and this is even more so for the African rabies-related lyssaviruses. There is no active surveillance for these viruses, and most laboratories in Africa do not have the capability to distinguish between RABV and rabies-related lyssaviruses. There have been only six positively identified encounters with DUVV to date. The first of these was in 1970 in a fatal human case from South Africa[60] after Mr. Duvenhage was bitten by a small bat. The bat was never positively identified. The next isolation of DUVV was only in 1981, again from South Africa. The virus was isolated directly from an insectivorous bat, but again the bat species was not positively identified.[93] In 1986 DUVV was isolated in Zimbabwe from the slit-faced bat (*Nycteris thebaica*).[31] The next isolation was only two decades later, again from South Africa, when a man died of rabies after contact with a small unidentified bat.[72] The first isolation outside southern Africa was made in 2007 from a human who died from a DUVV infection after being exposed in Kenya.[107] Recently (2012), DUVV was again isolated from an insectivorous bat in South Africa (personal communication). There is no active surveillance for DUVV in Africa, and the true incidence and specific reservoir species are unknown.

MOKV has never been isolated from bats but has been isolated from cats, dogs, shrews, and a rodent. The reservoir species is still unknown. The first isolation of the virus was made in 1968 from organ pools of shrews in Nigeria.[45,82] Thereafter, the virus was isolated again from a shrew (Cameroon, 1974),[52] from a rodent (Central African Republic, 1983),[78] from domestic cats and a dog (Zimbabwe, 1981/82),[29,30] again in 1989 from a cat in Ethiopia,[59] and in 1993 from a cat in Zimbabwe.[7] The first isolation of MOKV in South Africa was made from a domestic cat in 1970.[79] The next isolates were made only many years later (1995 to 1998), all from domestic cats[67,94] and then again in 2005/2006 from a cat and a dog.[76] Two reports of MOKV have been made from humans. The first was in 1969 from a 3.5-year-old Nigerian girl who recovered without neurological damage. In 1971 a six-year-old Nigerian girl died and MOKV was isolated from brain material.[26,27] In both these cases no material is available to verify these results.

LBV was first isolated from a pool of fruit bat (*Eidolon helvum*) brains in 1956 at Lagos Island in Nigeria,[10] but it was not until 1970 that it was identified as LBV.[82] Since then sporadic isolations of LBV have been made throughout Africa; from the Central African Republic (1974) from the frugivorous bat *Micropterus pusillus*[85] and then in June 1980 the first LBV isolation from South Africa was made from a frugivorous bat, *Epomophorus wahlbergi* (*E. wahlbergi*)[61,86] Another 13 bats from South Africa, all thought to be *E. wahlbergi*, were found positive for lyssavirus antigens during 1980/81.[21,47] In 1982 an isolation of LBV was made from a rabies-vaccinated domestic cat from in South Africa.[47] After this isolation, the virus was not encountered again in South Africa until 1990, when a further isolation was made from *E. wahlbergi*.[86] Two reports of LBV isolations in West Africa were made—one in 1985 from *Nycteris cambiensis* in Kindia, Guinea, and the other from *Eidolon helvum* in Dakar, Senegal.[42] An isolation was also made from a rabid cat in Zimbabwe in 1986[31] and from a rabid dog in Ethiopia in 1990.[59] From France, an isolate of LBV was reported in 1999[73] from *Rousettus aegyptiacus* that had been imported from an unknown location on the African continent.[2] Recently six new LBV isolations have been reported from *E. wahlbergi* in South Africa (2003, 2004, 2005, 2006, 2008, and 2013)[56–58] and in Kenya from frugivorous bats (2006, 2007, and 2008).[108] In addition, LBV was also isolated for the first time from a water mongoose (*Atilax paludinosis*)[56] and was diagnosed from a rabies vaccine failure case (domestic dog) in South Africa.[58]

In 2009 a new lyssavirus was isolated from a Commerson's leaf-nosed bat (*Hipposideros* spp) in Kenya,[106] and since then antibodies were detected indicating that this bat species is the possible reservoir host.[110] Although WCBV has only been isolated in the Caucasian region, antibodies that can neutralize this virus were also detected in insectivorous bats (*Miniopterus* spp) in Kenya.[109] The most recently proposed member of the *Lyssavirus* genus is IKOV. This virus was isolated from an African civet in the Serengeti National Park in Tanzania.[105]

11.4 Pathogenesis of Lyssaviruses

To be able to understand the principles of lyssavirus diagnostics, it is important to understand the infection dynamics of the virus. The transmission of a lyssavirus requires the deposition of the virus into the new host via saliva, and this is most commonly achieved through a bite wound.[75] Once the virus has entered the nerve it moves to the central nervous system,[92] after which it spreads through the spinal cord and between neurons. Incubation of the virus can be between 3 and 12 weeks, but longer incubation periods have also been reported,[36] and during this time no symptoms are observed. After reaching the brain stem the infection spreads rapidly in the brain and disease symptoms can be observed. Patterns of the virus spread within the central nervous system suggest that examination of the brain stem is critical to rabies diagnosis. Viral antigen is widespread in the brains of most animals positive for rabies, but because the spread may also be unilateral, especially in larger animals, a negative finding for rabies can be made only if a complete cross section of the brain stem is examined. Although the brain stem is the tissue most reliably found to contain viral antigen, the characteristic size and shape of intracytoplasmic inclusions produced as lyssavirus accumulates in the large neurons of the cerebellum are easily detected and recognized by diagnostic methods such as the FAT. Inclusion of this tissue with the brain stem yields a more confident diagnosis. After replication in the brain, the virus then spreads to other organs, such as salivary glands,[15] and is deposited throughout the body. The virus has been previously detected in skeletal and cardial muscle, adrenals, kidneys, the retina, the cornea, hair follicles, brown fat of bats, the myocardium, lungs, the bladder, milk, tears, epithelium of the tongue, the pancreas, and urine.[3,43,63] The virus is present in the saliva of an infected animal only after virus replication in the central nervous system and subsequent spread from the brain to the salivary glands. Because the virus may not always spread to salivary glands and may be present only intermittently in saliva, negative tests of salivary glands or the saliva cannot rule out rabies infection.

Immune responses to lyssavirus infections are complicated due to the neurotropism of the virus.[51] The nervous system is an immune-privileged site, and a virulent lyssavirus infection is able to suppress the peripheral immune system by producing a noncytopathic infection, and therefore no viral antigens are presented to lymphocytes in the periphery. Exposure to a lyssavirus does not always lead to a productive infection and may or may not lead to a detectable immune response.[69] The virus is able to escape the immune system and invade the nervous system. Virus-neutralizing antibodies (VNAs) are not detected until the final stages of clinical illness, and the presence of VNAs in the cerebrospinal fluid (CSF) is a reliable indication of present or past infection or recovery. The rate of specific acquired immunity in naturally exposed animals varies and may depend on the species and the virus variant.[70] The presence of lyssavirus antibodies is generally considered indicative of exposure to a lyssavirus that has not led to productive infection. Productive infection is only conclusively indicated by the presence of virus or viral RNA in appropriate samples, such as brain material. Rabies can present with variable and nonspecific symptoms in humans[75] and animals.[71] Clinical diagnosis of rabies is unreliable, and suspected rabies cases should be confirmed by laboratory diagnosis.

11.5 Lyssavirus Diagnostics

The approach used to diagnose a lyssavirus infection will differ in humans and animals due to the fact that lyssavirus diagnosis is most often performed postmortem on animals and intra vitam diagnosis is required in humans suspected to be infected with a lyssavirus. Brain material can be collected from animals suspected of being rabid and those that have been involved in potential human exposures, and diagnosis can be performed. In the case of a possible rabid animal, it is important that the diagnosis be quick and reliable in order to ensure correct postexposure prophylaxis (PEP) decisions to prevent fatal infections in exposed humans. It is also important to use reliable diagnostics in epidemiological studies in animals to be able to make informed decisions about prevention and control of the disease. In humans a diagnosis is required intra vitam to be able to provide improved patient care. The sensitivity of techniques for rabies diagnosis in humans varies greatly according to the stage of the disease, antibody status, intermittent nature of viral shedding, and the training of the technical staff. While a positive result is indicative of rabies, a negative result does not necessarily rule out the infection, and results should be confirmed postmortem from brain material, where possible. Lyssavirus diagnostic methods target either lyssaviral antigen (nucleoprotein), viral RNA, isolation of virus, or lyssavirus antibodies.

After positively diagnosing the presence of a lyssavirus it may be necessary to further characterize the virus. Most lyssavirus diagnostic methods used, such as the FAT, cannot discriminate between lyssavirus species and will only indicate that it is a lyssavirus. Antigenic and molecular techniques have permitted the classification of lyssaviruses into species as well as characterization of lyssavirus variants occurring in a specific location.[28]

11.5.1 *Detection of Negri Bodies*

Examination of brain tissue for Negri bodies[88] was the laboratory method of choice until immunofluorescent methods were introduced in the late 1950s and early 1960s. Negri bodies are round or oval inclusions within the cytoplasm of nerve cells of animals or humans infected with rabies. Staining with Mann's, Giemsa, or Sellers stain can permit differentiation of rabies inclusions from other intracellular inclusions since Negri bodies appear magenta in color and have small dark-blue interior basophilic granules. The presence of Negri bodies is, however, variable, and some experimentally infected cases of rabies display Negri bodies in brain tissue and others do not. Histological examination of tissues from clinically rabid animals show Negri bodies in about 50% of the samples, and in contrast, the FAT shows rabies antigen in nearly 100% of the samples.[101] Furthermore nonrabid tissues have shown inclusions indistinguishable from Negri bodies. However, several diagnostic laboratories in Africa still use this method as the primary diagnostic method for identifying lyssavirus infection since it requires simple staining techniques as well as only a light microscope. However, the presence or absence of Negri bodies should not be considered a preferred method for diagnosis of lyssaviruses.[101]

11.5.2 *Fluorescent Antibody Test*

The gold standard for lyssavirus diagnostics, the FAT,[22,101] is performed on brain tissue preparations. These preparations are treated with antilyssavirus serum (conjugate) that is labeled with fluorescein isothiocyanate (FITC), and results are viewed under a fluorescence microscope.[28] Depending on the conjugate used, various levels of positivity may be experienced for the rabies-related lyssaviruses.[101] The conjugate must be of a broad spectrum (detect all lyssaviruses) and used at the appropriate dilution to recognize all species and lyssavirus variants. Polyclonal conjugates are produced by the conjugation of FITC with purified globulins from hyperimmune serum that is prepared by vaccination of appropriate animals (e.g., goat, guinea pig) with purified RABV ribonucleocapsid proteins. Reagents are also available commercially, such as FITC-antirabies monoclonal globulin, which is a mixture of monoclonal antibodies. These commercially available conjugates have been shown to detect all lyssaviruses currently known. It is more economical for a diagnostic laboratory that tests high amounts of samples to produce conjugates, but this must be performed under strict quality control and the conjugates must be tested by an independent laboratory for sensitivity and specificity. Fresh, frozen, or glycerolized material may be examined in the FAT, but formalin fixation of test material may mask virus antigen and should be avoided. Formalin-fixed tissues can, however, be treated with trypsin prior to performing the FAT.[100] Viral antigen may be detected intra vitam in humans by using the FAT on skin biopsies from patients with clinical rabies.[101] Skin biopsies are usually taken from the nuchal area of the neck, with hair follicles containing peripheral nerves. Examination of at least 20 sections is required to detect rabies nucleocapsid inclusions around the base of hair follicles. Though sensitive, this technique may not be practical in all settings because of the need for a cryostat to prepare frozen tissue sections.

A limited factor of implementing and using the FAT in an African setting is the need for a fluorescent microscope to perform the test. Most laboratories in Africa do not have this apparatus available, and if it can be supplied it is difficult to service it regularly. This test also needs to be performed in a well-equipped diagnostic laboratory, and the quality of reagents is dependent on a cold chain. The test can therefore not be performed in field conditions.

Different panels of monoclonal antibodies were developed and can be used to identify the species of lyssaviruses involved or to identify the lyssavirus variants after positive identification with the FAT.[11,56,57,102] These panels of monoclonal antibodies are used in the FAT in the place of the broad spectrum lyssavirus conjugate. This method can be used in laboratories that are able to perform the FAT, since no additional equipment is needed to perform monoclonal antibody typing of lyssaviruses.

11.5.3 *Enzyme-Linked Immunosorbent Assay*

The enzyme-linked immunosorbent assay (ELISA) can be used for the detection of lyssavirus antigen, as described by Bourhy and Perrin,[13] who developed an ELISA called the rapid rabies enzyme immunodiagnosis (RREID) test that detects the rabies nucleocapsid antigen in brain tissue. The test, however, has not been commercially available since 2000. Another ELISA, named WELYSSA has also been described.[103] This test uses a cocktail of four mouse monoclonal antibodies (MAbs) to capture the antigen (viral nucleoprotein) and was able to detect most of the lyssavirus species, suggesting that this may be a potentially useful tool for rabies diagnosis.

11.5.4 *Direct Rapid Immunohistochemical Test*

A direct rapid immunohistochemical test (dRIT) has been developed by the Centers for Disease Control and Prevention (CDC), USA.[24,53] Like the FAT, the dRIT is performed on brain touch impressions, but the product of the reaction can be observed by light microscopy instead of fluorescence. Lyssavirus antigen appears as magenta inclusions against a blue neuronal background. The dRIT uses biotinylated antinucleocapsid monoclonal antibodies, and after incubation with a streptavidin- peroxidase complex, the antibody reagent is made visible with 3-amino-9-ethylcarbazole. A diagnosis can be made in less than an hour. When evaluated the dRIT showed a sensitivity and specificity equivalent to those of the FAT.[24,53] The test is simple, requires no specialized equipment (such as a fluorescent microscope), and can be successfully performed on samples preserved in glycerol solution for months or frozen for up to two years and in variable conditions of preservation. These qualities make it ideal for testing in developing countries, and its success has been indicated in Tanzania[53] and Chad.[24]

11.5.5 *Rapid Lateral Flow Immunochromatography*

Several rapid test kits relying on a lateral flow immunochromatography principle have been developed with varying degrees of success in terms of both sensitivity and specificity. These tests are based on gold-conjugated detector antibodies that interact with lyssavirus antigenic sites present in the sample. In one study this test has been evaluated against RABV representatives as well as against bacteria and other viruses, such as canine distemper, Japanese encephalitis, and pseudo-RABV, to determine cross reactivity with diseases that mimic rabies.[44] The dRIT was only 91.7% positive compared to an indirect FAT using brain samples and 80% positive using saliva samples compared to the RT-PCR method. The fact that this test produced an 8.3% false-negative rate with brain samples and a 20% false-negative rate with saliva samples indicates that interpretation of a negative result should be made with caution. Additional testing would be necessary, especially if human exposures occurred. The dRIT also demonstrated high specificity (100%) against representatives of African lyssavirus species.[111] This is a very simple test that can be performed in less than 10 minutes without any specialized equipment, infrastructure, technical expertise, or various chemicals involved. There are no critical points to field use, such as cold storage, since the test kit can be stored at room temperature. The FAT still remains the gold standard for lyssavirus diagnostics, but the dRIT can be used as a very useful additional tool in field surveillance for lyssaviruses where no other diagnostic method is available or in remote areas where samples cannot easily reach a diagnostic laboratory and would therefore not be tested.

11.5.6 *Detection of Lyssavirus RNA*

PCR, preceeded by reverse transcription, can be used to detect lyssavirus RNA.[8,35,64,90] Although such methods can have certain advantages and are highly sensitive, the FAT is still considered the most reliable primary diagnostic test—given the premise of lesser variability in antigen-antibody reactions, as opposed to nucleotide sequence characteristics.[101] Other factors, such as the risk of contamination when performing this method in laboratories where there is no strict quality control or the correct setup for performing very sensitive molecular methods is not available, should also be taken into account. However, in conjunction with conventional methods, PCR is an important application under some conditions,[91] especially in intra vitam diagnosis in humans where low amounts of virus RNA may be present.[20,65] Lyssavirus RNA can be detected in several biological fluids and samples (e.g., saliva, CSF, tears, and skin biopsy samples). Real-time RT-PCR is the latest of this class of method that can be used as a diagnostic tool for the confirmation of rabies.[77,84,95,96,112] Since real-time PCR does not require post-PCR manipulation it can be used in a high-throughput manner and is less prone to contamination.[65,96,97] The DNA sequence of PCR amplicons can also be determined and used to identify the specific lyssavirus species or virus variant involved by comparing the sequence obtained to other sequences available in the public domain (GenBank).[56,57] Most laboratories in Africa do not have the apparatus and infrastructure to perform these molecular techniques, and it is therefore most often not considered an option for diagnosis or characterization of lyssavirus infections in developing countries.

11.5.7 *Virus Isolation*

To confirm the results of the diagnosis or to enable more detailed characterization of a lyssavirus infection, it may be necessary to perform virus isolation. Lyssaviruses may be isolated from any sample containing lyssavirus antigen, such as brain suspension, salivary glands, saliva, tears, or CSF, with brain material being the preferred choice. Isolation may be based on the use of the rabies tissue culture infection test (RTCIT)[99] or the mouse inoculation test (MIT).[49,98]

Isolation of the virus can be achieved in 2–4 days in the RTCIT compared to 7–20 days using the MIT. Cell cultures of murine neuroblastoma (MNA) cells or brain material of infected mice can be examined for the presence of the virus by immunofluorescence staining of acetone fixed cells. It should be noted that specimens may produce no infectious virus even when it tested positive for lyssavirus infection by other diagnostic methods. This may be due to the presence of VNAs, very low virus titres, or the absence of a viable virus in the sample.[101] In developing countries the MIT may be easier to perform than the RTCIT since the latter requires expensive equipment and technical expertise. It should, however, be noted that a lyssavirus infection should be confirmed using the FAT on brain material of mice that succumbed to the disease or infected cells, and therefore laboratories performing the MIT should also be able to perform the FAT.

11.5.8 *Antibody Detection*

Antibodies indicative of a lyssavirus infection can only be detected very late in the disease, after symptoms have appeared, and therefore detection of antibodies is not routinely used as a diagnostic method to indicate lyssavirus infection. Methods to detect lyssavirus antibodies are most often used to measure exposure to a lyssavirus or to evaluate the efficiency of lyssavirus vaccines by measuring antibody titres. Lyssavirus antibodies can be detected by virus neutralization assays. These assays are designed to measure the presence of VNAs in serum samples by evaluation of the ability of serum samples to block virus infectivity. Serum dilutions are incubated with constant amounts of a challenge virus and the level of infectious virus determined by inoculating cell cultures. A challenge virus strain, depending on which lyssavirus antibodies are in question,[62] is adapted to cell culture and the virus titre of this challenge virus determined. The rapid fluorescent focus inhibition test (RFFIT)[84] or the fluorescent antibody virus neutralization (FAVN)[17] test are two methods used to measure lyssavirus-neutralizing antibodies. These methods determine the amount of infectious challenge virus still present after incubation with serum, by counting all the infectious foci (tissue culture infectious dose [TCID]) in a well or chamber after staining the cells using the FAT. Results can be reported as international units (IU) compared to a reference serum. It should be noted that a reference serum is only available for RABV and not for the rabies-related lyssaviruses, complicating the standardization of neutralization assays for these viruses. No statistical difference has been found between the results of these two tests,[14] and the main differences are the time to completion, volumes of test material used, and time to read results.[91] An advantage of the FAVN test is that it is performed in microtitre plates and can therefore be automated[39] and used for large-scale analysis. Both these tests require specialized equipment, technical expertise, and detection of infection by the FAT and are therefore not feasible to perform in most developing countries.

ELISA tests are also available that can be used to detect lyssavirus antibodies in serum samples.[18,19,81] These tests have been validated for the detection of RABV antibodies and can be used as an inexpensive and reliable method for detection of RABV antibodies[19] and are commercially available. Another advantage of using an ELISA is that it does not require handling of a live virus as with the virus neutralization tests. However sensitivity of rabies ELISA systems has been indicated to be lower than virus neutralization assays.[113]

11.6 Challenges for the Developing World

Human mortality due to canine endemic rabies is estimated to be 55, 000 deaths per year with around 24,200 deaths estimated to occur in Africa,[48,101] and more than 7 million people are treated worldwide each year due to exposure to RABV.[48] The exact numbers are unknown due to the absence of surveillance and proper lyssavirus diagnostics facilities available in Africa. It is widely accepted that the incidence of rabies cases is grossly underestimated, and it was shown that the number of human cases of rabies may be underestimated up to hundredfold in regions of Africa, for example, Tanzania.[16] Misdiagnosis is also common—for example, 11% of 133 cases diagnosed as encephalitic cerebral malaria in Malawi have retrospectively been identified as rabies.[54] Once rabies symptoms have occurred there is no cure, but the disease can be prevented by limiting exposures of humans to rabid animals by vaccinating the animals and, if an exposure occurs, by applying rabies postexposure prophylaxis.[74] Most human cases in less developed countries occur because of exposure to rabies-infected canines, and death occurs mainly due to a lack of access to biological agents for postexposure prophalaxis and/or poor recognition of the disease and awareness of treatment protocols. Even though cost-effective control programs exist for control of rabies in dogs, this disease has become an increasing problem in Africa.[101] In Africa and other developing countries lyssavirus diagnostics and surveillance is constrained due to a lack of diagnostic facilities and logistics of getting suspected samples to a diagnostic laboratory. Most suspect samples do not even reach a diagnostic facility due to remote areas and inaccessibility. The difficulties in obtaining a diagnostic result from field specimens have led to underreporting of the disease and ultimately resulted in a lack of commitment to control the disease. To be able to obtain commitment from governments and funding organizations to control rabies in animals by vaccination and prevent the disease in humans by postexposure prophylaxis, the true burden of the disease needs to be established. This can only be achieved if effective diagnostic methods are available that can be implemented in a developing-world setting.

References

1. Arai, Y. T., Kuzmin, I. V., Kameoka, Y., Botvinkin, A. D. (2003). New Lyssavirus genotype from the lesser mouse-eared bat (Myotis blythi), Kyrghyzstan. *Emerg. Infect. Dis.*, **9**, 333–337.
2. Aubert, F. A. (1999). Rabies in individual countries: France. Rabies Bulletin Europe, http://www.whorabies-bulletin.org/Aubert.
3. Balachandran, A., Charlton, K. (1994). Experimental rabies infection of non-nervous tissues in skunks (Mephitis mephitis) and foxes (Vulpes vulpes). *Vet. Pathol.*, **31**, 93–102.
4. Barnard, B. J. H., Hassel, R. H., Geyer, H. J., De Koker, W. C. (1982). Non-bite transmission of rabies in kudu (Tragelaphus Stresiceros). *Onderstepoort J. Vet. Res.*, **49**, 191–192.
5. Bingham, J. (2005). Canine rabies ecology in Southern Africa. *Emerg. Infect. Dis.*, **11**, 1337–1342.
6. Bingham, J., Foggin, C. M., Wandeler, A. I., Hill, F. W.G. (1999). The epidemiology of rabies in Zimbabwe. 1. Rabies in dogs (Canis familiaris). *Onderstepoort J. Vet. Res.*, **66**, 1–10.
7. Bingham, J., Javangwe, S., Sabeta, C. T., Wandeler, A. I., Nel, L. H. (2001). Report of isolations of unusual lyssaviruses (rabies and Mokola virus) identified retrospectively from Zimbabwe. *Tydskr. S. Afr. Vet. Ver.*, **72**, 92–94.
8. Black, E. M., Lowings, J. P., Smith, J., Heaton, P. R., and McElhinney, L. M. (2002). A rapid RT-PCR method to differentiate six established

genotypes of rabies and rabies-related viruses using TaqMan technology. *J. Virol. Methods,* **105**, 25–35.

9. Botvinkin, A. D., Poleschuk, E. M., Kuzmin, I. V., Borisova, T. I., Gazaryan, S. V., Yager, P., Rupprecht, C. E. (2003). Novel lyssavirus isolated from bats Russia. *Emerg. Infect. Dis.,* **9**, 1623–1625.
10. Boulger, L. R., Porterfield, J.S. (1958). Isolation of a virus from Nigerian fruit bats. *Trans. R. Soc. Trop. Med. Hyg.*, **52**, 421–424.
11. Bourhy, H., Kissi, B., Lafon, M., Sacramento, D.D., and Tordo, N. (1992). Antigenic and molecular characterization of bat rabies virus in Europe. *J. Clin. Microbiol.*, **30**, 2419–2426.
12. Bourhy, H., Kissi, B., Tordo, N. (1993). Molecular diversity of the lyssavirus genus. *Virology,* **194**, 70–81.
13. Bourhy, H., Perrin, P. (1996). Rapid rabies enzyme immunodiagnosis (RREID) for rabies antigen detection. In *Laboratory Techniques in Rabies.* Ed. Meslin, F. X., Kaplan, M. M., Koprowski, H., 105–113. Geneva: World Health Organization.
14. Briggs, D. J., Smith, J. S., Mueller, F. L., Schwenke, J., Davis. R. D., Gordon, C. R., Schweitzer, K., Orciaro, L. A., Yager, P. A., Rupprecht, C. E. (1998). A comparison of two serological methods for detecting the immune response after rabies vaccination in dogs and cats being exported to rabies-free areas. *Biologicals,* **26**, 347–355.
15. Charlton, K. M., Casey, G. A., Campbell, J. B. (1983). Experimental rabies in skunks: mechanisms of infection of the salivary glands. *Canad. J. Comp. Med.,* **47**, 363–369.
16. Cleaveland, S., Fèvre, E. M., Kaare, M., Coleman, P. G. (2002). Estimating human rabies mortality in the United Republic of Tanzania from dog bite injuries. *Bull. World Health Org.*, **80**, 304–310.
17. Cliquet, F., Aubert, M., Sagne, L. (1998). Development of a fluorescent antibody virus neutralization test (FAVN test) for the quantitation of rabies-neutralizing antibody. *J. Immunol. Methods,* **212**, 79–87.
18. Cliquet, F., McElhinney, L. M., Servat, A., Boucher, J. M., Lowings, J. P., Goddard, T., Mansfield, K. L., Fooks, A. R. (20040. Development of a qualitative indirect ELISA for the measurement of rabies virus-specific antibodies from vaccinated dogs and cats. *J. Virol. Methods,* **117**, 1–8.
19. Cliquet, F., Sagne, L., Schereffer, J. L., Aubert, M. F. A. (2000). ELISA test for rabies antibody titration in orally vaccinated foxes sampled in the fields. *Vaccine,* **18**, 3272–3279.
20. Crepin, P., Audry, L., Rotivel, Y., Cacoin, A., Caroff, C., Bourhy, H. (1998). Intravitam diagnosis of human rabies by PCR using saliva and cerebrospinal fluid. *J. Clin. Microbiol.*, **36**, 1117–1121.

21. Crick, J., Tignor, G. H., Moreno, K. 1982. A new isolate of Lagos Bat virus from the Republic of South Africa. *Trans. R. Soc. Trop. Med. Hyg.*, **76**, 211–213.

22. Dean, D. J., Abelseth, M. K., Atanasiu, P. (1996). The fluorescent antibody test. In *Laboratory Techniques in Rabies*. Ed. Meslin, F. X., Kaplan, M. M., Koprowski, H., 88–95, Geneva: World Health Organization.

23. Dietzschold, B., Rupprecht, C. E., Tollis, M., Lafon, M., Mattei, J., Wiktor, T. J., Koprowski, H. (1988). Antigenic diversity of the glycoprotein and nucleocapsid proteins of rabies and rabies-related viruses: Implications for epidemiology and control of rabies. *Rev. Infect. Dis.*, **10**, S785–S798.

24. Dürr, S., Naïssengar, S., Mindekem, R., Diguimbye, C., Niezgoda, M., Kuzmin, I., Rupprecht, C. E., Zinsstag, J. (2008). Rabies diagnosis for developing countries. *PLoS Negl. Trop. Dis.*, **2**(3), e206.

25. Durrheim, D. N., Speare, R., Petzer, M. (2002). Rabies post-exposure management in South Africa: a telephonic survey used as a rapid tool for operational research. *Trop. Med. Int. Health*, **7**, 459–461.

26. Familusi, J. B., Moore, D. L. (1972). Isolation of a rabies related virus from the CSF of a child with "aseptic meningitis". *Afr. J. Med. Sci.*, **3**, 93–96.

27. Familusi, J. B., Osunkoya, B. O., Moore, D. L., Kemp, G. E., Fabiyi, A. (1972). A fatal infection with Mokola virus. *Am. J. Trop. Med. Hyg.*, **21**, 959–963.

28. Flamand, A., Wiktor, T. J., Koprowski, H. (1980). Use of Hybridoma monoclonal antibodies in the detection of antigenic differences between rabies and rabies-related virus proteins: I. The nucleocapsid protein. *J. Gen. Virol.*, **48**, 97–104.

29. Foggin, C. M. (1982). Atypical rabies virus in cats and a dog in Zimbabwe. *Vet. Rec.*, **110**, 338.

30. Foggin, C. M. (1983). Mokola virus infection in cats and a dog in Zimbabwe. *Vet. Rec.*, **113**, 115.

31. Foggin, C. M. (1988). Rabies and rabies-related viruses In *Zimbabwe: Historical, Virological and Ecological Aspects*. PhD thesis. University of Zimbabwe. Harare.

32. Gould, A. R., Hyatt, A. D., Lunt, R., Kattenbelt, J. A., Hengstberger, S., Blacksell, S, D. (1998). Characterisation of a novel lyssavirus isolated from pteropid bats in Australia. *Virus Res.*, **54**, 165–187.

33. Hanlon, C. A., Kuzmin, I. V., Blanton, J. D., Weldon, W. C., Manangan, J. S., Rupprecht, C. E. (2005). Efficacy of rabies biologics against new lyssaviruses from Eurasia. *Virus Res.,* **111**, 44–54.
34. Haydon, D. T., Randall, D. A., Matthews, L., Knobel, D. L., Tallents, L. A., Gravenor, M. B., Williams, S. D., Pollinger, J. P., Cleaveland, S., Woolhouse, M. E., Sillero-Zubiri, C., Marino, J., Macdonald, D. W., Laurenson, M. K. (2006). Low-coverage vaccination strategies for the conservation of endangered species. *Nature,* **443**, 692–695.
35. Heaton, P. R., Johnstone, P., McElhinney, L. M., Cowley, R. O'Sullivan, E., Whitby, J. E. (1997). Heminested PCR assay for detection of six genotypes of rabies and rabies-related viruses. *J. Clin. Microbiol.,* **35**, 2762–2766.
36. Hemachudha, T. (1994). Human rabies: Clinical aspects, pathogenesis and potential therapy. In *Lyssaviruses.* Ed. Rupprecht, C. E., Dietzschold, B., Koprowski, H., 121–143. Berlin: Springer-Verlag.
37. Hofmeyr, M., Hofmeyr, D., Nel, L., Bingham, J. (2004). A second outbreak of rabies in African wild dogs (Lycaon pictus) in Madikwe Game Reserve, South Africa, demonstrating the efficacy of vaccination against natural rabies challenge. *Anim. Conserv.,* **7**, 193–198.
38. Hofmeyr, M., Bingham, J., Lane, E. P., Ide, A., Nel, L. H. (2000). Rabies in African wild dogs (Lycaon pictus) in the Madikwe Game Reserve, South Africa. *Vet. Rec.,* **146**, 50–52.
39. Hostnik, P. (2000). The modification of fluorescent antibody virus neutralization (FAVN) test for the detection of antibodies to rabies virus. *J. Vet. Med.,* **47**, 423–427.
40. Hubschle, O. J. (1988). Rabies in the kudu antelope (Tragelaphus strepsiceros). *Rev. Infect. Dis.,* **11**, S629–S633.
41. Hughes, G. J., Smith, J. S., Hanlon, C. A., Rupprecht, C. E. (2004). Evaluation of a TaqMan PCR assay to detect rabies virus RNA: influence of sequence variation and application to quantification of viral loads. *J. Clin. Microbiol.,* **42**, 299–306.
42. Institute Pasteur. (1985). Isolations of Lagos bat virus in West Africa. Internal reports of Centre Collaborateur OMS de Reference et Recherche Pour les Arbovirus. Dakar (Senegal), http://www.pasteur.fr/recherche/.
43. Jackson A. C. (2002). Pathogenesis. In *Rabies.* Ed. Jackson, A. C., Wunner, W. H., 246–282. London: Elsevier Science.
44. Kang, B., Oh, J., Lee, C., Park, B., Park, Y., Hong, K., Lee, K., Cho, B., Song, D. (2007). Evaluation of a rapid immunodiagnostic test kit for rabies virus. *J. Virol. Methods,* **145**, 30–36.

45. Kemp, G. E., Causey, O. R., Moore, D. L., Odelola, A., Fabiyi, A. (1972). Mokola virus: Further studies on Ibadan 27377, A new rabies-related etiological agent of zoonosis in Nigeria. *Am. J. Trop. Med. Hyg.*, **21**, 356–359.

46. King, A. A. (1996). Cell culture of rabies virus. In *Laboratory Techniques in Rabies.* Ed. Meslin, F. X., Kaplan, M. M., Koprowski, H., 114–130. Geneva: World Health Organization.

47. King, A., Crick, J. (1988). Rabies-related viruses. *In Rabies.* Ed. Campbell, J. B., Charlton, K. M., 177–199. Boston: Kluwer Academic.

48. Knobel, D. L., Cleaveland, S., Coleman, P. G., Fevre, E. M., Meltzer, M. I., Miranda, M. E., Shaw, A., Zinsstag, J., Meslin, F. X. (2005). Re-evaluating the burden of rabies in Africa and Asia. *Bull. World Health Org.*, **83**, 360–368.

49. Koprowski, H. (1996). The mouse inoculation test. In *Laboratory Techniques in Rabies.* Ed. Meslin, F. X., Kaplan, M. M., Koprowski, H., 80–87. Geneva: World Health Organization.

50. Kuzmin, I. V., Orciari, L. A., Arai, Y. T., Smith, J. S., Hanlon, C. A., Kameoka, Y., Rupprecht, C. E. (2003). Bat lyssaviruses (Aravan and Khujand) from Central Asia: phylogenetic relationships according to N, P and G gene sequence. *Virus Res.*, **97**, 65–79.

51. Lafon M. (2002). Immunology. In *Rabies.* Ed. Jackson, A. C., Wunner, W. H., 351–371. London: Elsevier Science.

52. Le Gonidec, G., Rickenbach, A., Robin, Y., Heme, G. (1978). Isolement d'une souche de virus Mokola au Cameroun. *Ann. Microbiol.* (Inst. Pasteur), **129A**, 245–249.

53. Lembo, T., Niezgoda, M., Velasco-Villa, A., Cleaveland, S., Ernest, E., Rupprecht, C. E. (2006). Evaluation of a direct, rapid immunohistochemical test for rabies diagnosis. *Emerg. Infect. Dis.*, **12**, 310–313.

54. Mallewa, M., Fooks, A. R., Banda, D., Chikungwa, P., Mankhambo, L., Molyneux, E., Molyneux, M. E., Solomon, T. (2007). Rabies encephalitis in malaria-endemic area, Malawi, Africa. *Emerg. Infect. Dis.*, **13**,136–139.

55. Mansfield, K., McElhinney, L., Hübschle, O., Mettler, F., Sabeta, C., Nel, L. H. and Fooks, A. R. (2006). A molecular epidemiological study of rabies epizootics in kudu (Tragelaphus strepsiceros) in Namibia. *BMC Veterinary Research* http://www.biomedcentral.com/1746-6148/2/2.

56. Markotter, W., Kuzmin, I., Rupprecht, C. E., Randles, J., Sabeta, C. T., Wandeler, A. I., Nel, L. H. (2006a). Isolation of Lagos bat virus from water mongoose. *Emerg. Infect. Dis.*, **12**, 1913–1918.

57. Markotter, W., Randles, J., Rupprecht, C. E., Sabeta, C. T., Wandeler, A. I., Taylor, P. J., Nel, L. H. (2006b). Recent Lagos bat virus isolations from bats (suborder Megachiroptera) in South Africa. *Emerg. Infect. Dis.*, **12**, 504–506.

58. Markotter, W., Van Eeden, C., Kuzmin, I., Rupprecht, C. E., Paweska, J. T., Swanepoel, R., Fooks, A. R., Sabeta, C. T., Cliquet, F., Nel, L. H. (2008). Epidemiology and pathogenicity of African bat lyssaviruses. *Dev. Biol. Basel Krager*, **131**, 317–325.

59. Mebatsion, T., Cox, J. H., Frost, J. W. (1992). Isolation and Characterization of 115 Street rabies virus isolates from Ethiopia by using monoclonal antibodies: Identification of 2 isolates as Mokola and Lagos Bat viruses. *J. Infect. Dis.*, **166**, 972–977.

60. Meredith, C. D. (1971). An unusual case of human rabies thought to be of chiropteran origin. *S. Afr. Med.*, **45**, 767–769.

61. Meredith, C. D., Standing, E. (1981). Lagos Bat virus in South Africa. *Lancet*, **11**, 832–833.

62. Moore, S. M., Ricke, T. A., Davis, R. D., Briggs, D. J. (2005). The influence of homologous vs. heterologous challenge virus strains on the serological test results of rabies virus neutralizing assays. *Biologicals*, **33**, 269–276.

63. Murphy, F. A., Harrison, A. K., Winn, W. C., Bauer, S. P. (1973). Comparative pathogenesis of rabies and rabies-like viruses: infection of the central nervous system and centrifugal spread of virus to peripheral tissues. *Lab. Invest.*, **29**, 1–16.

64. Nadin-Davis, S. A. (1998). Polymerase chain reaction protocols for rabies virus discrimination. *J. Virol. Methods*, **75**, 1–8.

65. Nagaraj, J. Vasanth, A. Desai, A. Kamat, S. Mad husudana, V. Ravi, T. (2006). Ante mortem diagnosis of human rabies using saliva samples: comparison of real time and conventional RT-PCR techniques. *J. Clin. Virol.*, **36**, 17–23.

66. Nel, L. H. (2005). Vaccines for lyssaviruses other than rabies. *Expert. Rev. Vaccines*, **4**, 533–540.

67. Nel, L., Jacobs, J., Jaftha, J., von Teichman, B., Bingham, J. (2000). New cases of mokola virus infection in South Africa: a genotypic comparison of Southern African virus isolates. *Virus Genes*, **20**, 103–106.

68. Nel, L. H., Rupprecht, C. E. (2007). Emergence of Lyssaviruses in the Old World: the case of Africa. In *Current Topics in Microbiology and Immunology: Wildlife and Emerging Zoonotic Diseases: The Biology, Circumstances and Consequences of Cross-Species Transmission*, 161–193. Heidelberg: Springer Berlin.

69. Niezgoda, M., Briggs, D. J., Shaddock, J., Dreesen, D. W., Rupprecht, C. E. (1997). Pathogenesis of experimentally induced rabies in domestic ferrets. *Am. J. Vet. Res.*, **58**, 1327–1331.

70. Niezgoda, M., Briggs, D. J., Shaddock, J., Rupprecht, C. E. (1998). Viral excretion in domestic ferrets (Mustela putorius furo) inoculated with a raccoon rabies isolate. *Am. J. Vet. Res.*, **59**, 1629–1632.

71. Niezgoda, M., Hanlon, C. A., Rupprecht, C. E. (2002). Animal rabies. In *Rabies*. Ed. Jackson, A. C., Wunner, W. H., 163–207. London: Elsevier Science.

72. Paweska, J. T., Blumberg, L., Liebenberg, C., Hewlett, R. H., Grobbelaar, A. A., Leman, P. A., Croft, J. E., Nel, L. H., Nutt, L., Swanepoel, R. (2006). Fatal human infection with rabies-related Duvenhage virus, South Africa. *Emerg. Infect. Dis.*, **12**, 1965–1967.

73. Picard-Meyer, E., Barrat, J., Wasniewski, M., Wandeler, A., Nadin-Davis, S., Lowings, J. P., Fooks, A. R., McElhinney, L., Bruyére, V., Cliquet, F. (2004). Epidemiology of rabid bats in France, 1989 to (2002). *Vet. Rec.*, **155**, 774–777.

74. Rupprecht, C., Gibbons R. (2004). Prophylaxis against rabies. *N. Engl. J. Med.*, **351**, 2626–2634.

75. Rupprecht, C. E., Hanlon, C. A., Hemachudha, T. (2002). Rabies re-examined. *Lancet Infect. Dis.*, **2**, 327–343.

76. Sabeta, C. T., Markotter, W., Mohale, D. K., Shumba, W., Wandeler, A. I., Nel, L. H. (2007). Mokola virus in domestic mammals, South Africa. *Emerg. Infect. Dis.*, **13**, 1371–1373.

77. Saengseesom, W., Mitmoonpitak, C., Kasempi molprn, S., Sitprija, V. (2007). Real-time PCR analysis of dog cerebrospinal fluid and saliva samples for ante-mortem diagnosis of rabies. *Souteast Asian J. Trop. Med. Public Health*, **38**, 53–57.

78. Saluzzo, J. F., Rollin, P. E., Dauguet, C., Digoutte, J. P., Georges, A. J., Sureau, P. (1984). Premier isolment du virus Mokola a partir D'un rongeur (Lophuromys sikapusi). *Ann. Virol.*, **135**, 57–66.

79. Schneider, L. G., Barnard, B. J. H., Schneider, H. P. (1985). Application of monoclonal antibodies for epidemiology investigations and oral vaccination studies: African viruses. In *Rabies in the Tropics*. Ed. Kuwert, E., Merieux, C., Koprowski, H., Bogel, K., 47–59. Heidelberg, New York, Berlin: Springer.

80. Schneider, L. G., Dietzschold, B., Dierks, R. E., Matthaeus, W., Enzmann, P. J., Strohmaier, K. (1973). Rabies group-specific ribonucleoprotein antigen and a test system for grouping and typing of rhabdoviruses. *J. Virol.*, **11**, 748–755.

81. Servat, A., Labadie, A., Hamen, A., Boue, F., Cliquet, F. Inter-laboratory trial to evaluate the reproducibility of a new ELISA to detect rabies antibodies in vaccinated domestic and wild carnivores. *Biologicals,* **36,** 19–26.

82. Shope, R. E., Murphy, F. A., Harrison, A. K., Causey, O. R., Kemp, G. E., Simpson, D. I. H., Moore, D. L. (1970). Two African viruses serologically and morphologically related to rabies virus. *J. Virol.*, **6**, 690–692.

83. Sillero-Zubiri, C., King, A. A., Macdonald, D. W. (1996). Rabies and mortality in Ethiopian wolves (Canis simensis). *J. Wildl. Dis.,* **32**, 80–86.

84. Smith, J. S., Yager, P. A., Baer, G. M. (1996). A rapid fluorescent focus inhibition test (RFFIT) for determining rabies virus-neutralizing antibody. In *Laboratory Techniques in Rabies.* Ed. Meslin, F. X., Kaplan, M. M., Koprowski, H., 181–192. Geneva: World Health Organization.

85. Sureau, P., Germain, M., Herve, J. P., Geoffrey, B., Cornet, J. P., Heme, G., Robin, Y. (1977). Isolement du virus Lagos bat en Empire Centrafricain. *Bull. Soc. Path. Exot.*, **70**, 467–470.

86. Swanepoel, R. (2004). *Infectious Diseases of Livestock: With Special Reference to Southern Africa.* 2nd edition. Ed. Coetzer, J. A. W., Tustin, R. C., 1123–1182. Cape Town, South Africa: Oxford University Press. Swart 1989.

87. Thomson, G. R. and Meredith, C. D. (1993). Rabies in bat-eared foxes in South Africa. *Onderstepoort J. Vet. Res.,* **60**, 399–403.

88. Tierkel, E. S., Atanasiu, P. (1996). Rapid microscopic examination for Negri bodies and preparation of specimens for biological tests. In *Laboratory techniques in rabies.* Ed. Meslin, F. X., Kaplan, M. M., Koprowski, H., 55–65. Geneva: World Health Organization.

89. Dietzgen, R., Calisher, C. H., Kurath, G., Kuzmin, I. V., Rodriguez, L. L., Stone, D. M., Tesh, R. B., et al. (2011). Rhabdoviridae. In A. M. Q. King, M. J. Adams, E. B. Carstens, & E. J. Lefkowitz (Eds.), *Virus Taxonomy: Ninth Report of the International Committee on Taxonomy of Viruses.* 9th edition, 654–681. San Diego, CA: Elsevier.

90. Tordo, N., Bourhy, H., Sacramento, D. (1995). Polymerase chain reaction technology for rabies virus. In *The Polymerase Chain Reaction (PCR) for HUMAN VIRAL DIAGNOSIS.* Ed. Clewley, J. P., 125–145. Boca Raton, FL: CRC Press.

91. Trimarchi, C. V., Smith, J. S. (2002). Diagnostic Evaluation. In *Rabies.* Ed. Jackson, A. C., Wunner, W. H., 307–349. London: Elsevier Science.

92. Tsiang, H., Ceccaldi, P. E., Lycke, E. 1991. Rabies virus infection and transport in human sensory dorsal root ganglia neurons. *J. Gen. Virol.*, **72**, 1191–1194.

93. Van der Merwe, M. (1982). Bats as vectors of rabies. *S. Afr. J. Sci.*, **78,** 421–422.

94. Von Teichman, B. F., De Koker, W. C., Bosch, S. J. E., Bishop, G. C., Meredith, C. D., Bingham, J. (1998). Mokola virus infection: description of recent South African cases and a review of the virus epidemiology. *J. S. Afr. Vet. Ass.*, **69**, 169–171.

95. Wacharapluesadee, S., Sutipanya, J., Damrongwatanapokin, S., Phumesin, P., Chamnanpood, P., Leowijuk, C., Hemachudha, T. (2008). Development of a Taqman real-time RT-PCR assay for the detection of rabies virus. *J. Virol. Methods,* (E-pub ahead of print).

96. Wakeley, P. R., Johnson, N., McElhinney, L. M., Marston, D., Sawyer, J., Fooks, A. R. (2005). Development of a real-time, Taqman reverse transcriptase-PCR assay for detection and differentiation of lyssavirus genotypes 1, 5, and 6. *J. Clin. Microbiol.*, **43**, 2786–2792.

97. Wakeley, P. R., Johnson, N., McElhinney, L. M., Marston, D., Sawyer, J., Fooks, A. R. (2006). Development of a real-time, differential RT-PCR Taq-Man assay for lyssavirus genotypes 1, 5 and 6. *Dev.Biol. (Basel),* **126**, 227–236.

98. Webster, L. T., Dawson, J. R. (1935). Early diagnosis of rabies by mouse inoculation: measurement of humoral immunity to rabies by mouse protection test. *Proc. Soc. Exp. Biol. Med.,* **32**, 570–573.

99. Webster, W. A., Casey, G. A. (1996).Virus isolation in neuroblastoma cell culture. In *Laboratory Techniques in Rabies.* Ed. Meslin, F. X., Kaplan, M. M., Koprowski, H., 96–104. Geneva: World Health Organization.

100. Whitfield, S. G., Fekadu, M., Shaddock, J. H., Niezgoda, M., Warner, C. K., Messenger S. L. (2001). A comparative study of the fluorescent antibody test for rabies diagnosis in fresh and formalin-fixed brain tissue specimens *J. Virol. Methods,* **95**, 145–151.

101. World Health Organization. (2005). *WHO Expert Consultation on Rabies. Technical Report Series 931.* Geneva: World Health Organization.

102. Wiktor, T. J., Flamand, A., Koprowski, H. (1980). Use of monoclonal antibodies in diagnosis of rabies virus infection and differentiation of rabies and rabies-related viruses. *J. Virol. Methods,* **1**, 33–46.

103. Xu, G., Weber, P., Hu, Q., Xue, H., Audry, L., Li, C., Wu, J., Bourhy, H. (2007). A simple sandwich ELISA (WELYSSA) for the detection of lyssavirus

nucleocapsid in rabies suspected specimens using mouse monoclonal antibodies. *Biologicals,* **35,** 297–302.

104. Freuling, C. M., Beer, M., Conraths, F. J., Finke, S., Hoffmann, B., Keller, B., Kliemt, J., Mettenleiter, T. C., Mühlbach, E., Teifke, J. P., Wohlsein, P., Müller, T. (2011). Novel Lyssavirus in Natterer's Bat, Germany. *Emerg. Infect. Dis.,* **17**(8), 1519–1522.

105. Marston, D. A., Horton, D. L., Ngeleja, C., Hampson, K., Mcelhinney, L. M., Banyard, A. C., Haydon, D., et al. (2012). Novel Lyssavirus in an African Civet 1. *Emerg. Infect. Dis.,* **18**(4), 664–667.

106. Kuzmin, I. V., Mayer, A. E., Niezgoda, M., Markotter, W., Agwanda, B., Breiman, R. F., Rupprecht, C. E. (2010). Shimoni bat virus, a new representative of the Lyssavirus genus. *Virus Res.* **149,** 197–210.

107. van Thiel P-PAM, de Bie RMA, Eftimov F, Tepaske R, Zaaijer HL, et al. (2009). Fatal human rabies due to Duvenhage virus from a bat in Kenya: failure of treatment with coma-induction, ketamine, and antiviral drugs. *PLoS Negl. Trop. Dis.,* **3**(7), e428.

108. Kuzmin, I. V., Niezgoda, M., Franka, R., Agwanda, B., Markotter, W., Beagley, J. C., Urazova, O. Y., Breiman, R. F., Rupprecht, C. E. (2008a). Lagos bat virus in Kenya. *J. Clin. Microbiol.,* **46,** 1451–1461.

109. Kuzmin, I. V., Niezgoda, M., Franka, R., Agwanda, B., Markotter, W., Beagley, J. C., Urazova, O. Y., Breiman, R. F., Rupprecht, C. E. (2008b). Seroprevalence to West Caucasian bat virus in Kenya. *Emerg. Infect. Dis.,* **14,** 1887–1889.

110. Kuzmin, I. V., Turmelle, A. S., Agwanda, B., Markotter, W., Niezgoda, M., Breiman, R. F., Rupprecht, C. E. (2011). Commerson's leaf-nosed bats (*Hipposideros commersonii*) is the likely reservoir of Shimoni bat virus. *Vector Borne Zoonotic Dis.,* **11,** 1465–1470.

111. Markotter, W., York, D., Sabeta, C. T., Shumba, W., Zulu, G., Le Roux, K., Nel, L. H. (2009). Evaluation of a rapid immunodiagnostic test kit for detection of African lyssaviruses from brain material. *Onderstepoort J. Vet. Res,* **76**(2), 257–262.

112. Coertse, J., Weyer, J., Nel, L. H., Markotter, W. (20100. Improved PCR methods for detection of African rabies and rabies-related lyssaviruses. *J. Clin. Microbiol.,* **48**(11), 3949–3955.

113. Ma, X., Niezgoda, M., Blanton, J. D., Recuenco, S., Rupprecht, C. E. (2012). Evaluation of a new serological technique for detecting rabies virus antibodies following vaccination. *Vaccine,* **3**(36), 5358–5362.

Chapter 12

Detection of Human Pathogens under Basic Laboratory Conditions by DNA Hybridization Arrays

Roman Wölfel

Bundeswehr Institute of Microbiology, Department for Medical Bio-Reconnaissance & Verification, Neuherbergstrasse 11, 80937 Munich, Germany

romanwoelfel@bundeswehr.org

Sensitive and affordable diagnostics even for rare diseases are of certain importance for public health labs in developing countries but are often not accessible to those who need them most. Recent interest highlights innovative technological approaches for overcoming the problem, but these must be balanced by the efforts and research needed to ensure effective implementation in the context of only basic laboratory equipment available in most countries in the developing world. This article discusses the use of chip-based hybridization arrays for diagnostic purposes under basic laboratory conditions. It describes a robust and field-usable chip array design for detection of human pathogens that couples conventional polymerase chain reaction (PCR) with a DNA hybridization array.

Viral Diagnostics: Advances and Applications

Edited by Robert S. Marks, Leslie Lobel, and Amadou Alpha Sall

ISBN 978-981-4364-43-0 (Hardcover), 978-981-4364-44-7 (eBook)

www.panstanford.com

During the last decade a variety of different microarray platforms have been described for identification and typing of bacteria and viruses.[1] Today, high-density, oligonucleotide-based whole-genome microarrays can be used for genomic analysis beyond simple gene expression profiling. But when looking at hundreds of thousands of analytes, the need to focus on subsets of those analytes is the basis for mid- and low-density panels built around disease states or genes of interest.[2,3]

So far, the microarray technology relies on well-equipped laboratories and well-trained personnel. Moreover, such assays are rather expensive, are often not appropriate for field-based outbreak investigations, and may be difficult to implement in those countries where the diseases are endemic.[4] Therefore research is directed to simplify this technique for use in the field and in laboratories with only basic lab equipment. To compete with so-called immunological-based point-of-care assays or to supplement them, microarray technology has to be as simple, reliable, and robust as possible.

A fast "low cost and density" (LCD) DNA microarray principle that could be used for direct identification and presumptive typing of pathogens is described here.

The working principle of the hybridization array is comparable to the robust protocols used for membrane-based reverse hybridization techniques, such as reverse line blots and line probe assays. The LCD microarray uses a so-called hybridization assay format, which reacts with biotin-labeled products of a (preperformed) specific conventional PCR (Fig. 12.1). It is a transparent, prestructured polymer support containing eight identical arrays in well-separated, individually addressable hybridization fields (Fig. 12.2A). The outer dimensions of 50 × 50 mm, combined with the nonfluorescent detection principle, based on the formation of a clearly visible substrate precipitate, allow the use of a very economical transmission-light film scanner to generate images for the data analysis. The pathogen-specific capture probes are between 15 and 25 nucleotides long and carry an immobilization tag at their 5′ end. They are spotted as duplicates in a 9 × 9 pattern, with average spot diameters of 300 μm (Fig. 12.2B). Each array is designed to detect pathogen-specific sequence motifs within PCR fragments. Four capture probes for the competitive internal control

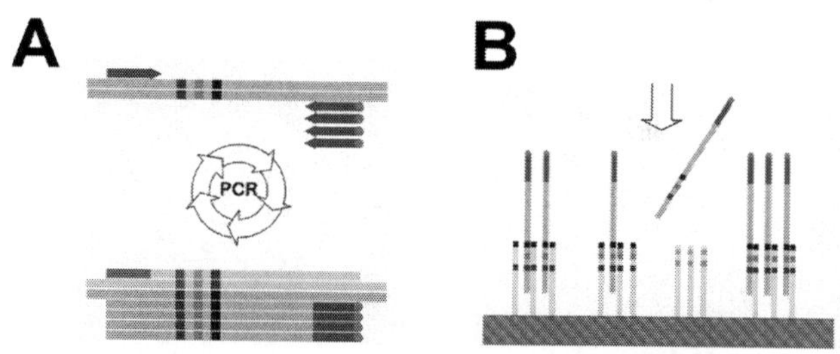

Figure 12.1 Schematic diagram of a hybridization array. (A) Principle of production of single-stranded DNA fragments by asymmetrical PCR conditions. Excess primers are labeled with biotin for subsequent detection by staining after specific hybridization on the array surface. (B) Amplified fragments are hybridized to capture probes, which are covalently bound on the surface of the hybridization array chip.

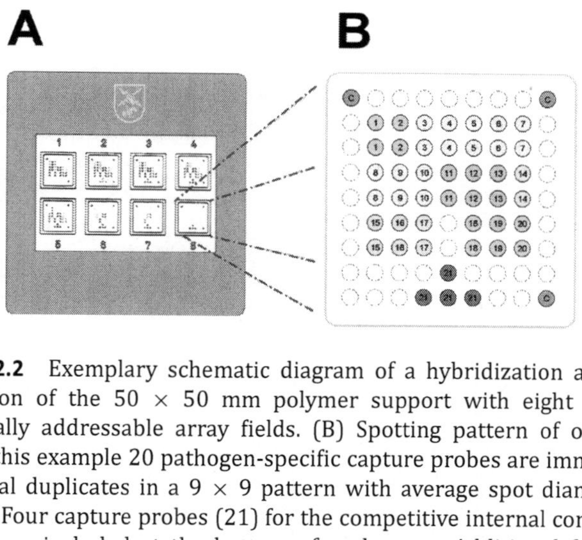

Figure 12.2 Exemplary schematic diagram of a hybridization array. (A) Illustration of the 50 × 50 mm polymer support with eight identical individually addressable array fields. (B) Spotting pattern of one array field. In this example 20 pathogen-specific capture probes are immobilized as vertical duplicates in a 9 × 9 pattern with average spot diameters of 300 μm. Four capture probes (21) for the competitive internal control PCR product are included at the bottom of each array. Additional functional control probes (C) to visualize successful hybridization and staining are immobilized in three corners of each field.

PCR product are included at the bottom of each array. Additional functional control probes with an unrelated sequence motif of fungal origin are immobilized in three corners of each field, to visualize successful hybridization and staining steps, and enable the user to position the array for analysis correctly.

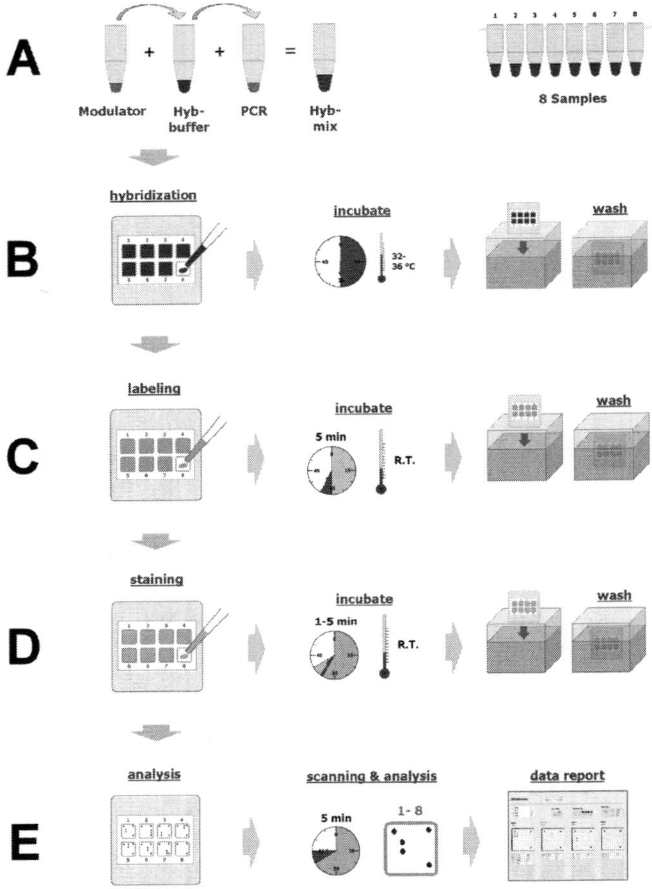

Figure 12.3 Work flow showing hybridization and visualization of prelabeled PCR products on an LCD microarray.

Nucleic acid extraction and PCR can be performed using a variety of commercially available kits and low-cost conventional PCR machines. Subsequently, hybridization and staining procedures are performed according to a simple and purely manual protocol: in brief, the biotin-labeled PCR product is combined with a formamide-based hybridization buffer (Fig. 12.3A). This mixture is applied to one of the individual fields of the arrays and hybridized (Figs. 12.1B and 12.3B) in a standard laboratory incubator. Following a short washing step in a wash buffer, the array is dried using an airstream from a simple compressed-air can purchased from a photography supply store. The specifically bound PCR products are further incubated with streptavidin peroxidase labeling at room temperature (Fig. 12.3C). Following an additional wash and drying cycle, the specific hybridization events are visualized by incubation with TMB (3,3′,5,5′-tetramethylbenzidin) as an enzyme substrate at room temperature, leading to the formation of dark blue precipitates at positions of label binding (Fig. 12.3D). Although these signals are already clearly visible to the human eye, in addition high-resolution images can be taken using an inexpensive transmission-light film-scanning device. These images allow the use of commercially available software for image analysis, data storage, or telemedicine transmission (Fig. 12.3E). The hybridization and staining technique provides a very fast (a total of 45 min with 15 min "hands-on" time only) economic, sensitive, and specific detection of PCR amplification products. Including the initial PCR step the whole protocol takes less than four hours, and up to 48 samples can easily be analyzed in parallel, owing to its simplicity. In addition to a conventional PCR cycler, only standard laboratory equipment as used for traditional enzyme-linked immunosorbent assay (ELISA) diagnostics is needed.

The LCD assay technique can be used for analysis of single-nucleotide polymorphism analysis, which is of advantage for differentiation of closely related microorganisms. In this case the LCD microarray method can compete with real-time PCR assays based on hybridization probes, because up to 30 variants can be detected in one single assay, thereby reducing time and avoiding a variety of costly hybridization probes. Such an application is shown in Fig. 12.4 using a Crimean-Congo hemorrhagic fever (CCHF) virus assay as an example: Though the genetic diversity of the CCHF virus, a member of the family *Bunyaviridae*, is extraordinarily high, this assay provides a simple method of detection and strain identification, obviating the need for sequencing. Its appropriateness on the clinical level has been proven by testing a large panel of virus strains as well as a sufficient number of original patient samples. Sensitivity in clinical specimen ranged around 26 copies of RNA per sample, which is equivalent to the sensitivities of diagnostic assays, which

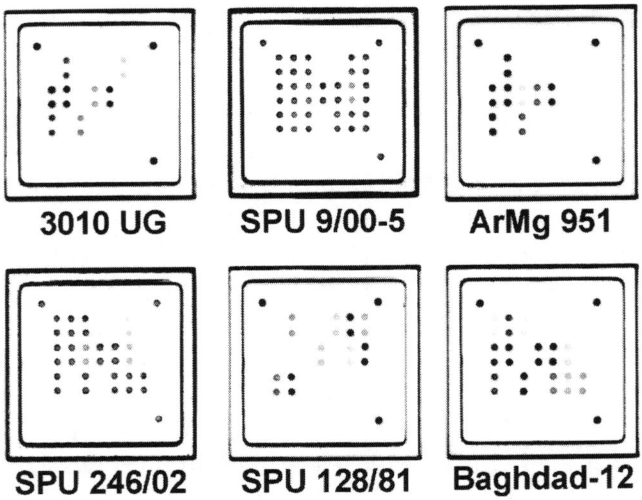

Figure 12.4 Discrimination capacities of an LCD microarray demonstrated for the CCHF virus as an example. Different strains of the CCHF virus (3010UG, Uganda; SPU 9/00-5, SPU246/02, SPU128/81, South Africa; ArMg951 Madagascar; Baghdad-12, Iraq) are detected by a CCHF-specific array chip, resulting in unique hybridization patterns. These patterns are based on sequence variability within an approximately 25-nucleotide region of the CCHF virus genome.

are targeted to one specific CCHF virus strain only. To determine whether an LCD array provides the same sensitivity as gel detection, the same amount of PCR product was analyzed by standard gel electrophoresis and LCD array hybridization. As shown in Fig. 12.5 even faint gel bands yielded corresponding hybridization signals on the array. Thus it can be concluded that the sensitivity of array detection was comparable to that of classical gel electrophoresis.

Due to the generic features of the LCD array technology, it should be applicable to many other fields in clinical microbiology, for example, for detecting whole virus families or for differentiating bacteria and fungi. Other examples, for genotyping *Mycobacterium tuberculosis* drug resistance,[5] for typing of human papilloma virus in squamous-cell cancer tissue,[6] and for differentiation of Coronavirus,[7] have recently been described. Further diagnostic assays for a variety of viral pathogens like the aforementioned

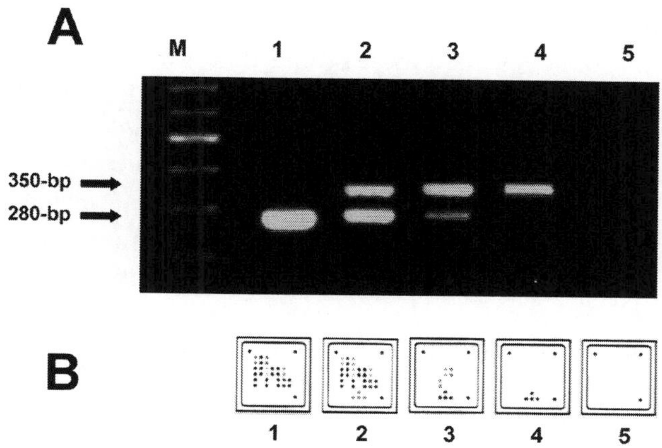

Figure 12.5 Comparison of conventional gel electrophoresis and LCD microarray hybridization as readout systems using the example of a CCHF virus-specific assay (280 bp amplicon; internal control 350 bp amplicon). Note that internal control amplification is suppressed by high concentrations of the CCHF virus target RNA. (A) 1.5% agarose gel analysis of the PCR products. Lane M: 100 bp molecular size ladder; lanes 1–3: CCHF virus strain Bagdad-12 in vitro–transcribed RNA (lane 1: 6×10^6 cps/rx, lane 2: 6×10^4 cps/rx, lane 3: 6 cps/rx); lane 4: internal control RNA only; lane 5: traces of ethanol added to the PCR mix to simulate PCR inhibition. (B) Corresponding specific hybridization pattern of both CCHF virus-specific amplicons and an internal control amplicon on the LCD microarray. Some CCHF-specific hybridization spots get lost at lower concentrations of viral RNA in the sample. Functional control spots are always visible at three angles. *Abbreviations*: bp, base pair; cps/rx, copies per reaction.

CCHF virus,[8] the Lassa virus, and the dengue virus have been developed, but most of them have not been published yet. In the future multiplex assays for identification of several independent pathogens causing a similar clinical entity, such as respiratory or neurological disorders, could be designed.

The availability of easy-to-use and accessible diagnostic tests is a key to the management and containment of infectious diseases in the developing world. Here the LCD microarray technique offers the potential for detecting the presence or absence of specific pathogens and makes it possible to screen a sample for different nucleic acid fragments in parallel. The ability to search a sample for genetic material corresponding to several different pathogens or pathogen strain variants is therefore greatly improved as compared to other techniques such as real-time PCR. The identification by hybridization is highly sequence based in contrast to the verification by amplicon size as done in conventional PCR assays.

Putting sophisticated reagents and detection systems into easy-to-use formats that deliver patient-empowering, point-of-care information is a goal that seems achievable by the simple hybridization assay described here. Applying this technical capacity to the medical needs of developing countries, especially for the diagnostics of rare but highly relevant diseases, is one of the challenges of the future.

References

1. Heller, M. J. DNA microarray technology: devices, systems, and applications. 2002. *Annu. Rev. Biomed. Eng.*, **4**, 129–153.
2. Hsia, C. C., V. E. Chizhikov, A. X. Yang, A. Selvapandiyan, I. Hewlett, R. Duncan, R. K. Puri, H. L. Nakhasi, and G. G. Kaplan. Microarray multiplex assay for the simultaneous detection and discrimination of hepatitis B, hepatitis C, and human immunodeficiency type-1 viruses in human blood samples. 2007. *Biochem. Biophys. Res. Commun.*, **356**, 1017–1023.
3. Khan, A. S. Genomics and microarray for detection and diagnostics. 2004. *Acta Microbiol. Immunol. Hung.*, **51**, 463–467.
4. Garner, P., A. Kiani, and A. Supachutikul. Diagnostics in developing countries. 1997. *BMJ*, **315**, 760–761.
5. Aragon, L. M., F. Navarro, V. Heiser, M. Garrigo, M. Espanol, and P. Coll. Rapid detection of specific gene mutations associated with isoniazid or rifampicin resistance in Mycobacterium tuberculosis clinical isolates using non-fluorescent low-density DNA microarrays. 2006. *J. Antimicrob. Chemother.*, **57**, 825–831.
6. Will, C., C. Schewe, K. Schluns, and I. Petersen. HPV typing and CGH analysis for the differentiation of primary and metastatic squamous cell carcinomas of the aerodigestive tract. 2006. *Cell Oncol.*, **28**, 97–105.
7. de Souza Luna, L. K., V. Heiser, N. Regamey, M. Panning, J. F. Drexler, S. Mulangu, L. Poon, S. Baumgarte, B.J. Haijema, L. Kaiser, and C. Drosten. Generic detection of coronaviruses and differentiation at the prototype strain level by reverse transcription-PCR and nonfluorescent low-density microarray. 2007. *J. Clin. Microbiol.*, **45**, 1049–1052.

8. Wölfel, R., J. T. Paweska, N. Petersen, A. Grobbelaar, P. A. Leman, R. Hewson, M. C. Georges-Courbot, A. Papa, V. Heiser, M. Panning, S. Gunther, and C. Drosten. Low-density macroarray for rapid detection and identification of Crimean-Congo hemorrhagic fever virus. 2009. *J. Clin. Microbiol.*, **47**(4), 1025–1030.

Chapter 13

Differentiation between Viral and Bacterial Respiratory Infections Using Chemiluminescence of Polymorphonuclear Leukocytes

Daria Prilutsky,[a,b,c] **Mark Last,**[b] **Leslie Lobel,**[a] **and Robert S. Marks**[c,d,e]

[a] *Department of Virology, Faculty of Health Science, Ben-Gurion University of the Negev, Beer-Sheva, Israel*
[b] *Department of Information Systems Engineering, Ben-Gurion University of the Negev, Beer-Sheva, Israel*
[c] *The National Institute of Biotechnology in the Negev, Ben-Gurion University of the Negev, Beer-Sheva, Israel*
[d] *Department of Biotechnology Engineering, Ben-Gurion University of the Negev, Beer-Sheva, Israel*
[e] *The Ilse Katz Center for Meso and Nanoscale Science and Technology, Ben-Gurion University of the Negev, Beer-Sheva, Israel*
dariap@bgu.ac.il, llobel@bgu.ac.il, rsmarks@bgu.ac.il

13.1 The Innate Immune System and Participating Cells

The environment contains a large array of potential pathogens, such as bacteria, viruses, parasites, fungi, chemicals, drugs, and pollen. To

Viral Diagnostics: Advances and Applications
Edited by Robert S. Marks, Leslie Lobel, and Amadou Alpha Sall

ISBN 978-981-4364-43-0 (Hardcover), 978-981-4364-44-7 (eBook)
www.panstanford.com

protect themselves from potentially nefast foreign invaders, humans have a laborious network of cells, molecules, and mechanisms that collectively are referred to as the immune system, both innate and adaptive. The direct aim of the immune system is to differentiate between self and nonself and to neutralize and remove the intruder.[1,2]

Of interest to this study is the innate immunity with which an individual is born and is always present and available at very short notice for protection from challenges by foreign invaders.[1,3] It accomplishes a wide variety of activities within minutes after invasion of the pathogen, including recognition and effector functions. The former involves the recognition of structures present in pathogens that are distinct from self. The effector functions include the release of substances exerting direct antimicrobial properties and production of cytokines and chemokines that recruit inflammatory cells to the infection site. The innate responses use cellular and molecular components, and the extent of the reactions remains constant regardless of previous exposure to the same pathogen. It was believed for many years that the innate immune response is nonspecific and nonselective. This dogma has been challenged by the discovery that innate immune responses are triggered by cellular pattern recognition receptors (PRRs) located on the plasma membrane or inside cells. They recognize specific conserved structures of invading pathogens called pathogen-associated molecular patterns (PAMPs).[3] The most important classes of PRRs are the toll-like receptors (TLRs) and the lectin-like receptors. TLRs recognize a wide variety of microbial structures, including lipopolysaccharide (LPS) of Gram-negative bacteria (TLR4), lipoteichoic acid and lipoproteins of Gram-positive bacteria (TLR2), bacterial DNA (TLR9), and viral RNA (TLR3, TLR7 and TLR8).[3,4] Lectin-like receptors are PRRs specialized in the recognition of polysaccharide chains of bacterial pathogens, including the mannans and glucans of pathogenic fungi.[5,6]

The cellular components of innate immunity comprise phagocytic cells, such as macrophages, monocytes, and neutrophils, as well as cells that release inflammatory mediators, such as basophils, mast cells, and eosinophils, in addition to natural killer cells. Phagocytic cells of the innate immune system consist of certain types of polymorphonuclear leukocytes (PMNs) and phagocytic monocytes (macrophages). Upon activation, these cells release cytokines, which have different effects on various cells. PMNs are a population of cells also referred to as granulocytes. About 98% of the PMNs are neutrophils, 1–2% eosinophils, and 0.1–0.2% basophils. The neutrophils are the most important circulating phagocytes. They form the body's first line of defense against invading particles, especially bacteria. PMNs migrate into the inflammatory site and kill the invading microorganisms by releasing toxic peroxide and superoxide radicals. Mononuclear phagocytes are cells giving rise to the circulating phagocytic monocytes, ultimately maturing into mobile or fixed macrophages, which, in turn, play an important role in phagocytosis in tissue.[1,2]

13.2 Phagocytosis as a First-Line Defense Mechanism against Pathogens

Phagocytosis is defined as a defense mechanism of innate immunity, mediated by specialized cells for ingestion and destruction of the invading foreign particles. This process is an evolutionarily ancient host cell endocytic response discovered by Elie Metchnikoff in the nineteenth century. The process of phagocytosis includes several steps: adhesion, or recognition of the particle as an initiation of ingestion; invagination of the outer cell membrane and enclosure of the particle within a phagocytic vacuole, the "phagosome"; and internalization, fusion with the granules, or lysosomes, forming a "phagolysosome," which contains a mixture of active agents.[7] In the phagolysosome, the pathogen is killed by exposure to enzymes, antimicrobial peptides, and reactive oxygen species (ROS).[2] Digestion products are then released from the cell into the environment. Phagocytosis is performed by blood monocytes, neutrophils (i.e., PMNs), and tissue macrophages. Phagocytosis may be enhanced by a variety of factors that make the foreign particle an easier target by a process called opsonization, or recognition of the foreign particle. Opsonization is the process whereby particles, such as microorganisms, become coated with molecules, allowing them to bind to receptors on phagocytes, facilitating recognition and full acceptance by such cells. Antibodies (especially IgG) and complement proteins, like C3b, can opsonize and are, therefore, referred to as opsonins.[1,2]

The systems participating in killing pathogens divide into two, oxygen-dependent and oxygen-independent mechanisms. The former mechanism of neutrophil killing is oxygen dependent. Phagocytozing neutrophils undergo a respiratory burst, during which ROS like superoxide anions, hydrogen peroxide (H_2O_2), hydroxyl radicals (OH^-), and hypochlorous acid (HOCl) are formed. The latter mechanisms act by using the contents of three neutrophil granule subsets, namely, azurophil, specific, and gelatinase granules, each containing enzymes, including proteases, antimicrobial proteins, and peptides.[8]

13.2.1 *Respiratory Burst: Mechanisms, Localization, and Techniques for Detection*

Phagocytic cells travel throughout the body in search of foreign pathogens or damaged tissue and respond to a wide variety of inflammatory factors. During the inflammatory response, neutrophils are mobilized to sites of inflammation, where they phagocytize pathogens and/or fragments of damaged tissue. Subsequently, a variety of reactive oxygen radical species and proteolytic enzymes are released, accompanied by increased consumption of the molecular oxygen, which directly destroy the engulfed material.[9]

13.2.1.1 Main mechanisms, products and enzymes of the respiratory burst

One of the most important antimicrobial systems of phagocytic cells is the reduced nicotinamide adenine dinucleotide phosphate (NADPH) phagocyte oxidase complex that is responsible for the generation of superoxide ($O_2\bullet$). $O_2\bullet$ and other oxygen-derived intermediates that can modify organic molecules are referred to as ROS.[10,11] The NADPH-oxidase complex is a plasma membrane–associated enzyme, dormant in resting cells and activated during phagocytosis of invading pathogens. It is activated to produce superoxide anions, with the secondary generation of other ROS, such as H_2O_2, OH^-, and HOCl, which function together in microbial killing in the phagosome.[12] Electrons are transferred from cytoplasmic NADPH to oxygen on the phagosomal side of the membrane, generating ROS. The second important enzyme participating in ROS production is myeloperoxidase (MPO), located in azurophil granules generating HOCl.[13]

The active NADPH oxidase is a multiprotein complex. The core enzyme comprises five components: $p40^{phox}$, $p47^{phox}$, $p67^{phox}$, $gp91^{phox}$, and $p22^{phox}$. These, in turn, are organized into either membrane subcomplexes or cytosolic, translocating to the plasma membrane units during the activation[14] (see Fig. 13.1D). The assembly of NADPH-oxidase components on the membrane of intracellular organelles leading to intracellular ROS production or on the plasma membrane causes a release of radicals into the phagosome or the extracellular space.[15] It is important to emphasize here that the generation of all of these ROS is dependent upon activation of the NADPH-oxidase complex. Because of the toxic nature of its products, it is not surprising that action of the oxidase must be tightly regulated and spatially controlled.[16]

13.2.1.2 Techniques used to measure reactive oxygen species

Numerous techniques have been developed over the years to follow the respiratory burst activity in phagocytes.[17–21] These techniques include microscopic visualization of intracellular oxidant production using stains, spectrophotometric techniques, fluorimetric techniques,[24,25] and chemiluminescence (CL) techniques.[26,27] The respiratory burst is associated with the emission of light, generated from reactive species, which can be amplified and detected as luminol-dependent chemiluminescence (LCL).[27,28] Several dyes have been developed that, after being excited by ROS, release energy in the form of light (CL). Among these, luminol (5-amino-2,3-dihydro-1,4-phthalazindione), with an emission at 460 nm, is the most intensively investigated and most frequently used in free radical research. To exhibit its luminescence, the luminol must be activated by an oxidant (ROS) in the presence of a catalyst (peroxidase, in this case MPO). The luminol-amplified CL reaction in neutrophils has three primary characteristics:[21,29] (1) a peroxidase

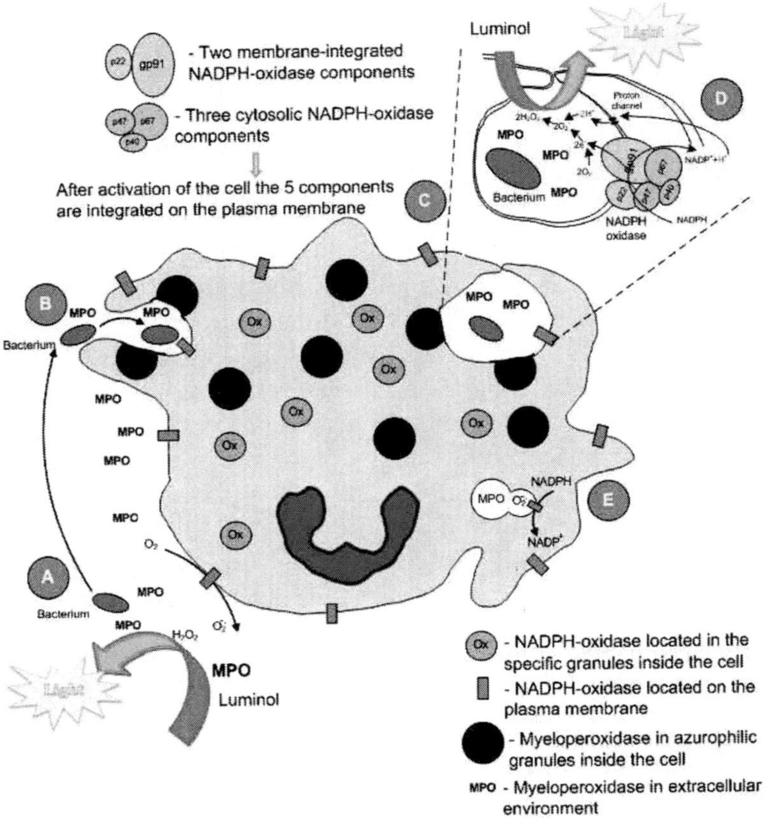

Figure 13.1 Schematic representation of mechanisms of phagocytosis and localization of the respiratory burst detected as a luminol CL. (A) Extracellular CL is produced by activating membrane-localized NADPH oxidase using MPO in the extracellular milieu. (B) Beginning of pseudopodia formation and enclosure of the bacterium into a phagosome. CL results from ROS produced by NADPH oxidase and extracellular MPO. (C) Formation of the phagosome and ingestion of the bacterium. (D) Close look at the phagosome: NADPH oxidase is integrated into the active enzyme by translocation of cytosolic components toward membrane components, and ROS is produced. CL in this site is referred to as intracellular in origin, while MPO is enclosed in the endocytic vesicle. (D) Intracellular CL reaction that is produced by fusion of azurophilic granule containing MPO and a specific granule containing NADPH oxidase. (Adapted from Ref. 31.)

(usually MPO, originating from azurophil granules), (2) oxygen metabolites produced by NADPH oxidase, and (3) both intracellular and extracellular reactions that are measurable.

The LCL technique is very sensitive and simple to perform and, as a consequence, is widely used to measure respiratory burst activity in phagocytic cells.[26,27]

Ideally, a technique designed to measure cellular production of ROS should:[17,21]

1. be specific for a particular oxygen metabolite and/or mechanism;
2. be highly sensitive;
3. not interfere with cellular function;
4. define the subcellular site of ROS generation by being able to measure both extracellularly and intracellularly released reactive oxygen metabolites;
5. not require expensive or complicated laboratory equipment; and
6. be easy to standardize.

At present, the most commonly used techniques for measuring release of ROS fail to satisfy the above criteria and appear to be insufficiently sensitive. Luminometry techniques, however, offer several advantages and are much more sensitive than other analytical techniques. The LCL technique allows measurement of intracellular as well as extracellular ROS production. However, more than one technique usually has to be included in the methodology for assessment of ROS production, that is, the LCL system used for intracellular assessment is different from the extracellular system and cannot be done simultaneously in one sample. Therefore, a direct simultaneous technique that can measure quantitatively the extracellular and intracellular ROS production is needed and proposed in this manuscript.

13.2.1.3 Localization of the luminol-dependent CL reaction

Upon stimulation with soluble or particulate matter in the presence of luminol, phagocytes respond by producing CL, which is linked to the increased production of oxidative metabolites. For an LCL reaction to take place, the different reagents must be present in the same subcellular compartment. Part of the CL reaction takes place extracellularly, where some of the oxygen metabolites produced are released to react with peroxidase in the extracellular milieu (see Fig. 13.1A,B). Extracellular peroxidase may have been released through a degranulation process or may originate from dead and disintegrated cells. The amount of extracellular peroxidase can, however, be a limiting factor in neutrophil CL[30–32] but can easily be fixed by adding extra peroxidase.

In addition, part of the CL reaction takes place intracellularly, in a subcellular compartment isolated from the surrounding environment (see Fig. 13.1C,E). As before, for this reaction the reactants necessary must be present also in the same subcellular compartment. Since the NADPH oxidase, producing oxygen radicals necessary for the reaction, lies in the membranes of endocytic vesicles (see Fig. 13.1C), or in the specific granules (see Fig. 13.1E), and the peroxidase (MPO) is localized into the azurophilic granules, granule fusion has to occur so as to bring the two enzymes together in the same compartment.[31] Alternatively, oxygen metabolites produced in one compartment could diffuse through two membranes, traversing the cytoplasm, to reach the MPO-containing compartment. Although the ability of H_2O_2 to diffuse through biological membranes is well known,[33] it is unlikely that this is the mechanism for the CL reaction, since the cytoplasm contains many oxygen radical scavengers that could remove the oxygen metabolites before they reach the azurophil granule.

The extracellular CL reaction can:[31]

1. result from an activation of the plasma membrane–localized NADPH oxidase, generating oxygen metabolites released to react with peroxidase (MPO) in the extracellular environment (see Fig. 13.1A); and
2. result from a reaction between oxygen metabolites and peroxidase in an endocytic vesicle during the steps of forming the phagosome but still in contact with the extracellular environment (see Fig. 13.1B).

The intracellular CL reaction can:[31]

1. result from a reaction between oxygen metabolites and peroxidase in a closed endocytic vesicle (phagosome) (see Fig. 13.1C,D); and

2. result from a reaction between oxygen metabolites and the peroxidase, combined in the same subcellular compartment through fusion of a granule containing MPO (azurophilic) and a granule containing NADPH oxidase (specific), respectively (see Fig. 13.1E).

13.2.1.4 Stimulation of the respiratory burst

Stimulants of phagocytic activity can be divided into different groups, depending on their solubility and interaction with membrane receptors. The first group, **soluble** receptor stimulants (e.g., *N*-formylmethionyl-leucyl-phenylalanine [fMLP], phorbol 12-myristate 13-acetate [PMA], conA, PAF, Caionophore A23187) interact with the phagocyte's membrane via a nonreceptor mechanism. Another important group of stimulants are those of a **particulate** nature like zymosan and latex beads.[21]

The most popular particle stimulant used to activate the immune system is zymosan, which is an insoluble cell wall polysaccharide of yeast (*Saccharomyces cerevisiae*). The preparation is composed primarily of β-glucans, the active component mediating the cellular effects, together with mannans, mannoproteins, and chitin, as well as small amounts of phosphorus, nitrogen, and magnesium.[34] Zymosan is used in many physiological and immunological studies and is recognized and phagocytozed principally by monocytes and macrophages, leading to cellular activation. In vitro, zymosan serves as a model for the study of the innate immune responses, activating the alternative complement cascade and becoming coated with C3b/C3bi, acting, therefore, as a convenient opsonized particle. It has been confirmed that activation ensues when the macrophage encounters the β-glucan molecule of zymosan, triggering the macrophage's full and powerful immune-stimulating capabilities.[35,36]

The inflammatory response triggered by zymosan is linked to its phagocytosis, a process mediated by a set of different receptors. The receptors participating in the zymosan recognition are Dectin-1, TLRs, and CD14 (TLR2, acting in collaboration with CD14 and TLR6). Together, these receptors facilitate inflammatory responses to the particle. Dectin-1 directly triggers phagocytosis and stimulates the production of ROS, thus contributing to microbial killing. TLRs induce signaling through NF-κB, leading to the production of inflammatory cytokines, responses enhanced by Dectin-1. Similarly, TLR signaling enhances responses, such as production of ROS, that are triggered by Dectin-1.[34,37,38]

In LCL reactions the response highly depends on the stimulus employed. If a particulate stimulus is employed, it becomes covered with opsonins (complement or/and immunoglobulins), the oxidative metabolites start to form, then phagocytosis takes place, then degranulation of azurophilic granules for release of MPO, and, finally, the formation of a phagosome. In this case, there is invagination of the phagocyte's outer membrane, which then proceeds to completely enclose the particle within a phagosome. If a liquid stimulant is used, it causes degranulation of specific granules with full activation of redox metabolism but little or no azurophilic degranulation. As a result, in this case, no phagosome is formed.[26]

Therefore, when different stimulants are used, different patterns of CL will be observed.

13.2.2 *Priming*

Understanding how a neutrophil undergoes functional transformation to facilitate an adequate host defense response has evolved over time. Only recently it has become clear that the neutrophil response is actually very finely modulated to provide an appropriate reaction to the invading pathogen. Factors implicated in this modulatory role are known as priming agents.

The initial standard for a neutrophil response was full activation. In this scenario, when presented with appropriate stimuli, a neutrophil quietly circulating through the bloodstream would be attracted to a site of infection and become fully activated. This process involved morphological polarization of the cell and subsequent chemotaxis, phagocytosis, production of ROS, and, perhaps, release of granule enzymes into the extracellular environment. An enormous amount of research has focused on understanding this process, addressing not only the type of stimuli causing activation and the consequences thereof but also the molecular mechanisms involved in the activation process. It soon became evident that activation was, in fact, not a process of all-or-none but a two-stage process. Thus, neutrophils would first encounter a stimulus that would not directly activate the cells but would leave the cells in a "primed" state. When the primed neutrophils encountered a second appropriate stimulus, a fully activated state would result. Given the two possibilities, a generally accepted consensus definition of priming has emerged: "Priming occurs when the neutrophil's functional response (e.g., ROS production, chemotaxis) to a stimulus is amplified by previous exposure of the cell to a priming agent." The priming agent itself would not, however, normally cause a noticeable functional response, except possibly when present in high concentrations.[39] Clearly, it is an oversimplification to think of neutrophils as occurring in only three distinct states, that is, resting, primed, and activated. Rather, the activation state of neutrophils seems to fall within a continuum, from nonresponsive to fully activated, in relation both to the functional response of the cells and to their morphological appearance.[40]

The level of neutrophil priming has been linked to the severity of disease and disease outcome, while a number of studies have suggested that priming may serve as a good clinical indicator of disease activity and the neutrophils circulate in an already pretuned state for future action.[41,42] Indeed, neutrophils with a primed phenotype have been observed in the blood of subjects where accelerated host defense responses would be appropriate, that is, enhanced neutrophil ROS production has been reported in patients with infections.[39,43,44]

In general, the overall effect of neutrophil priming is to prepare the cell for the appropriate level of response to a given stimulus. For example, exposure of neutrophils to an agent (LPS) was found to potentiate the oxidative burst upon exposure to subsequent stimulation by fMLP.[45] Many neutrophil functions can be enhanced by prior exposure to priming agents,[39] including:

1. the respiratory burst;
2. changes in the level of cell-surface adhesion molecule expression;
3. up-regulation of cell-surface receptors;
4. production and release of bioactive lipids;

5. degranulation and release of granule contents;
6. actin polymerization;
7. phosphorylation events;
8. increased phagocytosis;
9. activation of signal transduction enzymes; and
10. changes in the rate of neutrophil apoptosis.

It is becoming increasingly clear that priming agents can produce many effects in neutrophils, with one priming effect often influencing another. In addition, it is now evident that different priming agents can operate via distinctly dissimilar cellular mechanisms, possibly inducing variable combinations of priming responses.

13.2.3 *Characterization of the Dynamic Component Chemiluminescent Approach for Assessment of Functional States of Phagocytes*

Phagocytes in the body circulating in the "primed" state, that is a "pretuned for future tasks" state, reflecting the organism's readiness for defense, possess high predictive value that can be used as a sensitive infection marker to distinguish between different clinical states.

The approach presented herein corresponds to a dynamic assessment of phagocytic respiratory burst localization in a luminol-enhanced whole-blood CL system. This approach allows parallel measurement of intracellular, as well as extracellular, production of ROS from the same sample, in vitro. Performed after stimulation of a phagocyte respiratory burst, using data obtained from the same sample, the approach permits proper quantification of the relative contributions of intracellularly and extracellularly generated ROS to the total oxidative phagocyte response. The functional state of the phagocytes is determined by subjecting to a stimulating respiratory burst agent a first portion of a phagocyte-containing biological sample from a patient, following the reaction using a chemiluminescent probe. Upon exposing both a second and a third aliquot of the test biological sample from the patient to a phagocyte-priming agent and an aging process (that acts like a priming agent), respectively, a stimulating respiratory burst agent and a chemiluminescent probe, the relative contribution of both extracellularly and intracellularly produced reactive oxygen products can be assessed. The chemiluminescent response as a reflection of the first, second, and third aliquots of the sample measures the patient's dynamic phagocyte functional status.

Direct functional analysis of phagocytes in whole blood minimizes the contribution of artifacts introduced by cell purification, reduces the time required for testing, and maintains the in vivo cellular and molecular environment. This approach has been shown to be a sensitive, relatively simple, inexpensive, and time-saving procedure, enabling clinically relevant information from patients experiencing an infection via assessment of the functional status of the phagocytes to be obtained. The question remains, however, is this technique diagnostic?

13.2.4 *Components of Chemiluminescent Kinetics*

This approach for analyzing the activity of phagocytes, based on their CL response as a time-probabilistic process, originally enabled separation of the CL response into two components, assigned as either extracellular or intracellular.[46] Further development led to a more accurate analysis of the CL response of phagocytes, in the form of a triple-component resolution of the CL signal corresponding to three different mechanisms of the ROS formation.[47]

Since the events of light emission observed are relatively rare, Poisson-type distribution is used to describe processes whose probability of occurrence is low. Theoretically, the investigated LCL system of stimulated PMNs belongs to the probabilistic systems class, where the "luminous center" is defined as every space in and/or around the cell in which a sequence of biochemical processes leads to light emission and "luminous packages" are defined as the number of photons detectable as a single impulse by the luminometer.[46–50]

For a probabilistic description of light emission, Poisson-type distribution law can be used.[46] The LCL kinetics, $I(t)$, emitted by N luminous centers can thus be presented as:

$$I(t) = \frac{N\lambda^{m+1}}{(m-1)!} dt^{d(m-1)} e^{-\lambda t^{d}} \tag{13.1}$$

The LCL kinetics of PMN responses are described by four parameters:[47] N, λ, m and d, where:

$N=$ the number of luminous centers of the same registered type (with the same λ and m);
$\lambda =$ the average number of registered photons per time unit (e.g., for 1 min), reflecting a measure of the velocity of the processes leading to LCL;
$m =$ the quantity of light recorded by one luminous center during the whole time of measurement, thus representing the capacity of a luminous center to emit light; and
$d =$ a value that if differing from 1, reflects the cases when during the process of LCL, conditions for light emission are changed (e.g., the effects of stirring).

The registered LCL intensity expresses a complex combination of functionally different luminous centers, where the emission from all identical luminous centers forms one component out of the total LCL kinetics.[47]

Variations in the functional state of the cells alter the manifestation of the specific component, leading to changes in the parameters of those components resulting in different intensities of light. To describe the component, certain definitions should be made that are general concepts of this model:

Chemiluminescent capacity of one component (**S_i**): The whole quantity of light emitted during the response of that component. It is equal to the area under its chemiluminescent kinetics:[48–50]

$$S_i = N_i \times m_i (i = 1, 2, 3) \tag{13.2}$$

Chemiluminescent capacity of the whole response (**C**): The sum $\Sigma N_i \times m_i$, which is equal to the area under the whole model CL kinetics.[48–50]

The real CL kinetic data is modeled on the basis of Eq. 13.1, while each component contributes to the total intensity, depending on its own kinetics. In general, each recorded kinetic CL curve is presented as a sum of three components. The cellular-biochemical characteristics of the three components can be summarized as follows:[47,51]

(1) **The first component** represents processes of ROS production that take place near the plasma membrane. They are connected with phagocytosis and cause extracellular CL.
(2) **The second component** represents processes of ROS production located inside the cell. They are connected with phagocytosis and cause intracellular CL.
(3) **The third component** mainly represents processes that lead to intracellular CL. However, they are not directly connected with phagocytosis.

The time dependence of CL intensity, recorded after zymosan stimulation of PMNs, is represented in Fig. 13.2. The model components of the total CL kinetics are plotted in the same figure.

13.3 Functional States of Phagocytes

In order to assess the clinical state in the case of specific disorder, the evaluation of the functional state of defense-line phagocytes can be extremely useful information. There are three main parameters derived after decomposition of the total kinetics to its subcomponents that can be used in assessment of functional states of circulating phagocytes: capacity, effectiveness, and velocity of the respiratory burst occurring after particle stimulation.[52]

These parameters can be defined as follows:

Capacity (C): The total CL capacity of unit cells, reflecting their capability to generate ROS, calculated as an area under a specific CL curve: the total CL curve or the first-, second-, or third-component CL curve.

Effectiveness (E): The ratio of the capacity (area under the curve) of the second component to that of the first (Eq. 13.3):

$$\text{Effectiveness} = \frac{\text{Capacity of intracellular component}}{\text{Capacity of extracellular component}} \tag{13.3}$$

Indeed, there is a molecular explanation behind the formula, that is, to minimize neutrophil-mediated destruction of tissue components and be effective in killing invading pathogens, cellularly produced oxidative metabolites should be released inside a

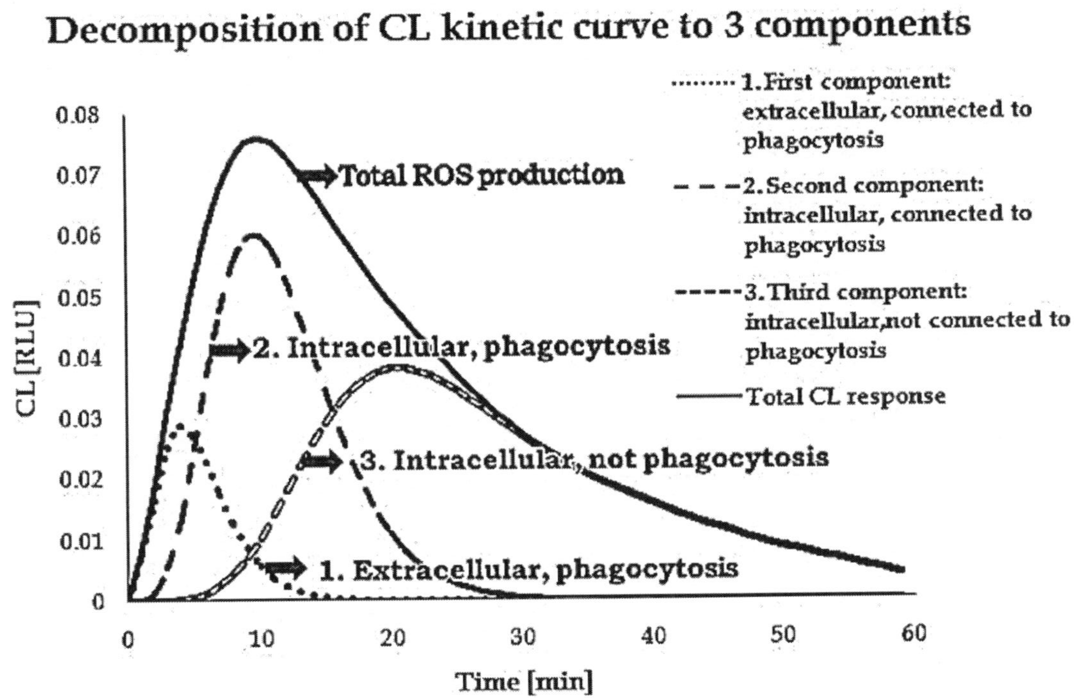

Figure 13.2 Representation of time dependence of CL intensity during stimulation with zymosan. CL response and its components resulted from the generation of a respiratory burst extracellularly and intracellularly by processes connected with phagocytosis and by intracellular processes that are not directly connected to phagocytosis. (Reprinted/adapted from Ref. 97. Copyright 2011, with permission from Elsevier.)

phagosome, contained within the cell, minimizing the ROS produced extracellularly.[31] As mentioned before, the capacities of the first and second components are related to extracellular and intracellular ROS generation during phagocytosis. Hence, the above ratio shows the effectiveness of ROS generation during phagocytosis.[52]

Velocity (V): The ratio of the sum of the capacities of the first and second components to the capacity of the third component of CL kinetics (Eq. 13.4).

$$V = \frac{\text{Intra} + \text{Extra components}}{\text{Third component}} \tag{13.4}$$

Indeed, also in definition of this parameter lies the molecular explanation of the phagocytosis phenomena, where CL kinetics, directly connected with phagocytosis (first and second components), is exhausted earlier than is CL not directly associated with phagocytosis (third component). This is why as the sum of the capacities of the first and second component (numerator) increases towards the total capacity of CL, the respiratory burst is achieved faster, the CL peaking earlier in time.

On the basis of these three parameters as coordinates of three-dimensional capacity-effectiveness-velocity (CEV) space and applying the "high" and "low" for quantifying these parameters, the particular functional state of circulating phagocytes can be assessed.[52] Some part of this space is considered as normal or "resting" (can be referred to as control), defined by low capacity, low velocity, and high effectiveness. The remaining CEV space is available to describe the effects of immunomodulating agents, like infection or a specific drug. On the basis of the three CL parameters (CEV) and a "high-low" scale, eight functional states were defined: fighting, effective, alternatively activated, frustatedly activated, restoring, standby, resting, and frustrated.[52] Using these states clinical approximation can be performed and the state of the immune system defined.

13.3.1 *Dynamic Assessment of Phagocytes' Functional States*

To describe not only the momentary state of the phagocyte during the test but also the reaction to modulating factors, the approach herein uses two additional systems, referred to as priming systems—those expressing the dynamic change of the phagocytes during a controlled functional shift. In general, these systems show us how much the phagocyte's functional state is unchangeable. If the phagocytes are too activated by possible infection in the body, the priming will not be able to change their functional status. If the phagocytes are in the resting state, priming will act in the same manner as an infection in the body and will show a signal increase. Two priming systems that are used in this approach are fMLP and storage of blood.

13.3.1.1 fMLP priming

N-formyl-methionyl-leucyl phenylalanine (fMLP) is a bacterial tripeptide that can act as a chemoattractant for neutrophils. Distinct receptors for these peptides have been detected on the membranes of human neutrophils, which are members of the G protein–coupled receptor (GPCR) superfamily, called formyl peptide receptors (FPRs).[53] In addition to having chemokinetic and chemotactic properties, fMLP also increases neutrophil adhesiveness and degranulation and stimulates an oxidative respiratory burst that is usually induced by phagocytosis.[54] It was shown in the previous study that fMLP priming causes an increase in velocity and a decrease in the effectiveness of the respiratory burst compared to a nonprimed sample from the same blood.[51]

13.3.1.2 Aging as a priming factor

Aging is achieved by postvenipuncture storage of blood in RT for an additional two hours and then recording the CL response. Blood storage has been described as modulating the phagocyte respiratory burst, as measured by CL. Because of the priming that takes place during the aging of blood, it was found that phagocytes undergo transition from states of "resting" (high effectiveness, low velocity) to "standby" (decreased effectiveness, higher velocity).[51,55]

13.3.2 *Functional States of Phagocytes Associated with Different Clinical States*

The killing of invading organisms by phagocytic cells forms a portion of the natural defense system available to humans and other animal species. Attempts have been made to correlate the primed activity of the phagocytes to the severity of the disease and its outcome.[42,56]

Cellular CL can be associated with changes from a normal state if the natural cellular defense mechanism is defective, increased, or decreased as a result of disease. Therefore, the quantification of light from human neutrophils has been found to be useful in the detection of genetic deficiencies and the study of inflammatory diseases, infection, degenerative disease, and cancer. The bulk of such efforts have addressed genetic diseases diagnosed by neutrophil abnormalities, for example, chronic granulomatous disease (CGD) and MPO deficiency. Studies related to cellular CL in inflammatory diseases include arthritis,[57,58] exercise or allergen-induced asthma,[59] and pollen-induced allergies.[60] Infection by bacteria,[61,62] viruses,[63] mycoplasma,[64] and yeast[62] can also be followed using the CL of human neutrophils. Diabetes,[65] renal dialysis,[66,67] and cancer[68] (including leukemias) have been addressed using cellular CL as well.[69]

The specific approach herein showed differences between patterns observed from the measurement of a healthy patient compared to a patient with urinary tract infection. During the infection the phagocytes were in an "activated" state compared to the "resting" state of the healthy patient.[51] Moreover, there is shown a clear ability to differentiate between clinical states associated with infections during peritoneal dialysis[70] and between viral and bacterial infections.[98]

Discrimination was possible between bacterial peritonitis, diabetes, and regular follow-up patients on peritoneal dialysis from control healthy personnel using the phagocyte's CL, the component approach, and data mining. Data implies that during development of any phagocyte-related disease, it is possible to monitor the overall change in the patient's humoral-immune-modulated phagocyte functional status, and as a result, this can serve as a diagnostic tool for prediction of the invading pathogen at early stages of infection.

13.3.3 *Phagocytic Function in Viral Infection*

Various viruses were studied for phagocytic function and CL response produced as a result of viral infection. It was found that patients with viral infections usually have reduced PMN function and, hence, decreased CL.[44] A variety of viruses have been shown to affect neutrophil functions in vitro. Influenza A virus, which has been most widely studied, inhibits chemotaxis and has been associated with a decreased function of phagocytic cells. The chemiluminescent responses of PMNs to various stimuli were inhibited after a very short time of incubation of the virus with PMNs.[71] Using CL assays, the respiratory burst response to influenza A virus was found to occur at an intracellular location.[72] Influenza virus was found to induce CL in the absence of antiviral antibodies, as demonstrated in mouse spleen cells and human neutrophils.[73,74] In phagocytes, influenza and paramyxo-viruses activated the generation of superoxide by a mechanism involving an interaction between viral surface glycoproteins and the phagocyte plasma membrane. Interestingly, viruses that activate this host defense mechanism were found to be toxic when injected into the bloodstream of animals.[75]

The viruses of herpes, measles, mumps, respiratory syncytial virus (RSV), and cytomegalovirus have also been shown to suppress chemotaxis, whereas mumps virus limits phagocytosis and RSV may interfere with intracellular killing of bacteria. Herpes simplex and vaccinia viruses were found to depress CL reactions.[63]

Pox virus has been shown to produce LCL when infecting mouse spleen cells, while the CL response reached a peak in 60–90 minutes postinfection and declined thereafter. Irradiated virus still produced CL, but heat-inactivated virus does not, indicating that a viral protein was important in producing light.[69]

Sendai virus, which causes respiratory tract infections in humans, mice, and swine has been shown to trigger CL in mouse spleen cells.[73]

Human immunodeficiency virus (HIV)-infected patients not only present changes in the number of circulating phagocytes but also abnormalities in phagocyte function, specifically altered oxidative metabolism in response to opsonin receptor-dependent stimulation.

Patients with early HIV infection exhibit increased CL per phagocyte in response to opsonized zymosan, whereas patients with advanced disease show significantly decreased CL per phagocyte.[76]

Human rhinoviruses (HRV) are the major cause of the common cold, one of the most frequent infectious diseases afflicting humans. Though HRV infections of the upper respiratory tract are usually harmless, there is increasing evidence that HRV sets the stage for more dangerous pathogens, asthmatic exacerbations, severe diseases in the lower respiratory tract, and even autoimmunity. The pathogenic mechanisms of HRV infection leading to such complications are still poorly understood.[77] Research on HRV has demonstrated that rhinovirus infections activate both eosinophils and neutrophils but they may not affect mast cell activity.[78]

In addition, it was found that stimulation of ROS by a viral challenge occurs at the cell surface, even in the absence of viral replication.[79]

13.3.4 *Phagocytic Function in Bacterial Infection*

Several substances, produced during the inflammatory reaction resulting from bacterial infection, may interact with the PMNs and cause functional changes in the cell. A number of reports have also shown altered metabolic activity in PMNs in response to bacterial infection. Changes in metabolic activity may be of consequence, not only in regard to microbicidal capacity, but also in relation to the potential tissue-damaging effect of the PMNs.[80] It was found that bacterial infections generally elevate host cellular CL, but not always, as noted in animal and human studies.[44,81] PMNs of the majority of patients with active bacterial infection are in an activated state, both functionally and metabolically, and produce increased CL. CL production remained increased with persistent infection but fell to control levels with appropriate therapy.[61] However, some patients display reduced PMN function, which may contribute to a fatal outcome of the disease.[44]

LCL has been used as an indicator of germ warfare, using *Franciella tularensis*. Virulent *Salmonella typhi* were found to depress cellular CL, so apparently increases are not always found associated with infection. This is the exception rather than the rule.[82]

13.4 Data Mining Techniques in Clinical Groups' Differentiation

Data mining is the process of sifting through large amounts of data and picking out relevant information and can be referred to otherwise as extracting or "mining" for knowledge. This is an essential process of observational data analysis for finding unsuspected relationships that summarize the data in novel ways that are both comprehensible and useful to the user. Data mining has been widely used in the computing industry and in society as a whole in recent years, due to the wide availability of huge amounts of data and the imminent need for turning it into useful information and knowledge.[83,84] The information gained can be used for various applications, like the medical field to include diagnosis of patient records, thus helping identify the best practices for uncovering gene expression signatures in different diseases (genomics)[85,86] and proteomics[87,88] and many other nonmedical applications, such as improving food and drug product safety and detection of terrorists or criminals.[83,89] Traditional statistical analysis involves an approach that is usually directed, in that a specific set of expected outcomes exists (hypothesis development and testing). However, there is more to data mining than the technical tools used. It involves the goal of knowledge discovery (learning new and useful things).[89] The relationships and summaries induced by data mining algorithms are often referred to as models or patterns. Data mining tasks include the discovery of concept/class descriptions, associations and correlations, classification, prediction, clustering, trend analysis, outlier and deviation analysis, and similarity analysis.[83] A variety of analytic models have been used in data mining. The standard model types include regression (normal for prediction, logistic for classification), artificial neural networks, and decision trees.[89]

In this work, decision tree classification algorithm C4.5 (J48) was used for finding a model that describes and distinguishes data classes, which represent medical conditions/diagnosis of patients.

13.5 Differentiation between Viral and Bacterial Respiratory Infections Using a Chemiluminescent Approach

Infectious diseases are still a major health problem worldwide, being associated with significant morbidity and mortality. Among infectious diseases, lower respiratory infections are a primary cause of death for children under the age of five. Community-acquired pneumonia is one of the most common infectious diseases addressed by clinicians and is associated with a broad range of causative agents. Despite medical advances, differentiation between viral and bacterial etiology based on clinical symptoms is currently very difficult and is a challenge to clinicians.[90,91] While the viral infection generally resolves with supportive care only, bacterial infection usually necessitates the administration of antibiotics. Moreover, treating viral infections with antibiotics is ineffective and contributes to the development of antibiotic resistance, allergic reactions, toxicity, and greater health care costs.[92–94] The most common ways to diagnose viral and bacterial infections are time consuming and often have poor sensitivity and specificity. Hence, the treatment of community-acquired pneumonia is often empirical. An alternative rapid, sensitive, and commercially available diagnostic tool, describing whether the infection is viral or bacterial in origin, is needed for the manifestation of proper treatment on the admission of the patient.

On the basis of the PMNs' functional differences in the case of viral and bacterial infections, the approach here of dynamic assessment of phagocytic respiratory burst localization in a luminol-enhanced whole-blood CL system could be used for building a distinctive pattern between these two clinical groups.

13.5.1 *Description of an Experiment*

To check the ability of this method to build a diagnostic tool for differentiating viral versus bacterial infections, 69 patients with different kinds of infections, diagnosed as viral or bacterial in origin, were recruited.[98] In this study, respiratory as well as urinary tract infections were analyzed. Diagnosis was made according to specific clinical parameters, and it was assumed that the main source of illness was bacterial or viral, and the mixed infections diagnosis was minimized for clarification of the border between viral and bacterial etiology. Healthy medical personnel served as the control group.

Chemiluminescent assays were performed using whole blood, zymosan as a stimulant, and luminol as an amplifier of the signal. Three measurements for every patient, that is, standard, primed with fMLP as a priming agent and aged blood, where aging served as a priming agent, were performed.

After performing the measurements, the analysis of data was done in consecutive steps. The kinetic curve was decomposed to three biological components using the component method described earlier. The final number of kinetic parameters calculated for every patient stood at 82. The list of basic parameters is shown in Table 13.1. The next step was the assignment of patients into groups according to their clinical states. The patients recruited formed three groups: bacterial infections (33 cases), viral infections (36 cases), and controls (6 cases). As a last step, the data mining decision tree algorithm[95] J48 (C4.5) was used for inducing a classification model that discriminated between clinical groups using the selected combinations of chemiluminescent parameters for group characterization and differentiation. The final decision tree was previously published.[98]

Moreover, the functional states of the phagocytes could be assessed using average chemiluminescent behavior that could provide information on the state of the immune system in specific infection.

13.6 Data Mining Algorithms and CL Information Can Differentiate between Clinical Groups and Assess Functional States of Phagocytes

Owing to the experimentation performed, the patients with different infections could be grouped according to their chemiluminescent behavior and the functional states of phagocytes could be characterized. Figure 13.3A presents the CL curves in the case of different infections and Fig. 13.3B the main parameters (capacity,

Table 13.1 List of the parameters used to construct a classification model for differentiating clinical groups (*also applies to priming and aging systems) (reprinted and adapted with permission from Ref. 70. Copyright 2008 American Chemical Society)

Abbreviation	Definition	Reference
S	Standard sample	—
SP	Primed sample	—
SA	Aged sample	—
Extra	Extracellular phagocytosis emission	47
Intra	Intracellular phagocytosis emission	47
nonPhago	Non-phagocytosis-related emission	47
Parameter	**Definition**	**Reference**
PtimeS	Peak time of standard* sample	96
CapS	Capacity(x) of standard* sample	52, 96
AmpS	Amplitude of standard* sample	96
SlopeS	Peak of standard* sample divided by time required to reach it	21, 48–50
RelPtimeSP	Peak time of primed* sample divided by peak time of standard sample	21, 48–50
RelCapSP	Capacity of primed* sample divided by capacity of standard sample	21, 48-50
S_eff	Effectiveness(xx) of standard* sample	52
S_vel	Velocity(xxx) of standard* sample	52
RelEff_SP	Effectiveness of primed* sample divided by effectiveness of standard sample	50
RelVel_SP	Velocity of primed* sample divided by the velocity of standard sample	50
RelPhagS	Phagocytosis capacity of standard* sample divided by total capacity of standard sample	50
TimeP	Time peak of specific peak of specific sample	50, 96
CapP	Capacity of specific peak of specific sample	50, 96
ExtraS	Extra-cellular phagocytosis related emission of standard* sample	21, 48–50
TimeExS	Time peak of extra-phago emission of standard* sample	50, 96
nonPhagoS	Non-phago-related CL of standard* sample	21, 48–50
RelTimeExSP	Time of extra-cellular phagocytosis-related emission of primed* sample divided by time of extra-cellular phagocytosis-related emission of standard sample	50
RelExtraSP	Extra-cellular capacity of primed sample divided by extra-cellular capacity of standard sample	50
RelIntrano-PhagSP	Intracellular non-phagocytosis capacity of primed* sample divided by intracellular non-phagocytosis capacity of standard sample	21, 48–50

x: Capacity is presented by the total CL capacity of unit cells, which reflects their ability to generate ROS, calculated as the total area under the CL curve.[52]

xx: Effectiveness is presented as a ratio of the capacity of the second component to that of the first and shows the effectiveness of ROS generated during phagocytosis.[52]

xxx: Velocity is presented as a ratio of the sum of the capacities of the first and second components to the capacity of the third component of CL kinetics and shows the velocity to achieve a respiratory burst.[52]

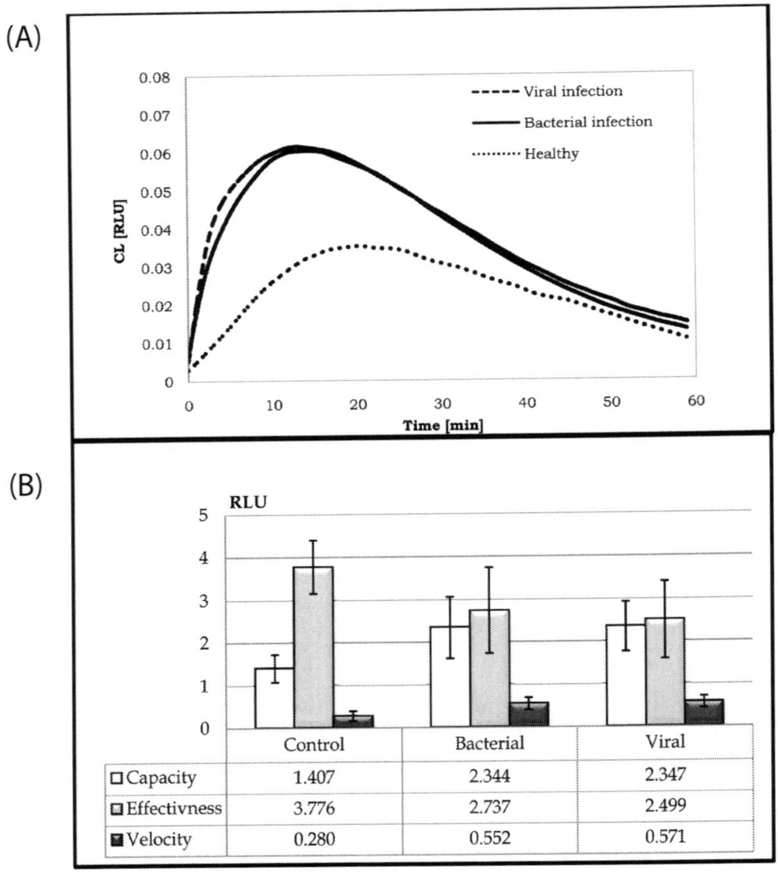

Figure 13.3 (A) Representative CL kinetic curves recorded using a standard system for patients with viral or bacterial infections as compared to healthy personnel. Two curves in the case of infection are elevated as compared to the control because of the activated immune system. (B) Comparison of derived respiratory burst parameters (capacity, effectiveness and velocity) in the case of control, bacterial and viral infections. The capacity and velocity of ROS production in the case of two infections were higher than in control cases, showing that the immune system is much more activated in the presence of infection. (Reprinted and adapted with permission from Ref. 98. Copyright 2011 American Chemical Society)

effectiveness, and velocity) that are used in estimation of functionality of the immune system. The capacity and velocity of ROS production in either type of infection were higher than in control cases, meaning that the immune system is much more activated in the presence of infection and the CL response of the cells is elevated (see Fig. 13.3B). The effectiveness of the phagocytosis in both infections was lower than in controls. In the case of bacterial infection, the ROS produced during phagocytosis occurred predominantly inside the cells and mainly generated in processes connected to phagocytosis. As it was shown in studies before,[44,61,81] PMNs of the majority of patients with active bacterial infection were in an activated state, both functionally and metabolically, and produced elevated CL, as was also shown herein. This study presented the patients with possible mixed viral infection (influenza and HRV), and it was difficult to characterize the behavior of a specific virus. The response of the phagocytes in the case of viral infection was elevated. Compared to previous studies, influenza viral infection has been associated with decreased phagocytic cell function and the inhibition of chemiluminescent responses of PMNs to various stimuli,[44,71,81] and HRV virus was shown to activate neutrophils.[78] The response of the phagocytes in the case of viral infections using this approach herein was elevated and showed a more extracellular origin of ROS production. The patterns of the CL curves associated with two different etiological agents are both elevated (see Fig. 13.3A), with lower than the controls' effectiveness in the bacterial infections and also emphasis on the extracellular part being higher in the case of viral infections. These characteristic CL shape differences were used for a description of different clinical groups by component analysis. The functional states of phagocytes were assigned as "resting" in the case of control and as "effective" or "fighting" in the case of viral or bacterial infections.[98] This was assigned according to the capacity, effectiveness, and velocity.[52]

After the evaluation of the functional status of the immune system, classification models were induced and three clinical groups (i.e., control, viral, and bacterial infections) were classified with 94.7% accuracy in the training set and 88.9% in a testing set as previously described.[98] In general, data mining succeeded in extracting specific differentiating mathematical combinations of the chosen kinetic parameters and actually proving that the pattern can be determined according to the infection involved. This model can now serve for evaluation of the blind unknown cases. Of course, the extension of the study population should increase the precision of the diagnosis and cover all possible spectra of the phagocyte activity during a specific kind of infection.

The system shows that the level of activation of circulating phagocytes can serve as a diagnostic marker and can evaluate the immune status of the patient upon admission to the hospital. The combination of whole-blood CL, kinetic signature, and data mining can lead to a fast, sensitive, inexpensive, and informative diagnostic immunologic tool.

13.7 Prospects

The present diagnostic enables immediate intervention as a first line of protection against a biowarfare weapon, which is crucial at the early stages of the infection, before symptoms appear or before conventional techniques such as ELISA (takes one week to obtain antibodies after infection) or polymerase chain reaction (PCR) (requires the presence of microorganism and a sophisticated lab) may be able to do the job. As the innate nonspecific immune system response is being activated immediately after an infection and phagocytes being the "warriors" in this battle, the approach presented here of evaluating phagocytic functional status using a dynamic component chemiluminescent technique and data mining algorithms can be envisioned to be deployed in clinics or HAZMAT teams, serving as an on-site real-time diagnostic tool. Indeed, if a bioterror case scenario occurs, investigating a putatively exposed population would enable the determination of a sudden statistically significant increase in exposure to microorganisms, thus determining whether the bioterror event was real or a hoax.

The system is relatively inexpensive and can be developed to be sensitive and specific by collecting a bigger population and constructing more accurate classification models. These models can serve in the future as a "black box" automated technology for diagnostics of different infectious diseases and more. The differentiation can be done on the level of the background pathology, clinical etiology, and, probably, classification to the groups of the organisms involved. This system can be developed as the tool for diagnosing infections by different groups of viruses and different kinds of bacteria.

The main objectives for improvement of the proposed method are extending the data set by checking more patients with specific diseases for the development of stronger classification rules and improving technique accuracy by increasing the number of parameters involved in the analysis.

Acknowledgments

We thank Dr. Moni Magrisso for his participation in the study at its early stage. We thank Dr. Ohad Etzion and Dr. Boris Rogachev for providing clinical information and blood samples. Daria Prilutsky thanks the Council for Higher Education for Converging Technologies for its fellowship support.

References

1. Benjamini, E., R. Coico, and G. Sunshine, *Immunology: A Short Course*. 4th ed. 2000, New York: Wiley-Liss.
2. Goldsby, R.A., et al., *Immunology*. 5th ed. 2003, W. H. Freeman.
3. Blasi, F., P. Tarsia, and S. Aliberti, Strategic targets of essential host-pathogen interactions. *Respiration*, 2005. **72**(1), 9–25.
4. Akira, S., and H. Hemmi, Recognition of pathogenassociated molecular patterns by TLR family. *Immunol Lett*, 2003. **85**(2), 85–95.
5. Netea, M.G., J.W. Van der Meer, and B.J. Kullberg, Role of the dual interaction of fungal pathogens with pattern recognition receptors in the activation and modulation of host defence. *Clin Microbiol Infect*, 2006. **12**(5), 404–409.
6. Gow, N.A., A.J. Brown, and F.C. Odds, Fungal morphogenesis and host invasion. *Curr Opin Microbiol*, 2002. **5**(4), 366–371.
7. van Oss, C.J., Phagocytosis: an overview. *Methods Enzymol*, 1986. **132**, 3–15.

8. Mayer-Scholl, A., P. Averhoff, and A. Zychlinsky, How do neutrophils and pathogens interact? *Curr Opin Microbiol*, 2004. **7**(1), 62–66.
9. DeLeo, F.R., and M.T. Quinn, Assembly of the phagocyte NADPH oxidase: molecular interaction of oxidase proteins. *J Leukoc Biol*, 1996. **60**(6), 677–691.
10. Fang, F.C., Antimicrobial reactive oxygen and nitrogen species: concepts and controversies. *Nat Rev Microbiol*, 2004. **2**(10), 820–832.
11. Fialkow, L., Y. Wang, and G.P. Downey, Reactive oxygen and nitrogen species as signaling molecules regulating neutrophil function. *Free Radic Biol Med*, 2007. **42**(2), 153–164.
12. Sumimoto, H., K. Miyano, and R. Takeya, Molecular composition and regulation of the Nox family NAD(P) H oxidases. *Biochem Biophys Res Commun*, 2005. **338**(1), 677–686.
13. Segal, A.W., How neutrophils kill microbes? *Annu Rev Immunol*, 2005. **23**, 197–223.
14. Babior, B.M., NADPH oxidase: an update. *Blood*, 1999. **93**(5), 1464–1476.
15. Bengtsson, T., K. Orselius, and J. Wettero, Role of the actin cytoskeleton during respiratory burst in chemoattractant-stimulated neutrophils. *Cell Biol Int*, 2006. **30**(2), 154–163.
16. Karlsson, A., and C. Dahlgren, Assembly and activation of the neutrophil NADPH oxidase in granule membranes. *Antioxid Redox Signal*, 2002. **4**(1), 49–60.
17. Dahlgren, C., and A. Karlsson, Respiratory burst in human neutrophils. *J Immunol Methods*, 1999. **232**(1–2), 3–14.
18. Cohen, H.J., and M.E. Chovaniec, Superoxide generation by digitonin-stimulated guinea pig granulocytes. A basis for a continuous assay for monitoring superoxide production and for the study of the activation of the generating system. *J Clin Invest*, 1978. **61**(4), 1081–1087.
19. Hyslop, P.A., and L.A. Sklar, A quantitative fluorimetric assay for the determination of oxidant production by polymorphonuclear leukocytes: its use in the simultaneous fluorimetric assay of cellular activation processes. *Anal Biochem*, 1984. **141**(1), 280–286.
20. Tarpey, M.M., and I. Fridovich, Methods of detection of vascular reactive species: nitric oxide, superoxide, hydrogen peroxide, and peroxynitrite. *Circ Res*, 2001. **89**(3), 224–236.
21. Magrisso, M., and R.S. Marks, Phagocyte luminescent sensor, in *Handbook of Biosensors and Biochips*, R.S. Marks, et al., Editors. 2007, John Wiley & Sons, 511–529.

22. Model, M.A., L.S. Ganelina, and R.F. Todd, 3rd, A microscopic study of Fc gamma RIII-mediated respiratory burst in neutrophils. *Immunobiology*, 1998. **199**(1), 39–50.

23. van der Valk, P., and C.J. Herman, Leukocyte functions. *Lab Invest*, 1987. **56**(2), 127–137.

24. Emmendörffer, A., et al., A fast and easy method to determine the production of reactive oxygen intermediates by human and murine phagocytes using dihydrorhodamine 123. *J Immunol Methods*, 1990. **131**(2), 269–275.

25. Bass, D.A., et al., Flow cytometric studies of oxidative product formation by neutrophils: a graded response to membrane stimulation. *J Immunol*, 1983. **130**(4), 1910–1917.

26. Allen, R.C., Phagocytic leukocyte oxygenation activities and chemiluminescence: a kinetic approach to analysis. *Methods Enzymol*, 1986. **133**, 449–493.

27. Allen, R.C., and L.D. Loose, Phagocytic activation of a luminol-dependent chemiluminescence in rabbit alveolar and peritoneal macrophage. *Biochem Biophys Res Commun*, 1976. **69**(1), 245–252.

28. Allen, R.C., R.L. Stjernholm, and R.H. Steele, Evidence for the generation of an electronic excitation state(s) in human polymorphonuclear leukocytes and its participation in bactericidal activity. *Biochem Biophys Res Commun*, 1972. **47**(4), 679–684.

29. Lundqvist, H., and C. Dahlgren, Isoluminol-enhanced chemiluminescence: a sensitive method to study the release of superoxide anion from human neutrophils. *Free Radic Biol Med*, 1996. **20**(6), 785–792.

30. Edwards, S.W., Luminol- and lucigenin-dependent chemiluminescence of neutrophils: role of degranulation. *J Clin Lab Immunol*, 1987. **22**(1), 35–39.

31. Dahlgren, C. Is lysosomal fusion required for the granulocyte chemiluminescence reaction? 1989. *Free Radic. Biol. Med*, **6**(4), 399–403.

32. Dahlgren, C., Difference in extracellular radical release after chemotactic factor and calcium ionophore activation of the oxygen radical-generating system in human neutrophils. *Biochim Biophys Acta*, 1987. **930**(1), 33–38.

33. Ohno, Y., and J.I. Gallin, Diffusion of extracellular hydrogen peroxide into intracellular compartments of human neutrophils. Studies utilizing the inactivation of myeloperoxidase by hydrogen peroxide and azide. *J Biol Chem*, 1985. **260**(14), 8438–8446.

34. Gantner, B.N., et al., Collaborative induction of inflammatory responses by dectin-1 and Toll-like receptor 2. *J Exp Med*, 2003. **197**(9), 1107–1117.

35. Cabib, E., et al., Carbohydrates as structural constituents of yeast cell wall and septum. *Pure Appl Chem*, 1991. **63**(4), 483–489.

36. Kollar, R., et al., Architecture of the yeast cell wall. Beta(1–>6)-glucan interconnects mannoprotein, beta(1–>)3-glucan, and chitin. *J Biol Chem*, 1997. **272**(28), 17762–17775.

37. Frasnelli, M.E., et al., TLR2 modulates inflammation in zymosan-induced arthritis in mice. *Arthritis Res Ther*, 2005. **7**(2), R370–R379.

38. Underhill, D.M., Toll-like receptors: networking for success. *Eur J Immunol*, 2003. **33**(7), 1767–1775.

39. Swain, S.D., T.T. Rohn, and M.T. Quinn, Neutrophil priming in host defense: role of oxidants as priming agents. *Antioxid Redox Signal*, 2002. **4**(1), 69–83.

40. Kitchen, E., et al., Demonstration of reversible priming of human neutrophils using platelet-activating factor. *Blood*, 1996. **88**(11), 4330–4337.

41. Rivkind, A.I., et al., Neutrophil oxidative burst activation and the pattern of respiratory physiologic abnormalities in the fulminant post-traumatic adult respiratory distress syndrome. *Circ Shock*, 1991. **33**(1), 48–62.

42. Wakefield, C.H., et al., Polymorphonuclear leukocyte activation. An early marker of the postsurgical sepsis response. *Arch Surg*, 1993. **128**(4), 390–395.

43. Bass, D.A., et al., Subpopulations of neutrophils with increased oxidative product formation in blood of patients with infection. *J Immunol*, 1986. **136**(3), 860–866.

44. Solberg, C.O., et al., Polymorphonuclear leukocyte function in bacterial and viral infections. *Scand J Infect Dis*, 1982. **14**(1), 11–18.

45. Guthrie, L.A., et al., Priming of neutrophils for enhanced release of oxygen metabolites by bacterial lipopolysaccharide. Evidence for increased activity of the superoxide-producing enzyme. *J Exp Med*, 1984. **160**(6), 1656–1671.

46. Magrisso, M.J., et al., A new approach for analysis of chemiluminescent kinetics of activated phagocytes in blood. *J Biolumin Chemilumin*, 1995. **10**(2), 77–84.

47. Magrisso, M.J., et al., Model components of luminal chemiluminescence generated by PMNL. *J Biochem Biophys Methods*, 1995. **30**(4), 257–269.

48. Magrisso, M., and R.S. Marks, Chemiluminescent method and device for evaluating the in vivo functional state of phagocytes. Publication No.: WO/2006/092787, International Application No.: PCT/IL2006/00027. 2005.

49. Magrisso, M., and R.S. Marks, Chemiluminescent method and device for a single-test assessment of the in vivo functional state of phagocytes. Publication No.: WO/2006/092788, International Application No.: PCT/IL2006/00027. 2006.

50. Magrisso, M., and R.S. Marks, Chemiluminescent method and device for identifying respiratory infections of different origins. Patent No. in Submission: 22611-WO-07. 2007.

51. Magrisso, M., et al., Fiber-optic biosensor to assess circulating phagocyte activity by chemiluminescence. *Biosens Bioelectron*, 2006. **21**(7), 1210–1218.

52. Magrisso, M.Y., et al., Functional states of polymorphonuclear leukocytes determined by chemiluminescent kinetic analysis. *Luminescence*, 2000. **15**(3), 143–151.

53. Fu, H., et al., Ligand recognition and activation of formyl peptide receptors in neutrophils. *J Leukoc Biol*, 2006. **79**(2), 247–256.

54. McCall, C.E., et al., In vitro responses of human neutrophils to N-formyl-methionyl-leucyl-phenylalanine: correlation with effects of acute bacterial infection. *J Infect Dis*, 1979. **140**(3), 277–286.

55. Zgliczynski, J.M., et al., Functional states of neutrophils as suggested by whole blood chemiluminescence. *Acta Biochim Pol*, 1988. **35**(4), 331-342.

56. Stevens, D.L., et al., Analysis of circulating phagocyte activity measured by whole blood luminescence: correlations with clinical status. *J Infect Dis*, 1994. **170**(6), 1463–1472.

57. James, D.W., W.H. Betts, and L.G. Cleland, Chemiluminescence of polymorphonuclear leukocytes from rheumatoid joints. *J Rheumatol*, 1983. **10**(2), 184–189.

58. Chou, C.T., et al., Neutrophil chemiluminescence and superoxide production in patients with rheumatoid arthritis: the effect of zymosan, phorbol myristate acetate and platelet-activating factor. *Zhonghua Min Guo Wei Sheng Wu Ji Mian Yi Xue Za Zhi*, 1991. **24**(3), 281–288.

59. Vachier, I., et al., Increased oxygen species generation in blood monocytes of asthmatic patients. *Am Rev Respir Dis*, 1992. **146**(5 Pt 1), 1161–1166.

60. Lindberg, R.E., J.L. Pinnas, and J.F. Jones, Polleninduced chemiluminescence: inhibition by serum from allergic individuals. *J Allergy Clin Immunol*, 1982. **69**(4), 388–396.

61. Barbour, A.G., et al., Chemiluminescence by polymorphonuclear leukocytes from patients with active bacterial infection. *J Infect Dis*, 1980. **141**(1), 14–26.

62. Robinson, P., et al., Chemiluminescent response to pathogenic organisms: normal human polymorphonuclear leukocytes. *Infect Immun*, 1984. **43**(2), 744–752.

63. Faden, H., P. Sutyla, and P.L. Ogra, Effect of viruses on luminol-dependent chemiluminescence of human neutrophils. *Infect Immun*, 1979. **24**(3), 673–678.

64. Krausse, R., U. Ullmann, and C. Wagener, In vitro influence of Mycoplasma species on the stimulation of human polymorphonuclear granulocytes. *Zentralbl Bakteriol Mikrobiol Hyg [A]*, 1988. **270**(1–2), 228–236.

65. Shah, S.V., J.D. Wallin, and S.D. Eilen, Chemiluminescence and superoxide anion production by leukocytes from diabetic patients. *J Clin Endocrinol Metab*, 1983. **57**(2), 402–409.

66. Ritchey, E.E., J.D. Wallin, and S.V. Shah, Chemiluminescence and superoxide anion production by leukocytes from chronic hemodialysis patients. *Kidney Int*, 1981. **19**(2), 349–358.

67. Duwe, A.K., S.I. Vas, and J.W. Weatherhead, Effects of the composition of peritoneal dialysis fluid on chemiluminescence, phagocytosis, and bactericidal activity in vitro. *Infect Immun*, 1981. **33**(1), 130–135.

68. Heberer, M., et al., Chemiluminescence of granulocytes and monocytes in diluted whole blood samples: a tumor marker? *Cancer Detect Prev*, 1983. **6**(1–2), 273–280.

69. Van Dyke, K., and C. Van Dyke, Cellular chemiluminescence associated with disease states. *Methods Enzymol*, 1986. **133**, 493–507.

70. Prilutsky, D., et al., Dynamic component chemiluminescent sensor for assessing circulating polymorphonuclear leukocyte activity of peritoneal dialysis patients. *Anal Chem*, 2008. **80**(13), 5131–5138.

71. Abramson, J.S., et al., Influenza A virus-induced polymorphonuclear leukocyte dysfunction. *Infect Immun*, 1982. **37**(2), 794–799.

72. Kazhdan, M., et al., Human neutrophil respiratory burst response to influenza A virus occurs at an intracellular location. *J Leukoc Biol*, 1994. **56**(1), 59–64.

73. Peterhans, E., Chemiluminescence: an early event in the interaction of Sendai and influenza viruses with mouse spleen cells. I. The role of the envelope glycoproteins in the stimulation of chemiluminescence. *Virology,* 1980. **105**(2), 445–455.

74. Mills, E.L., et al., Initiation of the respiratory burst of human neutrophils by influenza virus. *Infect Immun,* 1981. **32**(3), 1200–1205.

75. Peterhans, E., Reactive oxygen species and nitric oxide in viral diseases. *Biol Trace Elem Res,* 1997. **56**(1), 107–116.

76. Pitrak, D.L., et al., Impaired phagocyte oxidative capacity in patients with human immunodeficiency virus infection. *J Lab Clin Med,* 1998. **132**(4), 284–293.

77. Kirchberger, S., O. Majdic, and J. Stockl, Modulation of the immune system by human rhinoviruses. *Int Arch Allergy Immunol,* 2007. **142**(1), 1–10.

78. Greiff, L., et al., Effects of rhinovirus-induced common colds on granulocyte activity in allergic rhinitis. *J Infect,* 2002. **45**(4), 227–232.

79. Kaul, P., et al., Rhinovirus-induced oxidative stress and interleukin-8 elaboration involves p47-phox but is independent of attachment to intercellular adhesion molecule-1 and viral replication. *J Infect Dis,* 2000. **181**(6), 1885–1890.

80. Briheim, G., O. Stendahl, and C. Dahlgren, Simulaion of a metabolic response in vitro between neutrophils and a meningococcal culture filtrate. *J Infect Dis,* 1987. **156**(4), 676–680.

81. McCarthy, J.P., et al., Differential alterations in host peripheral polymorphonuclear leukocyte chemiluminescence during the course of bacterial and viral infections. *Infect Immun,* 1980. **30**(3), 824–831.

82. Kossack, R.E., et al., Diminished neutrophil oxidative metabolism after phagocytosis of virulent Salmonella typhi. *Infect Immun,* 1981. **31**(2), 674–678.

83. Han, J., and M. Kamber, *Data Mining: Concepts and Techniques.* 2nd ed. 2006, Morgan Kaufmann.

84. Hand, D., H. Mannila, and P. Smyth, *Principles of Data Mining.* 2001, Cambridge, MA; London: MIT Press.

85. Nevins, J.R., and A. Potti, Mining gene expression profiles: expression signatures as cancer phenotypes. *Nat Rev Genet,* 2007. **8**(8), 601–609.

86. Dudley, J., and A.J. Butte, Enabling integrative genomic analysis of high-impact human diseases through text mining. *Pac Symp Biocomput,* 2008, 580–591.

87. Phan, J.H., C.F. Quo, and M.D. Wang, Functional genomics and proteomics in the clinical neurosciences: data mining and bioinformatics. *Prog Brain Res*, 2006. **158**, 83–108.
88. Fung, E.T., et al., Bioinformatics approaches in clinical proteomics. *Expert Rev Proteomics*, 2005. **2**(6), 847–862.
89. Olson, D.L., and D. Delen, *Advanced Data Mining Techniques*. 2008, Springer.
90. File, T.M., Community-acquired pneumonia. *Lancet*, 2003. **362**(9400), 1991–2001.
91. Virkki, R., et al., Differentiation of bacterial and viral pneumonia in children. *Thorax*, 2002. **57**(5), 438–441.
92. Nuutila, J., and E.M. Lilius, Distinction between bacterial and viral infections. *Curr Opin Infect Dis*, 2007. **20**(3), 304–310.
93. Livermore, D.M., Minimising antibiotic resistance. *Lancet Infect Dis*, 2005. **5**(7), 450–459.
94. French, G.L., Clinical impact and relevance of antibiotic resistance. *Adv Drug Deliv Rev*, 2005. **57**(10), 1514–1527.
95. Quinlan, J.R., *C4.5: Programs for Machine Learning*. 1993, San Francisco, CA: Morgan Kaufmann.
96. Giltinan, D.M., et al., Design and analysis considerations in evaluating the chemiluminescence response of mouse spleen cells. *J Clin Microbiol*, 1986. **23**(3), 531–535.
97. Prilutsky, D., et al. Classification of infectious diseases based on chemiluminescent signatures of phagocytes in whole blood. *Artif Intell Med*, 2011. **52**(3), 153–163.
98. Prilutsky, D., et al. Differentiation between viral and bacterial acute infections using chemiluminescent signatures of circulating phagocytes. *Anal Chem*, 2011. **83**(11), 4258–4265.

Chapter 14

Phage Display for Viral Diagnostics

Danit Atias,[a,b] **Leslie Lobel,**[a] **and Robert S. Marks**[b,c]

[a] *Department of Virology, Ben-Gurion University of the Negev, Beer-Sheva, Israel*
[b] *Department of Biotechnology Engineering, Ben-Gurion University of the Negev, Beer-Sheva, Israel*
[c] *The National Institute of Biotechnology in the Negev, Ben-Gurion University of the Negev, Beer-Sheva, Israel*
danitatias@gmail.com, atiasdan@bgu.ac.il, rsmarks@bgu.ac.il, llobel@bgu.ac.il

14.1 Phage Display for Advanced Diagnostics

Attempts of small countries and terrorist groups to obtain biological warfare agents have escalated the need to provide the armed forces and civilians with miniature, easy-to-use, disposable instruments for detection and identification of potentially hazardous biological agents. Traditional methods for the detection and identification of pathogens lack the speed and sensitivity to be of use in the field, since they are not real time or even typically completed in a single day. One way of dealing with this problem is to improve the bioreceptor that binds the analyte, which is one of the essential elements of any immuno-based detector device.[1] Molecular recognition systems that can be used for rapid identification can improve response time and thus avert or reduce the number of casualties associated with

Viral Diagnostics: Advances and Applications
Edited by Robert S. Marks, Leslie Lobel, and Amadou Alpha Sall

ISBN 978-981-4364-43-0 (Hardcover), 978-981-4364-44-7 (eBook)
www.panstanford.com

a potential bioterrorism or biowarfare event.[2] One of the newer techniques being developed for construction of bioreceptors for pathogen detection involves phage display technology.

"Phage display," first introduced by George Smith in 1985,[3] is a powerful technique that allows expression and presentation of peptides or proteins on the phage surface. According to this method, a coding domain of interest is fused to that of a bacteriophage coat protein, resulting in phage particles that display the encoded protein as a chimeric protein.[4,5] This procedure can also be performed with an ensemble of coding domains, resulting in a phage library that can contain potentially billions of phage variants (up to 10^{10}).[5] In general, the DNA that encodes the displayed protein is encapsulated within the same virion, therefore providing a direct link between phenotype and genotype.[4,6] This enables rapid amplification and characterization of the desired clone through DNA sequence analysis of the insert. Phage display has many advantages over other methods for recombinant peptide and protein expression, including ease of manipulation, low production costs, reproducibility, production on a large scale with proper protein folding, and ability to analyze a collection of recombinant molecules to identify those with the highest affinity.[1,5,7] Thus phage display peptides can be a good alternative to the common method for chemical synthesis of short peptides, the solid-phase peptide synthesis (SPPS), in the case where the presence of the phage itself does not interfere with the system (see Table 14.1).[8]

For example, on a 96-well enzyme-linked immunosorbent assay (ELISA) microplate, phage display allows a higher number of particles to be coated and also provides better accessibility to the peptide.[9,10] Several studies have demonstrated the potential of using phage display immunogenic peptides or proteins as bases for vaccination.[11–13]

Phage display is not limited to short peptides, as it can be used to express polypeptides of a variety of sizes and with varied biological activity (e.g., cytokines, antibodies, receptors, enzymes, and DNA-binding proteins).[6]

In particular, phage display can solve some of the problems that result from using antibodies as recognition reagents. In some cases,

Table 14.1 Comparison between SPPS and phage display peptides

	SPPS	Phage display
Production cost	The instrument and the reagent are very expensive (to achieve high yield)	Low
Working time	Time consuming	Time consuming (but shorter than manual SPPS)
Large scale production	Hard to produce in large scale	Large scale
Mutation	Can create sequence of the same amino acid or shorter from the planned one	Reproducible
Washing	Washing after each amino acid supplement	No need
Cleaning	There is need to clean the peptides from the resin	Short centrifuge with PEG
Fraction checking	To check the received fraction in HPLC	To check the received fraction in Mass Spectrometry (first time only)

Abbreviation: HPLC, high-performance liquid chromatography.

antibodies are difficult to isolate due to the nonantigenic nature of the analyte or the target to be analyzed consists of a special matrix that is not compatible with antibody function. Additionally, conventional polyclonal antibody technology is time consuming and requires much manpower to obtain antisera from animals and the production of monoclonal antibodies is also complicated and time consuming.[14,15]

This chapter will focus on the biology of phages in general and the filamentous and the T7 phage in particular, phage display technology, and applications. It will also highlight the different diagnostic methods that can utilize the phage display systems as bioreceptors for viral diagnostics.

14.2 Biology of Phages

Bacteriophages, like animal viruses, can be divided into those containing RNA genomes (typically small and single stranded), those with small DNA genomes (generally less than 10 kilobases [kb], usually single stranded), and those with medium to large DNA genomes (30 to 200 kb, double stranded).[16] Bacteriophages can be classified further according to three primary types of life cycles:

lytic, temperate, and filamentous. The lytic or virulent phages can multiply only in bacteria and kill the cell by lysis at the end of the life cycle. Examples of this group are the tailed phages T7, T3, and T4 that were isolated originally as members of the seven-type phage that grows in *Escherichia coli* (*E. coli*).[17] Temperate or lysogenic phages are those that can multiply either via the lytic cycle or enter a quiescent state in the cell. Whereas most of the phage genes are not transcribed in the quiescent state, the phage genome exists in a repressed state. The phage DNA, in this repressed state, is called a prophage because it is not a phage but it has the potential to produce phages. In most cases the phage DNA actually integrates into the host chromosome and is replicated along with its host. The lysogenic state may persist indefinitely. The cell harboring a prophage is termed a lysogen. The λ phage is one notable example of this group of phages.[16,18]

The filamentous phage consists of a group of bacteriophages that do not lyse their host cells but rather are secreted from the bacterium.[13,19] More importantly, capsid proteins that are displayed on the membrane of the phage are secreted into the periplasmic space of the bacterium, where they fold prior to secretion of the phage. The periplasmic space of *E. coli* provides a favorable redox potential (as compared to the bacterial cytoplasm) for proper folding of disulfide-linked proteins. As such, recombinant proteins tethered to filamentous phage capsid proteins tend to fold in their native conformation when expressed for display on the surface of the phage (Fig. 14.1).

This overview on phage display systems will focus on two common phages, filamentous and T7. Both have been utilized for display of recombinant proteins on the surface of phages and have mature vector systems. In addition, they each have their own unique advantages for incorporation into biosensors that will be examined later.

14.3 Filamentous Phages

The filamentous phage constitutes a large family of bacterial viruses that infect mainly Gram-negative bacteria. Their defining charac-

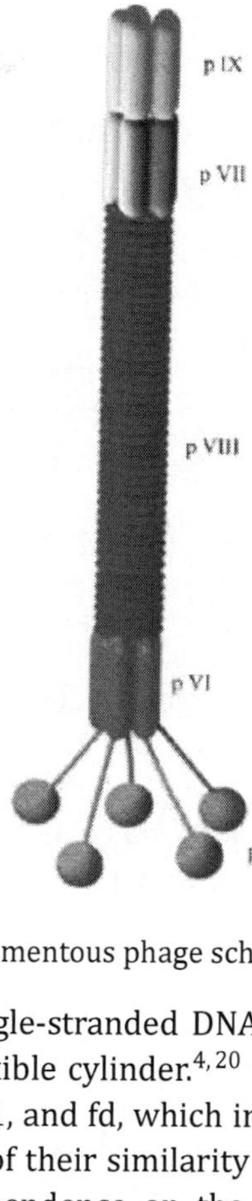

Figure 14.1 Filamentous phage schematic structure.[5]

teristic is a circular, single-stranded DNA (ssDNA) genome, which is encased in a long, flexible cylinder.[4,20] The best characterized of these phages are M13, f1, and fd, which infect *E. coli* containing the F' ion factor.[21] Because of their similarity in structure, homologous genomes, and their dependence on the F' element for infection, M13, f1, and fd are collectively referred to as the Ff phages ("Ff", F-specific filamentous). Unless otherwise specified, the properties of filamentous phages described next refer to the Ff phage. Unlike most bacterial viruses, the life cycle of the filamentous phage ends with secretion from the infected bacteria in the form of phage particles without cell deathor lysis.[4,19]

14.3.1 *Structure of the Filamentous Phage Virion*

Phage particles are normally composed of five coat proteins (see Fig. 14.2) arranged in a long, flexible cylinder approximately 7 nm

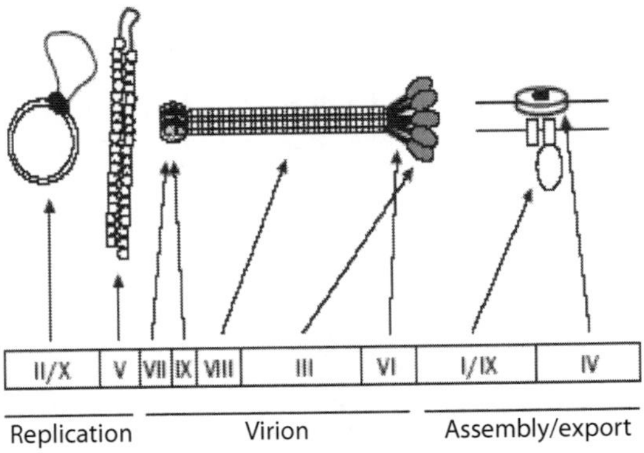

Figure 14.2 Filamentous phage genes and gene products.[4]

wide by 900 to 2,000 nm long, depending upon the size of the recombinant genome, which is normally 6–7 kb. Cloning DNA into a nonessential region of the genome can create longer phage particles, although the longer they are, the more sensitive the particles are to breakage.[4,19] The 6400-nucleotide ssDNA of the Ff phage is encapsidated in a 930 nm particle, whereas a 221-nucleotide "microphage" variant is 50 nm long.[22]

The viral particles contain approximately 2,700 copies of a single major coat protein, pVIII, oriented at a 20° angle from the particle axis and overlapped like fish scales to form a right-handed helix.[15,23] The blunt end of the phage contains three to five copies each of pVII and pIX capsid proteins.[4,19] Phage assembly begins at the pVII–pIX end, and in the absence of either protein, no particle is formed.[24] The other end of the particle is pointed, and lollipop-like knobs can be seen, by electron microscopy, extending from the tip in certain preparations.[25] This end contains about five copies each of pIII and pVI, both of which are needed in order for the phage to enter and detach from the cell membrane; without these two proteins, assembly cannot terminate.[4,19] The N-terminal domain of pIII binds to the bacterial cell receptor, the tip of the F'-encoded pilus, and the C-terminal domain of TolA.[26] The pIII coat protein can only be incorporated into the particle if pVI is present.[27] The 406-residue pIII capsid protein is responsible for the knob structure at the end of the phage particle.[28]

The phage genome is oriented within the phage particle. Its orientation is determined by the packaging signal (PS), which is located in the noncoding IG region of the genome (IG located near gene IV). The PS is positioned at the pVII–pIX end of the particle and is necessary for efficient encapsidation of circular ssDNA into the phage particle.[29]

14.3.2 *Life Cycle of the Filamentous Phage*

The life cycle of the filamentous bacteriophage involves three major stages: 1) infection of the bacterial cell, 2) replication and gene expression, and 3) assembly and secretion from the infected bacteria. In the first stage the bacteriophage infects the bacterial cell by utilizing the F' element-encoded bacterial pilus (conjugative pilus) as a primary receptor for infection of bacterial cells (see Fig. 14.3). Infection normally begins when the N2 (or D2) domain of the capsid pIII protein binds the tip of an F pilus.[19,30] The interaction between N2 and the pilus releases the N1 (or D1) domain from its

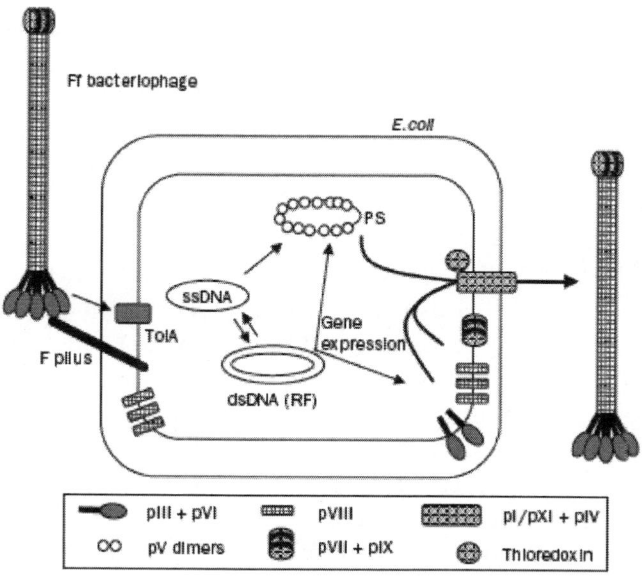

Figure 14.3 Filamentous phage life cycle.[4]

normal interaction with N2, making it available to bind a discrete domain (D3) of the bacterial TolA protein.[30–32]

Three Tol proteins (Q, R, and A) are required for phage infection.[33–36] They mediate depolymerization of the phage coat protein into the cytoplasmic membrane and the translocation of the viral ssDNA into the bacterial cytoplasm, although the details of how this process is accomplished remain unclear.

Once the viral ssDNA enters the cytoplasm, the second stage begins, consisting of replication of the genome and gene expression. Host enzymes (RNA and DNA polymerases, topoisomerase, and gyrase) convert it to a double-stranded, supercoiled molecule called the replicative form (RF). The RF serves as a template for phage gene expression and rolling circle replication, which generates ssDNA molecules. Early after infection, when the concentration of the phage-encoded ssDNA-binding protein (pV) is low, newly synthesized ssDNAs are converted to RFs and both RF and phage protein increase exponentially. When sufficient pV has accumulated, pV dimers cover the ssDNA and prevent polymerase access, thereby blocking their conversion to RF. The pII (or pX) protein is required for the stable accumulation of ssDNA at this stage, but the mechanism by which it acts is not known.[37,38] The DNA is oriented in pV-ssDNA complexes, with the PS hairpin protruding from one end; this complex is a substrate for phage assembly.[39]

The Ff genome contains nine closely packed genes and one major noncoding region (IG), which contains the origin of replication for + and – strand synthesis and the PS. Two of the phage genes (I and II) have internal start sites from which in-frame restart proteins are produced. Both kinds of proteins are necessary for successful phage production. The phage genome encodes for 11 proteins: three (pII, pX, and pV) are needed to generate ssDNA, three (pI, pXI, and pIV) are required for phage assembly, and five (pIII, pVI, pVII, pVIII, and pIX) are part of the phage particle. All phage proteins are synthesized concurrently but at an appropriate rate.[4,19]

Filamentous phages assemble and secrete by a unique process that does not kill bacterial cells. Instead, newly formed virions are secreted from the cell during assembly. The eight phage-encoded proteins that are directly involved in assembly are integral membrane proteins. The five viral coat proteins (pIII, pVI, pVII, pVIII, pIX) reside in the cytoplasmic membrane prior to their incorporation into the phage.[40] Two of these proteins, pIII and pVIII, are synthesized as precursors. Assembly sites are composed of the three morphogenetic proteins, pI, pXI, and pIV, which interact via their periplasmic domains. pI and pXI are located on the cytoplasmic membrane, and pIV is located on the outer membrane.[4,41,42]

Phage assembly begins when pI recognizes the PS, which protrudes from one end of the pV-ssDNA complex. Initiation takes place only if the two minor coat proteins (pVII and pIX) are located at the tip of the particle and the ssDNA is present.[43] It is believed that pVII and pIX interact with the PS.[44] During elongation, pV is removed from the ssDNA and several thousand copies of pVIII are added to the ssDNA. The elongation requires host thioredoxin (TrxA) and adenosine triphosphate (ATP) hydrolysis, presumably by pI. When the end of the DNA is reached, assembly is terminated by the addition of pIII and pVI, and a conformational change in pIII, which detaches the nascent particles from the membrane through the pIV channel. The first progeny phage particles can be found in the culture supernatant about 10 minutes after infection (at 37°C). The phage numbers increase exponentially for about 40 minutes, after which the rate becomes linear. About 1,000 phages per cell are produced during the first hour. Under optimal conditions, the infected cells can continue to grow and divide and thus produce phages indefinitely.

14.3.3 *Filamentous Phage Display*

All five coat proteins have been used to display proteins or peptides in varying degrees (see Table 14.2). However, by far the most commonly used virion proteins for phage display are pVIII and pIII. pVI has been shown useful for display through a C-terminal fusion.[45] pVII and pIX have been used for fusion of antibody fragments to their N-termini.[46] pIII is commonly used for phage display of functional proteins because of its tolerance for large insertions, its compatibility with monovalent display, and the broad availability of appropriate vectors. pIII fusions are typically constructed at the N-terminal, but pIII C-terminal fusions are also possible through a linker peptide.[47] pVIII phage display systems are mainly useful for display of short peptide sequences tethered to every copy of the protein on the phage capsid. These sequences are inserted typically at the N- terminal of the pVIII capsid protein.[4] The main difference between pIII and pVIII phage display systems is in the capacity of the display and the display copy number—pIII can display large proteins, and pVIII optimally displays peptides in many copies on the capsid.[21,48–50] Large proteins can also be displayed on pVIII; however, the total copy number of these molecules on the phage is limited strictly by molecular hindrance to a few copies. In this case a wild-type copy of the pVIII gene must also be supplied for proper phage assembly.[4,21,49] Small peptides (6–10 residues) tethered to pVIII can be expressed in approximately 2,700 copies on the surface of the phage particle, and in this case, wild-type copies of pVIII are not required.[4,51]

In addition to display of single proteins and peptides, filamentous phages have also been utilized for production of phage display libraries that express all open reading frames (ORFs) within a given cDNA library. Libraries of this type have proven useful for identification of bacterial proteins as well as domains that interact with a range of target molecules.[52–54] In some cases recombinant proteins within these libraries can be toxic to filamentous phages or interfere with release of the recombinant capsid protein. Therefore, special phagemid systems have been developed to facilitate toxic-protein display.[55] The main advantage of filamentous phage display over the lytic system is its long, proven history of displaying peptides or proteins in the proper conformation. In addition, it is much easier to purify filamentous phages since they are secreted from bacteria and can readily be purified by filtration. As such, readily purified phage particles can, in many cases, be directly used for diagnostic applications.

14.4 T7 Phage: Structure of the Virion

The structures of T7 virions (see Fig. 14.4) consist of an icosahedral capsid (the head) with a diameter of about 60 nm, with a volume of about 10^5 nm^3.[17,76–79] The capsid shell consists of 415 molecules in two forms of the gene 10 protein (gp10A and gp10B).[76,80–82]

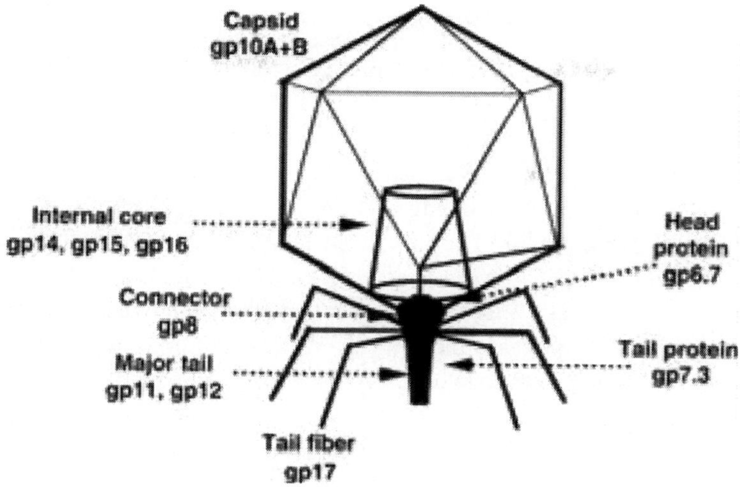

Figure 14.4 Schematic diagram of the T7 virion.[76]

Inserted at one vertex of the capsid is the head–tail connector, composed of 12 gp8 molecules.[83,84] The connector has a 12-lobed wide domain inserted into the head cavity and a narrower domain that interacts with the tail.[85] Inside the head, and attached to the head–tail connector in the coaxial orientation, is a 26 nm × 21 nm cylindrical structure that is usually referred to as the internal core.[86]

The core, which exhibits eightfold symmetry,[87] consists of three distinct proteins: gp14, gp15, and gp16.[76] The genome is spooled around the internal core in six coaxial shells.[77,88] The T7 phage genome contains 39,936 base pairs (bp) of linear double-stranded DNA.[16] The DNA has a center-to-center helix spacing of 2.4 nm in β-form.[88] The stubby tail is 23 nm long, tapering from a diameter of 21 nm at the connector to 9 nm at its distal end, and is known to consist of several proteins gp11, gp12, and gp 7.3. The exact position of gp 7.3 has not yet been determined, but particles that were assembled in the absence of gp 7.3 failed to adsorb to cells.[17,77]

Attached to the tail are six symmetrically positioned tail fibers. Each fiber is composed of a trimer of gp17 that forms a kinked structure.[89,90] The gp17 consists of a linear array of four globules that are thought to bind directly to the bacterial cell.[76,89]

14.4.1 *Life Cycle of T7*

The life cycle of the T7 bacteriophage involves four major stages: 1) adsorption and penetration to the bacteria cell, 2) replication, 3) assembly, and 4) lysis of the bacteria in order to exit from the bacteria. In the first stage, the bacteriophages infect an *E. coli* bacterium by the interaction of its tail fibers with the lipopolysaccharide (LPS) on the cell surface. However, the stubby T7 tail is too short to cross the *E. coli* cell envelope into the cytoplasm. So it was proposed that virion proteins are injected into the cell to function as an extensible tail to form a channel across the cell envelope enabling passage of the genome.[76,77]

Immediately after infection the second stage begins, consisting of replication of the genome and gene expression. The phage nucleic acid takes over the host biosynthetic machinery and phage-specific mRNAs, and early proteins are made. Transcription is performed by the host RNA polymerase from a promoter near the end of the molecule, whose products include an antirestriction factor, a protein kinase that phosphorylates and inactivates *E. coli* RNA polymerase, a T7-specific RNA polymerase, and a DNA ligase. After a few minutes, the T7 polymerase replaces the host polymerase and all transcription takes place on 81% of the genome. Early proteins are needed for phage DNA synthesis, shutting off host DNA and protein biosynthesis. The first genes transcribed by the T7 polymerase include a DNA polymerase and recombination genes. Later, after replication ensues, the late genes encoding virion components and at least one of the lysis genes are expressed. Late gene expression does not require replication, as it does in T4, and the host polymerase is replaced rather than reprogrammed. The first cycle of DNA replication proceeds bidirectionally from a unique origin on the linear monomer.[16,91] Because DNA polymerase moves only in one direction, unit-size DNA has no means of replicating the 3′ terminal gaps, which arise by excision of primer RNA. To solve this problem, T7 replication generates concatemers formed by hybridization of the unreplicated 3′ redundant regions of newly synthesized DNA.[16,91] To complete replication T7 completes the ends of the genome, a task facilitated by the terminal redundancy of the T7 genome; the first and the last 160 bp are identical.[16,91]

After the replication of the genome and gene expression the third stage begins. At this stage the nucleic acid and structural proteins that have been made are assembled and infectious phage particles accumulate within the cell. Assembly starts with formation of proheads, which contain the proteins gp8, gp10, gp14, gp15, gp16 and the scaffolding protein gp9.[17,91,92] Then the DNA is packaged in the presence of gp18 and gp19. The proteins gp9, gp18, and gp19 are not present in the mature phage. The tail proteins (gp11, gp12) collectively assemble on the full head, and subsequently the tail fibers are bound to the conical tail.[91] Alternatively, the assembly can start with the formation of incomplete prohead shells that contain the scaffold and capsid proteins, which are then closed by insertion of a connector-core complex.[17,93]

Following assembly of phage particles, the fourth stage begins with bacterial host lysis due to the accumulation of phages and lysis proteins followed by release of phages into the medium. The only T7 proteins known to function in lysis are the products of genes 3.5 (lysozyme) and 17.5 (holing).[91,94] Gene 3.5 was initially thought to be responsible only for cell lysis, but it has been demonstrated that gene 3.5's most important function is in control of initiation and termination by T7 RNA polymerase, which, in turn, affects DNA replication and packaging.[94,95] T7 lysozyme is also an *N*-acetylmuramyl-*L*-alanine amidase, which breaks up the peptidoglycan wall. The lysozyme is blocked from access to the cell wall by the inner membrane. It has been suggested that at some point the phage-encoded holing protien (gp17.5) enlarges membrane penetrability, thereby exposing the peptidoglycan wall to the lysozyme, with rapid lysis of the cells ensuing.[94,96]

14.4.2 *T7 Phage Display*

T7 phage display is possible on the capsid proteins, gp10A, and gp10B, which are present in 415 copies. The fact that functional T7 particles can be made from either of the two capsid proteins, including a range of mixtures of two different protomers, prompted the exploration of the T7 capsid protein as a C-terminal display platform (see Table 14.2). The T7 phage is released by cell lysis, and the translocation of fusion proteins through the cell membrane

Table 14.2 Comparison between filamentous and T7 phage display systems[4,21]

Type of phage	Protein type	Size (aa)	Type of display	Type of fusion	References
Filamentous	pIII	406 (mature protein without signal sequence)	Peptides and functional proteins (large insertions), cDNA cloning, Antibodies, Cytokines, Receptores, Enzymes, Enzymes inhibitors, DNA binding proteins, Cellulose, binding proteins	C and N- terminal	47, 56–68
	pVI	112	Polypeptides encoded by cDNA, Enzymes, Enzymes inhibitors	C -terminal	45, 69
	pVII	33	Peptides, fusion of antibody fragments, Enzymes, Enzymes inhibitors	N-terminal	46
	pVIII	50(mature protein without signal sequence)	Short peptides	C and N-terminal	4, 19, 70–73
	pIX	32	Fusion of antibody fragments	N-terminal	46
T7	gp10 (A and B)	gp10A- 344, gp10B- 397	Fusion of peptides and cDNA	C-terminal	9, 74, 75

is thus avoided. The biology of T7 is well understood and offers a variety of host-vector systems for a range of applications.[17] It is reported that the vector T7Select415-1b, Novagen, is capable of accepting peptides of up to 40–50 amino acids in length for display at 415 copies per particle.[10,74,97,98] The vector T7Select1-1b, from Novagen, is designed for display of peptides and proteins at approximately one copy per phage particle and can tolerate fusions of up to 900–1,200 amino acids at that display density.[21] cDNA libraries constructed with the T7Select1-1b system or its derivatives have been used for display and panning against a variety of different targets, including small-molecule chemical probes.[75,99–102] The T7 display (select 1–2 series) system has also been used to select for RNA-binding proteins from cDNA libraries.[103] Lehmann et al. reported that in an indirect ELISA, C-terminal T7 phage display systems have some major advantages in comparison with N-terminal pVII or pIII filamentous phage display systems.[10]

14.5 Principles and Applications of Phage Display

One of the most impressive aspects of phage display is the variety of uses for this technology.[13] The primary applications for phage display involve the use of phage display libraries, which are screened and enriched by a process known as biopanning. In this technique a phage library is incubated with an immobilized target. Those few phages in the population with an affinity for the target attach to the receptor molecule, while the remaining phages, which do not bind, are washed away. Bound phages are subsequently eluted following thorough washing. Infection of bacteria with the eluted phages results in amplification of the recombinant phages. After several rounds of such enrichment, a phage population with high affinity and specificity to the target can thereby be obtained. The chosen primary structure of the peptide expressed by a given clone can be deduced by means of DNA sequencing of the original insert.[15]

The main applications of phage display are discussed next.

14.5.1 *Phage Display of Natural Peptides*

Display of natural peptides, from known proteins, on a phage particle is mainly used for two applications.[13] The first application involves mapping epitopes of monoclonal and polyclonal antibodies. In this application, phage display is used as a means to identify clones that were expressing epitopes recognized by a given antibody. DNA segments encoding approximately 10 amino-acid-long peptides are inserted into the coding domain of a capsid protein and thus are expressed on the phage capsid. These short peptides consist of linear epitopes from the original antigenic protein. A library of these phage-displayed peptides can then be screened against a given antibody to identify the specific epitope to which it binds.[104–107] When an epitope is found it can be used for development of probes for the detection of biological threat agents,[1] vaccines, diagnostics, and therapy of diseases such as cancer.[108] In the second application, phage display is used for the purpose of generating immunogens. Short peptides derived from various pathogen coat proteins can be expressed on a phage and used to elicit or identify antibodies against these pathogenic epitopes.[109,110] This method might be used for vaccination purposes and in diagnostic technologies.[9,111,112]

14.5.2 *Phage Display of Random Peptides*

Random peptide phage display libraries, in general, consist of a combinatorial representation of all linear peptide epitopes of a defined length. Production of these segments is achieved through degenerate synthetic oligonucleotides of a specified length that are cloned into the capsid protein of a phage. Like natural peptide libraries, random peptide libraries have been invaluable for mapping of monoclonal and polyclonal antibody-binding domains.[5,107] It has been successfully used to identify immunodominant peptide sequences of antigens, generate peptide competitors of antigen-antibody interactions, and map accessible and/or functional sites of numerous antigens.[13,15,107,113,114]

Random peptide phage display libraries are also used for identifying ligands. A great variety of potential ligands for a vast array of receptors is present in a random peptide library.[115] In some cases, the peptides resemble the receptor's ligand in terms of sequence. In other cases, the ligand mimics the binding of a nonprotein ligand.[13,116–118] The development of a "substrate phage" is another important application of this technology. In this case, libraries are not used for simply identifying a ligand for a target molecule but rather for defining substrate specificity. Many different posttranslational modifications can also be mapped using this approach.[119,120]

14.5.3 *Phage Display of Proteins or Protein Domains*

In this approach, whole proteins or specific domains are displayed on the phage. Even though these proteins and domains are tethered to a phage protein, they usually retain their normal binding and enzymatic activity.[13,108] One of the most promising applications of this technology is the ability to isolate recombinant antibodies (expressed as single-chain variable fragments) to a wide variety of antigens. The antibody library is generated by a random combination of variable light (VL) and variable heavy (VH) chain coding domains produced as single-chain variable antibody fragments (scFvs).[13,121,122] Once the library is constructed it can be screened against any desirable antigen by the biopanning technique. This method has helped overcome some of the current limitations in generating human monoclonal antibodies or humanizing mouse antibodies. It is cost effective, less time consuming, and relatively easy.[15] However, once a suitable scFv is isolated by phage display it needs to be reengineered into a human antibody backbone and tuned for optimal specificity and sensitivity. The reengineered human soluble recombinant antibodies can then be produced in cell culture systems and be used as in vitro diagnostic reagents.[15] Many recombinant antibodies that were isolated by phage display showed high affinity and specificity.[123–125]

Another possible application for phage display of proteins or protein domains is the screening of libraries of mutant proteins to improve affinity for a target.[5,13,15]

Novel enzyme inhibitors have been identified by this technology.[106,126] Protein–protein interactions have also been investigated with phage-displayed libraries.

For this purpose we do not wish to explore all combinatorial combinations of amino acids as defined by random peptide libraries but merely the amino acid combinations defined by the genome under study. To this end, it is possible to express cDNA encoded proteins on the surface of phages, which can then be tested against a particular immobilized target, in vitro, using biopanning enrichment.[13]

14.5.4 *Multiple-Display Phages*

The ability to simultaneously display two different peptides on the surface of filamentous (fd) bacteriophages was demonstrated in 1997. Cells that contain a plasmid with a modified pVIII gene were infected with an engineered bacteriophage carrying a second and different copy of a modified pVIII gene. Hybrid virions were thereby created in which the pVIII capsid protein consisted of a mixture of the wild-type and two modified coat proteins carrying different peptide inserts.[127] The ability to display different peptides on the surface of the T4 phage was recently demonstrated in 2006.[12] Multiple epitopes of human immunodeficiency virus (HIV) were displayed on the capsid protein of the T4 phage to construct multicomponent HIV vaccines. The assembly system facilitated display of three different HIV antigens on the capsid surface. This system offers new direction and insights for HIV vaccine development, with the potential to increase the breadth of both cellular and humoral immune responses.[12]

Recently an innovative technique has been developed that utilizes a phage display system to simultaneously display multiple distinct peptides on a phage, each of them in multiple copies. This system is based on filamentous (M13) phages, in which five peptides were expressed and displayed on different pVIII capsid proteins. Calibration of the system was based on peptides from West Nile virus (WNV) that were designed using in silico prediction of B-cell linear epitopes. A phagemid encoding five different epitopes was introduced into *E. coli* JM109(DE3) (Promega). These bacterial cells were infected with helper phages (VCSM13) to enable production of wild-type proteins. The recombinant phages were produced with pVIII proteins, displaying the various epitopes and wild-type pVIII

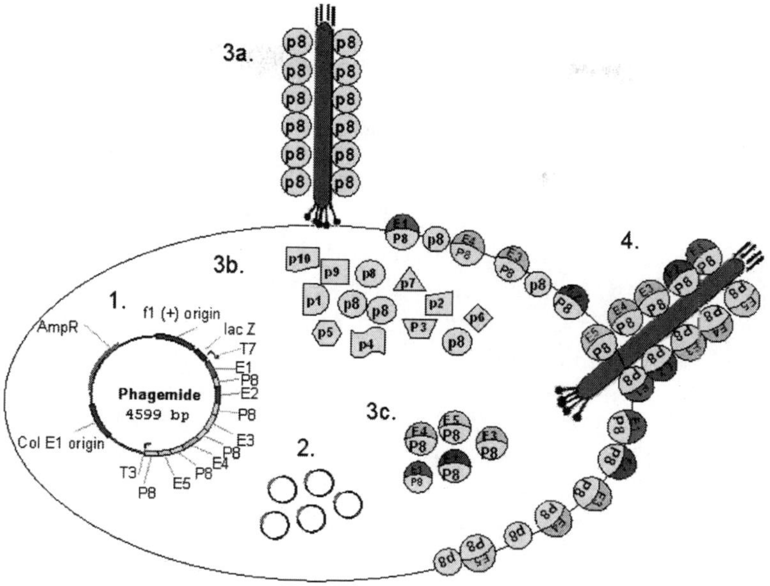

Figure 14.5 The multiple-display phage system: (1) Phagemid transformation into competent *E. coli* cells, JM109(DE3. The phagemid encodes five different epitopes, each one fused to the pVIII protein. (2) Growth of cells. During this time the phagmid multiples. (3a) Infecting the cells with helper phages (VCSM13) (3b) Production of wild-type proteins (3c) Production of pVIII with various epitope fusion peptides. (4) Assembly and secretion of recombinant phage-displayed multiepitopes.[128]

protein (see Fig. 14.5).[128] This offers an efficient solution when simultaneous analysis of binding to multiple epitopes is required. Furthermore, it affords a simple solution for incorporation of multiple peptides into a simpler diagnostic system and is especially suited for inclusion in nanoscale-based diagnostics such as in lab-on-chip technology. This system also has the potential to be the basis for multivalent vaccines.

14.6 Use of Phage-Displayed Epitopes for Viral Diagnostics

There is a growing use for phage display systems for developing diagnostic tools. This section will focus on different diagnostic techniques that employ phage-displayed epitopes as bioreceptors for detecting viruses or specific antibodies against viruses.

14.6.1 *ELISA and Phage Display*

ELISA is an established technique used primarily for serological analysis to detect the presenceof an antibody or an antigen in a sample. A detectable signal is generated by an enzyme conjugate that serves to detect a specific analyte. The enzyme conjugate can be used in either a chemiluminescent or a chromogenic ELISA assay, which is the most commonly used.[129] Herrmann et al. demonstrated the first phage-displayed peptide-based ELISA for detection of WNV immunoglobulin G (IgG). The Ep15 epitope, derived from the WNV E protein DIII, was cloned into a T7 phage display system that was then used as a recombinant antigen in a chemiluminescent ELISA format. Their system showed a limit of detection at a serum titer of 1:51,200 and a dynamic range from 1:100 to 1:2000. In a screen of a panel of 66 human sera samples using this technology, a sensitivity of 67% and a specificity of 100% were achieved. Considering the ease of antigen preparation, its stability, and the optimum display properties of the T7 bacteriophage, it is apparent that this approach can be useful for the preparation of highly sensitive and specific anti-WNV immunoglobulin diagnostic kits.[9]

Indirect ELISA based on a filamentous M13 phage was used in the work of Ramanujam et al. They used a phage-displayed peptide (sequence TLTTKLY), isolated from phage-displayed peptide library, from the highly virulent Newcastle disease virus (NDV) strain AF2240. Thy established an ELISA that could be used to differentiate virulent NDV strains from avirulent strains.[130]

Newcastle disease is a highly contagious zoonotic bird disease affecting many domestic and wild avian species. Exposure of humans to infected birds can cause mild conjunctivitis and influenza-like symptoms.

14.6.2 *Dot Blot Assay and Phage Display*

The dot blot is a common technique in molecular biology used to detect biomolecules. It represents a simplification of the Northern blot (RNA detection), Southern blot (DNA detection), or Western blot (protein detection) methods. In a dot blot the biomolecules to be detected are not first separated by chromatography. Instead, a mixture containing the molecules to be detected is applied directly on a membrane as a dot. This is then followed by specific detection with either nucleotide probes (for a Northern blot and a Southern blot) or antibodies (for a Western blot). Since separation of the samples is not required, it offers significant savings in time. However, if two molecules of different sizes are detected, they will still appear as a single dot. Dot blots therefore can only confirm the presence or absence of a biomolecule or biomolecules.

Lee et al. developed a dot blot assay that detected only virulent NDV strains in order to differentiate the virulent strain from the avirulent strain. They isolated a M13 filamentous phage clone displaying the peptide sequence SWGEYDM that was capable of distinguishing virulent from avirulent NDV strains.[131] This work was based on previous work of this group using phage display peptides in an ELISA format, which is described before.[130]

14.6.3 *PCR, Immuno-PCR, and Phage Display*

Polymerase chain reaction (PCR) is a technique widely used in molecular biology. With PCR it is possible to amplify a single or a few copies of a piece of DNA across several orders of magnitude, generating millions or more copies of the DNA. PCR can be extensively modified to perform a wide array of genetic manipulations. Immuno-PCR (IPCR) is a highly promising technique for the ultrasensitive analysis of antigens,[132] combining the versatility of ELISAs with the amplification power and sensitivity of PCR. On the basis of this technology a new approach was developed involving a phage display system that was called phage display mediated immuno-PCR (PD-IP-CR) (see Fig. 14.6).[133] Instead of utilization of a monoclonal antibody (mAb) and chemically bonded DNA that is required in the conventional IPCR, a recombinant M13 filamentous phage particle is applied as a ready reagent for an IPCR experiment. The phage-displayed single-chain variable fragment (scFv) specifically recognizes the analyte and phage DNA directly serves as a PCR template, respectively. The idea has been

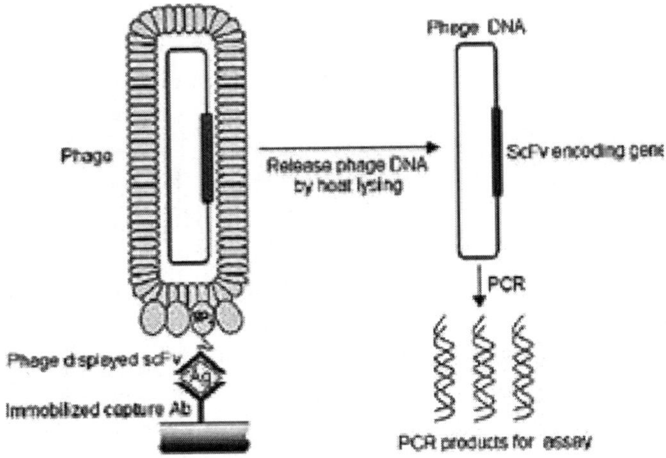

Figure 14.6 Schematic diagram of phage display–mediated IPCR.[133]

demonstrated by applying Hantaan virus nucleocapsid protein (NP) and prion protein (PrP) as detection targets. The detection sensitivity was increased 1,000- to 10,000-fold as compared with conventional colorimetric ELISA.

14.6.4 *Electrochemical Phage Immunosensors*

An electrochemical biosensor is a self-contained integrated device capable of providing specific or semiquantitative analytical information using a biological recognition element (biochemical receptor), which is retained in direct spatial contact with an electrochemical transduction element.[134]

Recently an amperometric immunosensor for the detection of anti-WNV IgG antibodies was described. The immunosensor is based on the immobilization of phage-displayed WNV peptides incorporated into a poly(pyrrole-alkyl ammonium) film electrogenerated at the surface of a glassy carbon electrode. The designed immunosensor has proven to have high analytical performance in terms of a low limit of detection (10^7), a fast response time (5–20 sec.), and great reproducibility (see Fig. 14.7).[135]

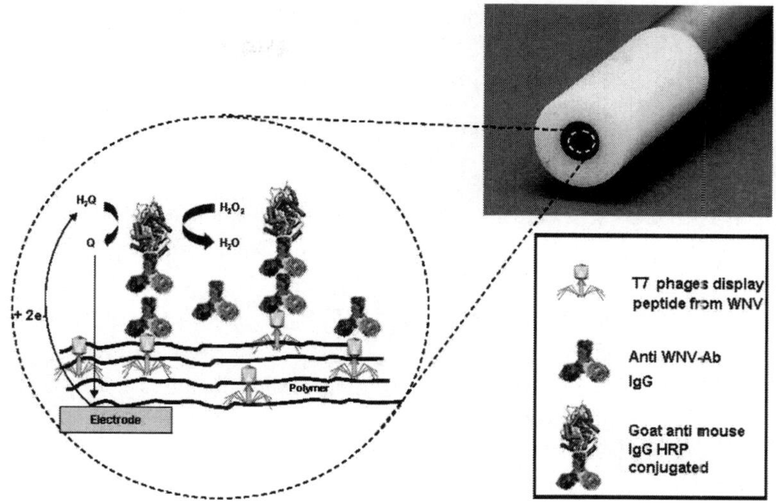

Figure 14.7 Schematic representation of the functioning principle of the WNV amperometric immunosensor through HRP-conjugated secondary antibody that catalyzed the formation of electroactive quinone species (Q) in the presence of hydroquinone (HQ) as a substrate.[128,135] *Abbreviation*: HRP, horseradish peroxidase.

14.7 Prospects for Use of Phage Display in Biosensors and Biochips

Although phage display was first introduced over 20 years ago, the development and application of this technology is still being explored and new applications are constantly being identified. Exploitation of phage display technology will lead to the isolation and production of a broad range of bioreceptors, including recombinant antibodies and peptides, with predefined specificities. In addition, emergent technologies based on phage display will benefit diagnostics by producing molecules that are otherwise unobtainable by traditional approaches. Exploitation of the advantages of a phage display system can serve as convenient sources of proteins/peptides. Phage-displayed epitopes for pathogen diagnostics can also improve the detection specificity, in particular for viruses. Phage display can also be the base for rapid identification and thus avert or reduce the number of casualties associated with a potential bioterrorism or biowarfare event.

References

1. Petrenko, V.A., and V.J. Vodyanoy, Phage display for detection of biological threat agents. 2003. *J Microbiol Methods,* **53**(2), 253–262.
2. Iqbal, S.S., et al. A review of molecular recognition technologies for detection of biological threat agents. 2000. *Biosens Bioelectron,* **15**(11–12), 549–578.
3. Smith, G.P., Filamentous fusion phage: novel expression vectors that display cloned antigens on the virion surface. 1985. *Science,* **228**(4705), 1315–1317.
4. Russel, M., and H.B.L., Tim Clackson, Introduction to phage biology and phage display, in *Phage Display: A Practical Approach* (2004), H.B.L. Tim Clackson, Editor. Oxford University Press. 1–26.
5. Arap, M.A., Phage display technology: applications and innovations. 2005. *Genetics and molecular biology,* **28,** 1–9.
6. Benhar, I., Biotechnological applications of phage and cell display. 2001. *Biotechnol Adv,* **19**(1), 1–33.
7. Cesareni, G., L. Castagnoli, and G. Cestra, Phage displayed peptide libraries. 1999. *Comb Chem High Throughput Screen,* **2**(1), 1–17.
8. Miranda, L.P., and P.F. Alewood, Challenges for protein chemical synthesis in the 21st century: bridging genomics and proteomics. 2000. *Biopolymers,* **55**(3), 217–226.
9. Herrmann, S.L., B. Leslie, L. Bin, H. Mendelson, E. Ben-Nathan, D. Dussart, P. Porgador, A. Ragerzisman, B and Marks, R.S., T7 phage display of Ep15 peptide for the detection of WNV IgG. 2006. *Virol Methods.*
10. Lehmann, D., R. Sodoyer, and S. Leterme, Characterization of BoHV-1 gE envelope glycoprotein mimo topes obtained by phage display. 2004. *Vet Microbiol,* **104**(1–2), 1–17.
11. Willis, A.E., R.N. Perham, and D. Wraith, Immunological properties of foreign peptides in multiple display on a filamentous bacteriophage. 1993. *Gene,* **128**(1), 79–83.
12. Sathaliyawala, T., et al. Assembly of human immunodeficiency virus (HIV. antigens on bacteriophage T4: a novel in vitro approach to construct multicomponent HIV vaccines. 2006. *J. Virol,* **80**(15), 7688–7698.

13. Brian, K., and R.H.H. Kay, Principles and applications of phage display, in *Phage Display of Peptides and Proteins* (1996), J.W. Beian K. Kay, John McCafferty, Editors. Academic Press. 21–28.
14. Goldman, E.R., et al. Phage-displayed peptides as biosensor reagents. 2000. *J Mol Recognit,* **13**(6), 382–387.
15. Azzazy, H.M., and W.E. Highsmith, Jr., Phage display technology: clinical applications and recent innovations. 2002. *Clin Biochem,* **35**(6), 425–445.
16. Campbell, A.M., Bacteriophage, in *Fields Virology* (1996), D.M.k. Bernard N. Fields, Peter M, Howley, Editors. Lippincott-Raven.
17. Molineux, I.J., The T7 group, in *The Bacteriophage* (2006), R. Calendar, Editor. Oxford University Press. 277–301.
18. Hendrix, W.R., and Casjens, S., Lambda and its genetic neighborhood, in *The Bacteriophage* (2006), R. Calendar, Editor. Oxford University Press. 409–447.
19. Russel, M., and Model, P., Filamentus phage, in *The Bacteriophage* (2006), R. Calendar, Editor. Oxford University Press. 146–160.
20. Chopin, M.C., et al. Filamentous phage active on the gram-positive bacterium Propionibacterium freudenreichii. 2002. *J Bacteriol,* **184**(7), 2030–2033.
21. Lindqvist, B.H., Phage in display, in *The Bacteriophage* (2006), R. Calendar, Editor. Oxford University Press. 686–694.
22. Specthrie, L., et al. Construction of a microphage variant of filamentous bacteriophage. 1992. *J Mol Biol,* **228**(3), 720–724.
23. Marvin, D.A., Filamentous phage structure, infection and assembly. 1998. *Curr Opin Struct Biol,* **8**(2), 150–158.
24. Lopez, J., and R.E. Webster, Morphogenesis of filamentous bacteriophage f1: orientation of extrusion and production of polyphage. 1983. *Virology,* **127**(1), 177–193.
25. Gray, C.W., R.S. Brown, and D.A. Marvin, Adsorption complex of filamentous fd virus. 1981. *J Mol Biol,* **146**(4), 621–627.
26. Bradley, D.E., and J. Whelan, Escherichia coli tolQ mutants are resistant to filamentous bacteriophage that adsorb to the tips, not the shafts, of conjugative pili. 1989. *J Gen Microbiol,* **135**(7), 1857–1863.
27. Rakonjac, J., J. Feng, and P. Model, Filamentous phage are released from the bacterial membrane by a two-step mechanism involving a short C-terminal fragment of pIII. 1999. *J Mol Biol,* **289**(5), 1253–1265.

28. Armstrong, J., R.N. Perham, and J.E. Walker, Domain structure of bacteriophage fd adsorption protein. 1981. *FEBS Lett,* **135**(1), 167–172.

29. Griffith, J., M. Manning, and K. Dunn, Filamentous bacteriophage contract into hollow spherical particles upon exposure to a chloroform-water interface. 1981. *Cell,* **23**(3), 747–753.

30. Deng, L.W., P. Malik, and R.N. Perham, Interaction of the globular domains of pIII protein of filamentous bacteriophage fd with the F-pilus of Escherichia coli. 1999. *Virology,* **253**(2), 271–277.

31. Riechmann, L., and P. Holliger, The C-terminal domain of TolA is the coreceptor for filamentous phage infection of *E. coli.* 1997. *Cell,* **90**(2), 351–360.

32. Stengele, I., et al. Dissection of functional domains in phage fd adsorption protein. Discrimination between attachment and penetration sites. 1990. *J Mol Biol,* **212**(1), 143–149.

33. Sun, T.P., and R.E. Webster, fii, a bacterial locus required for filamentous phage infection and its relation to colicin-tolerant tolA and tolB. 1986. *J Bacteriol,* **165**(1), 107–115.

34. Sun, T.P., and R.E. Webster, Nucleotide sequence of a gene cluster involved in entry of E colicins and single-stranded DNA of infecting filamentous bacteriophage into Escherichia coli. 1987. *J Bacteriol,* **169**(6), 2667–2674.

35. Click, E.M., and R.E. Webster, Filamentous phage infection: required interactions with the TolA protein. 1997. *J Bacteriol,* **179**(20), 6464–6471.

36. Click, E.M. and R.E. Webster, The TolQRA proteins are required for membrane insertion of the major capsid protein of the filamentous phage f1 during infection. 1998. *J Bacteriol,* **180**(7), 1723–1728.

37. Fulford, W., and P. Model, Gene X of bacteriophage f1 is required for phage DNA synthesis. Mutagenesis of in-frame overlapping genes. 1984. *J Mol Biol,* **178**(2), 137–153.

38. Fulford, W., and P. Model, Regulation of bacteriophage f1 DNA replication. I. New functions for genes II and X. 1988. *J Mol Biol,* **203**(1), 49–62.

39. Bauer, M., and G.P. Smith, Filamentous phage morphogenetic signal sequence and orientation of DNA in the virion and gene-V protein complex. 1988. *Virology,* **167**(1), 166–175.

40. Endemann, H., and P. Model, Location of filamentous phage minor coat proteins in phage and in infected cells. 1995. *J Mol Biol,* **250**(4), 496–506.

41. Horabin, J.I., and R.E. Webster, An amino acid sequence which directs membrane insertion causes loss of membrane potential. 1988. *J Biol Chem,* **263**(23), 11575–11583.

42. Brissette, J.L., and M. Russel, Secretion and membrane integration of a filamentous phage-encoded morphogenetic protein. 1990. *J Mol Biol,* **211**(3), 565–580.

43. Feng, J.N., P. Model, and M. Russel, A trans-envelope protein complex needed for filamentous phage assembly and export. 1999. *Mol Microbiol,* **34**(4), 745–755.

44. Russel, M., and P. Model, Genetic analysis of the filamentous bacteriophage packaging signal and of the proteins that interact with it. 1989. *J Virol,* **63**(8), 3284–3295.

45. Jespers, L.S., et al. Surface expression and ligand-based selection of cDNAs fused to filamentous phage gene VI. 1995. *Biotechnology (N Y),* **13**(4), 378–382.

46. Gao, C., et al. Making artificial antibodies: a format for phage display of combinatorial heterodimeric arrays. 1999. *Proc Natl Acad Sci U S A,* **96**(11), 6025–6030.

47. Fuh, G., and S.S. Sidhu, Efficient phage display of polypeptides fused to the carboxy-terminus of the M13 gene-3 minor coat protein. 2000. *FEBS Lett,* **480**(2–3), 231–234.

48. Nakayama, G.R., et al. Improving the copy numbers of antibody fragments expressed on the major coat protein of bacteriophage M13. 1996. *Immunotechnology,* **2**(3), 197–207.

49. Sidhu, S.S., G.A. Weiss, and J.A. Wells, High copy display of large proteins on phage for functional selections. 2000. *J Mol Biol,* **296**(2), 487–495.

50. Jestin, J.L., G. Volioti, and G. Winter, Improving the display of proteins on filamentous phage. 2001. *Res Microbiol,* **152**(2), 187–191.

51. Zeri, A.C., et al. Structure of the coat protein in fd filamentous bacteriophage particles determined by sol id-state NMR spectroscopy. 2003. *Proc Natl Acad Sci U S A,* **100**(11), 6458–6463.

52. Jacobsson, K., and L. Frykberg, Shotgun phage display cloning. 2001. *Comb Chem High Throughput Screen,* **4**(2), 135–143.

53. Djojonegoro, B.M., M.J. Benedik, and R.C. Willson, Bacteriophage surface display of an immunoglobulin-binding domain of Staphylococcus aureus protein A. 1994. *Biotechnology (N Y),* **12**(2), 169–172.

54. Rudgers, G.W., and T. Palzkill, Protein minimization by random fragmentation and selection. 2001. *Protein Eng,* **14**(7), 487–492.

55. Beekwilder, J., et al. A phagemid vector using the *E. coli* phage shock promoter facilitates phage display of toxic proteins. 1999. *Gene,* **228**(1–2), 23–31.
56. Daniels, D.A., and D.P. Lane, Phage Peptide Libraries. 1996. *Methods,* **9**(3), 494–507.
57. Cabilly, S., The basic structure of filamentous phage and its use in the display of combinatorial peptide libraries. 1999. *Mol Biotechnol,* **12**(2), 143–148.
58. Rodi, D.J., and L. Makowski, Phage-display technology–finding a needle in a vast molecular haystack. 1999. *Curr Opin Biotechnol,* **10**(1), 87–93.
59. Hoogenboom, H.R., et al. Antibody phage display technology and its applications. 1998. *Immunotechnology,* **4**(1), 1–20.
60. Griffiths, A.D., and A.R. Duncan, Strategies for selection of antibodies by phage display. 1998. *Curr Opin Biotechnol,* **9**(1), 102–108.
61. Buchli, P.J., Z. Wu, and T.L. Ciardelli, The functional display of interleukin-2 on filamentous phage. 1997. *Arch Biochem Biophys,* **339**(1), 79–84.
62. Scarselli, E., G. Esposito, and C. Traboni, Receptor phage. Display of functional domains of the human high affinity IgE receptor on the M13 phage surface. 1993. *FEBS Lett,* **329**(1–2), 223–226.
63. Forrer, P., S. Jung, and A. Pluckthun, Beyond binding: using phage display to select for structure, folding and enzymatic activity in proteins. 1999. *Curr Opin Struct Biol,* **9**(4), 514–520.
64. Huang, W., J. Petrosino, and T. Palzkill, Display of functional beta-lactamase inhibitory protein on the surface of M13 bacteriophage. 1998. *Antimicrob Agents Chemother,* **42**(11), 2893–2897.
65. Fujii, I., et al., Evolving catalytic antibodies in a phage-displayed combinatorial library. 1998. *Nat Biotechnol,* **16**(5), 463–467.
66. Rebar, E.J., and C.O. Pabo, Zinc finger phage: affinity selection of fingers with new DNA-binding specificities. 1994. *Science,* **263**(5147), 671–673.
67. Choo, Y., et al. Promoter-specific activation of gene expression directed by bacteriophage-selected zinc fingers. 1997. *J Mol Biol,* **273**(3), 525–532.
68. Berdichevsky, Y., et al. Phage display of a cellulose binding domain from Clostridium thermocellum and its application as a tool for antibody engineering. 1999. *J Immunol Methods,* **228**(1–2), 151–162.

69. Hufton, S.E., et al. Phage display of cDNA repertoires: the pVI display system and its applications for the selection of immunogenic ligands. 1999. *J Immunol Methods,* **231**(1–2), 39–51.

70. Wilson, D.R., and B.B. Finlay, Phage display: applications, innovations, and issues in phage and host biology. 1998. *Can J Microbiol,* **44**(4), 313–329.

71. Corey, D.R., et al. Trypsin display on the surface of bacteriophage. 1993. *Gene,* **128**(1), 129–134.

72. Wang, L., et al. Cloning of anti-Gal Fabs from combinatorial phage display libraries: structural analysis and comparison of Fab expression in pComb3H and pComb8 phage. 1997. *Mol Immunol,* **34**(8–9), 609–618.

73. Markland, W., et al. Design, construction and function of a multicopy display vector using fusions to the major coat protein of bacteriophage M13. 1991. *Gene,* **109**(1), 13–19.

74. Houshmand, H., G. Froman, and G. Magnusson, Use of bacteriophage T7 displayed peptides for determination of monoclonal antibody specificity and biosensor analysis of the binding reaction. 1999. *Anal Biochem,* **268**(2), 363–370.

75. Yamamoto, M., Y. Kominato, and F. Yamamoto, Phage display cDNA cloning of protein with carbohydrate affinity. 1999. *Biochem Biophys Res Commun,* **255**(2), 194–199.

76. Kemp, P., L.R. Garcia, and I.J. Molineux, Changes in bacteriophage T7 virion structure at the initiation of infection. 2005. *Virology,* **340**(2), 307–317.

77. Molineux, I.J., No syringes please, ejection of phage T7 DNA from the virion is enzyme driven. 2001. *Mol Microbiol,* **40**(1), 1–8.

78. Stroud, R.M., P. Serwer, and M.J. Ross, Assembly of bacteriophage T7. Dimensions of the bacteriophage and its capsids. 1981. *Biophys J.* **36**(3), 743–57.

79. Ronto, G., et al. Structure of bacteriophage T7. Small-angle X-ray and neutron scattering study. 1983. *Biophys J,* **43**(3), 309–314.

80. Condreay, J.P., S.E. Wright, and I.J. Molineux, Nucleotide sequence and complementation studies of the gene 10 region of bacteriophage T3. 1989. *J Mol Biol,* **207**(3), 555–561.

81. Condron, B.G., J.F. Atkins, and R.F. Gesteland, Frameshifting in gene 10 of bacteriophage T7. 1991. *J Bacteriol,* **173**(21), 6998–7003.

82. Condron, B.G., R.F. Gesteland, and J.F. Atkins, An analysis of sequences stimulating frameshifting in the decoding of gene 10 of bacteriophage T7. 1991. *Nucleic Acids Res,* **19**(20), 5607–5612.
83. Cerritelli, M.E., and F.W. Studier, Purification and characterization of T7 head-tail connectors expressed from the cloned gene. 1996. *J Mol Biol,* **258**(2), 299–307.
84. Kocsis, E., et al. Improved methods for determination of rotational symmetries in macromolecules. 1995. *Ultramicroscopy,* **60**(2), 219–228.
85. Valpuesta, J.M., et al. Structural analysis of the bacteriophage T3 head-to-tail connector. 2000. *J Struct Biol,* **131**(2), 146–155.
86. Serwer, P., et al. The conformation of packaged bacteriophage T7 DNA: informative images of negatively stained T7. 1997. *J Struct Biol,* **120**(1), 32–43.
87. Cerritelli, M.E., et al. A second symmetry mismatch at the portal vertex of bacteriophage T7: 8-fold symmetry in the procapsid core. 2003. *J Mol Biol,* **327**(1), 1–6.
88. Cerritelli, M.E., et al. Encapsidated conformation of bacteriophage T7 DNA. 1997. *Cell,* **91**(2), 271–280.
89. Steven, A.C., et al. Molecular substructure of a viral receptor-recognition protein. The gp17 tail-fiber of bacteriophage T7. 1988. *J Mol Biol,* **200**(2), 351–365.
90. Kato, H., H. Fujisawa, and T. Minagawa, Subunit arrangement of the tail fiber of bacteriophage T3. 1986. *Virology,* 153(1), 80–86.
91. Kruger, D.H., and C. Schroeder, Bacteriophage T3 and bacteriophage T7 virus-host cell interactions. 1981. *Microbiol Rev,* **45**(1), 9–51.
92. Serwer, P., R.H. Watson, and S.J. Hayes, Detection and characterization of agarose-binding, capsid-like particles produced during assembly of a bacteriophage T7 procapsid. 1982. *J Virol,* **42**(2), 583–594.
93. Cerritelli, M.E., and F.W. Studier, Assembly of T7 capsids from independently expressed and purified head protein and scaffolding protein. 1996. *J Mol Biol,* **258**(2), 286–298.
94. Heineman, R.H., I.J. Molineux, and J.J. Bull, Evolutionary robustness of an optimal phenotype: re-evolution of lysis in a bacteriophage deleted for its lysin gene. 2005. *J Mol Evol,* **61**(2), 181–191.
95. Zhang, X., and F.W. Studier, Mechanism of inhibition of bacteriophage T7 RNA polymerase by T7 lysozyme. 1997. *J Mol Biol,* **269**(1), 10–27.

96. Young, R., Bacteriophage lysis: mechanism and regulation. 1992. *Microbiol Rev,* **56**(3), 430–481.

97. Castillo, J., B. Goodson, and J. Winter, T7 displayed peptides as targets for selecting peptide specific scFvs from M13 scFv display libraries. 2001. *J Immunol Methods,* **257**(1–2), 117–122.

98. Sokoloff, A.V., et al. The interactions of peptides with the innate immune system studied with use of T7 phage peptide display. 2000. *Mol Ther,* **2**(2), 131–139.

99. Hansen, M.H., B. Ostenstad, and M. Sioud, Identification of immunogenic antigens using a phage-displayed cDNA library from an invasive ductal breast carcinoma tumour. 2001. *Int J Oncol,* **19**(6), 1303–1309.

100. Savinov, S.N., and D.J. Austin, The cloning of human genes using cDNA phage display and small-molecule chemical probes. 2001. *Comb Chem High Throughput Screen,* **4**(7), 593–597.

101. Sche, P.P., et al. Corrigendum to: "Display cloning: functional identification of natural product receptors using cDNA-phage display" [Chemistry & Biology 6 (1999. 707–716]. 2001. *Chem Biol,* **8**(4), 399–400.

102. Kataoka, K., et al. A set of Hox proteins interact with the Maf oncoprotein to inhibit its DNA binding, transactivation, and transforming activities. 2001. *J Biol Chem,* **276**(1), 819–826.

103. Danner, S., and J.G. Belasco, T7 phage display: a novel genetic selection system for cloning RNA-binding proteins from cDNA libraries. 2001. *Proc Natl Acad Sci U S A,* **98**(23), 12954–12959.

104. Tsunetsugu-Yokota, Y., et al. Expression of an immunogenic region of HIV by a filamentous bacteriophage vector. 1991. *Gene,* **99**(2), 261–265.

105. Wang, L.F., et al. Use of a gene-targeted phage display random epitope library to map an antigenic determinant on the bluetongue virus outer capsid protein VP5. 1995. *J Immunol Methods,* **178**(1), 1–12.

106. Valuev, V.P., et al. ASPD (Artificially Selected Proteins/Peptides Database): a database of proteins and peptides evolved in vitro. 2002. *Nucleic Acids Res,* **30**(1), 200–202.

107. Wang, L.F., and M. Yu, Epitope identification and discovery using phage display libraries: applications in vaccine development and diagnostics. 2004. *Curr Drug Targets,* **5**(1), 1–15.

108. Yip, Y.L., and R.L. Ward, Application of phage display technology to cancer research. 2002. *Curr Pharm Biotechnol,* **3**(1), 29–43.

109. de la Cruz, V.F., A.A. Lal, and T.F. McCutchan, Immunogenicity and epitope mapping of foreign sequences via genetically engineered filamentous phage. 1988. *J Biol Chem,* **263**(9), 4318–4322.

110. Greenwood, J., A.E. Willis, and R.N. Perham, Multiple display of foreign peptides on a filamentous bacteriophage. Peptides from Plasmodium falciparum circumsporozoite protein as antigens. 1991. *J Mol Biol,* **220**(4), 821–827.

111. Minenkova, O.O., et al. Design of specific immunogens using filamentous phage as the carrier. 1993. *Gene,* **128**(1), 85–88.

112. di Marzo Veronese, F., et al. Structural mimicry and enhanced immunogenicity of peptide epitopes displayed on filamentous bacteriophage. The V3 loop of HIV-1 gp120. 1994. *J Mol Biol,* **243**(2), 167–172.

113. Bottger, V., et al. Comprehensive epitope analysis of monoclonal anti-proenkephalin antibodies using phage display libraries and synthetic peptides: revelation of antibody fine specificities caused by somatic mutations in the variable region genes. 1995. *J Mol Biol,* **247**(5), 932–946.

114. Wright, R.M., et al. Binding epitope of somatostatin defined by phage-displayed peptide libraries. 1995. *Biotechnology (N Y),* **13**(2), 165–169.

115. Cortese, R., et al., Identification of biologically active peptides using random libraries displayed on phage. 1995. *Curr Opin Biotechnol,* **6**(1), 73–80.

116. Dyson, M.R., and K. Murray, Selection of peptide inhibitors of interactions involved in complex protein assemblies: association of the core and surface antigens of hepatitis B virus. 1995. *Proc Natl Acad Sci U S A,* **92**(6), 2194–2198.

117. Oldenburg, K.R., et al. Peptide ligands for a sugar-binding protein isolated from a random peptide library. 1992. *Proc Natl Acad Sci U S A,* **89**(12), 5393–5397.

118. Daniels, D.A., and D.P. Lane, The characterisation of p53 binding phage isolated from phage peptide display libraries. 1994. *J Mol Biol,* **243**(4), 639–652.

119. Westendorf, J.M., P.N. Rao, and L. Gerace, Cloning of cDNAs for M-phase phosphoproteins recognized by the MPM2 monoclonal antibody and determination of the phosphorylated epitope. 1994. *Proc Natl Acad Sci U S A,* **91**(2), 714–718.

120. Smith, M.M., L. Shi, and M. Navre, Rapid identification of highly active and selective substrates for stromelysin and matrilysin using

bacteriophage peptide display libraries. 1995. *J Biol Chem,* **270**(12), 6440–6449.

121. Dantas-Barbosa, C., M.M. Brigido, and A.Q. Maranhao, Construction of a human Fab phage display library from antibody repertoires of osteosarcoma patients. 2005. *Genet Mol Res,* **4**(2), 126–140.

122. Maranhao, A.Q., and M.M. Brigido, Expression of anti-Z-DNA single chain antibody variable fragment on the filamentous phage surface. 2000. *Braz J Med Biol Res,* **33**(5), 569–579.

123. McCafferty, J., et al. Phage antibodies: filamentous phage displaying antibody variable domains. 1990. *Nature,* **348**(6301), 552–554

124. Barbas, C.F., 3rd, et al. Assembly of combinatorial antibody libraries on phage surfaces: the gene III site. 1991. *Proc Natl Acad Sci U S A,* **88**(18), 7978–7982.

125. Clackson, T., et al. Making antibody fragments using phage display libraries. 1991. *Nature,* **352**(6336), 624–628.

126. Roberts, B.L., et al. Directed evolution of a protein: selection of potent neutrophil elastase inhibitors displayed on M13 fusion phage. 1992. *Proc Natl Acad Sci U S A,* **89**(6), 2429–2433.

127. Malik, P., and R.N. Perham, Simultaneous display of different peptides on the surface of filamentous bacteriophage. 1997. *Nucleic Acids Res,* **25**(4), 915–916.

128. Atias, D., Lobel, L., Virta, M., and Marks, R.S., Phage-displayed epitopes as bioreceptors for biosensors, in *Handbook of Biosensors and Biochips* (2007), D.C.C. Robert S. Marks, Isao Karube, Christopher R. Lowe and Howard H. Weetall, Editors. John Wiley and Sons. 115–131.

129. Herrmann, S.L., B. Landes, S. Rager-Zisman, B and Marks, R.S., Chemiluminescent optical fiber immunosensor for the detection of anti-West Nile virus IgG. 2005. *Talanta,* **66**, 6–14.

130. Ramanujam, P., et al. Pathotyping of Newcastle disease virus with a filamentous bacteriophage. 2004. *Biotechniques,* **36**(2), 296–300, 302.

131. Lee, T.C., et al. Detection of virulent Newcastle disease virus using a phage-capturing dot blot assay. 2006. *J Virol Methods,* **136**(1–2), 224–229.

132. Sano, T., C.L. Smith, and C.R. Cantor, Immuno-PCR: very sensitive antigen detection by means of specific antibody-DNA conjugates. 1992. *Science,* **258**(5079), 120–122.

133. Guo, Y.C., et al. Phage display mediated immuno-PCR. 2006. *Nucleic Acids Res,* **34**(8), e62.

134. Thevenot, D.R., et al. Electrochemical biosensors: recommended definitions and classification. 2001. *Biosens Bioelectron*, **16**(1–2), 121–131.

135. Ionescu, R.E., et al. Amperometric immunosensor for the detection of anti-West Nile virus IgG. 2007. *Anal Chem*, **79**(22), 8662–8668.

Chapter 15

Nanolithography and Biochips' Role in Viral Detection

Inbal Tsarfati-BarAd and Levi A. Gheber

Department of Biotechnology Engineering, Ben-Gurion University of the Negev, Beer-Sheva, Israel

Inbaltsa@bgu.ac.il, glevi@bgu.ac.il

15.1 The Need for Portable Biochips for Viral Detection

A viral outbreak may cause heavy repercussions to human life as well as the economy and the environment. For this reason health services rely on fast diagnosis and disease surveillance to minimize consequences of such outbreaks.[1] Whether the epidemic is due to natural causes or a bioterrorism act, a main issue in controlling viral outbreaks is the ability to detect and identify the virus as early as possible. Only after the type and location of the virus is known, proper measures can be taken in order to contain the outbreak.[2] However, detection methods are currently expensive and time consuming and require complex laboratory equipment, for example, polymerase chain reaction (PCR) and enzyme-linked immunosorbent assay (ELISA). There is therefore a pressing need for deployable, *on-site* detection devices able to simultaneously

Viral Diagnostics: Advances and Applications
Edited by Robert S. Marks, Leslie Lobel, and Amadou Alpha Sall

ISBN 978-981-4364-43-0 (Hardcover), 978-981-4364-44-7 (eBook)
www.panstanford.com

screen a number of candidates and quickly identify and quantify the threat. Various degrees of deployment are conceivable, such as handled, point-of-care devices, monitoring devices for high-exposure institutions such as hospitals, or on a battlefield.[3] One of the most appealing approaches to high-throughput parallel screening is the arrayed biosensor, or *biochip*. The current microarray technology, however, produces large biochips which require heavy and expensive separate readout systems (scanners) and thus are used predominantly in large facilities such as research institutes and hospitals. Miniaturization of current microarrays by orders of magnitude is the first step in achieving portable, deployable, arrayed biosensors.

15.2 Arrayed Biosensors: Biochips

A biosensor array (biochip) is a device enabling the detection of target molecules in a sample. The target molecules may be of a biological nature, and/or the detection method is based on a biological capture molecule. It consists of an array of biological test sites, immobilized to a solid support. The test sites are dots of various biological molecules (termed "spots"), such as DNA, protein, and antibodies. These molecules serve as selective capture agents. Upon application of a solution on the surface, the target molecule will bind to the capture molecule if it is present in the solution.

In a direct assay, the target molecule itself is (nonspecifically) labeled. In an indirect assay, a secondary labeled specific detection molecule is subsequently applied. In both cases, the binding of the target protein to a certain spot is indicated by the appearance of a signal (typically a fluorescent signal). Spots consisting of various molecules on the same surface enable multiple tests simultaneously, thereby improving throughput and specificity.[4] This technology has launched a considerable advancement in the field of both proteomics and genomics research, as well as medical diagnostics.[5]

15.3 The Need for Miniaturization

Current microarray technology produces spots $\sim$100 μm in diameter, with a spacing of 300–400 μm. Reading of the assay results are based mostly on fluorescence; thus detection is done using a microscope objective with a typical viewfield ~1.5 mm in diameter. The spot size and spacing lead to imaging of only ~9 spots simultaneously in one such viewfield; therefore, the microarray is rasterscanned in front of the objective to allow collection of the whole array image. Scanning is performed over several centimeters in the horizontal directions. However, the *Z* (vertical) direction must be controlled with micrometer precision in order to avoid defocusing of the sample. To achieve this precision, large and complex machines—scanners—are employed, which use highprecision robotic components. This fact makes scanners heavy, large, and expensive and thus not portable. Miniaturization of the active spots by two or three orders of magnitude would allow the observation of an array consisting of hundreds of spots in one view field, thus avoiding the need for large scanning machines. This alone can reduce considerably the biochip's basic investment and constitutes a big step toward portable chips.

Nanoarrays will furthermore reduce the amount of biomaterials used as capture molecules (which constitute a disproportionately large component of a biochip's cost) and thus reduce the biochip's production price. Moreover, since the reactions take place in small volumes, mass (and heat) transfer durations are dramatically reduced, leading to shortening of the incubation. This will lead to more rapid diagnostics.[4]

Due to these considerations, there is a search for the most practical way to fabricate nanostructures of biological molecules, using several nanolithography techniques explained later

15.4 Nanolithography

Lithography can be classified in a number of ways:

1. **Positive/negative**: Positive lithography involves deposition of materials that accumulate on the surface. Negative lithography is the removal of material or the creating of depressions in the surface.
2. **Direct/indirect**: Indirect lithography is a multistep procedure, in which patterns are first inscribed on the surface and the

structures are fabricated after a number of subsequent steps, according to the "blueprint" previously inscribed. With direct lithography, the features are directly fabricated on to the surface in the desired locations.

Nanolithography is the application of these approaches, resulting in structures with typical dimensions of nanometers. All combinations between (1) and (2) are possible, and the choice depends on the type of desired structures, their composition, size, and precision of fabrication and the ultimate ease of production and cost.

These techniques can be used either to chemically modify the chip's surface in preparation for the molecules' immobilization (indirect) or to print the capture molecules themselves (direct). They can be used to fabricate libraries of different viruses and viruses' constituents or capture molecules recognizing these. These libraries will constitute a major source of knowledge and can promote a big step in biomedical research.[5]

One example is the use of biologically active virus particles that can be used in the monitoring of singlecell infectivity, for example, how the number of virus particles, their orientation, and their chemical immobilization and presentation affect cellular infectivity for single and small collections of virus particles.[6] Another is directly detecting viral particles by specific capture molecules such as antibodies against specific viral proteins. Obviously the reverse approach is possible: immobilization of viral constituents as capture molecules for detection of antibodies in patient samples.

15.5 SPM-Based Nanolithography Methods

The majority of classical lithography methods, borrowed from microelectronics, are unsuitable for biological materials. Most of them demand vacuum, etching with strong acids, or irradiation with UV light, all of which can be harmful to biomaterials such as DNA and proteins. Thus, nanobiolithography must use a method that operates under biological conditions: atmospheric pressure, room temperature, and moderate pH.[7]

To create nanostructures an accurate and highresolution device is required. The device must be able to operate in ambient conditions and in liquid, thus making the scanning probe microscope (SPM) an excellent candidate. During the last decade a number of SPMbased nanolithography methods were developed. They use various modifications of the atomic force microscope (AFM), to create nanostructures of biological material on a substrate, with nanometer precision (nanobiolithography). The principal variations are listed later

15.5.1 *Nanografting*

A thin molecular layer is formed on the surface, typically a self-assembled monolayer (SAM), which serves as the "resist" thus preventing the spontaneous adsorption of the biomolecules on the surface. In the next step, an AFM probe is used to "shave" molecules from the protective layer, revealing the substrate only in precisely controlled positions. The patterned surface is then incubated with a solution containing the desired molecules, which are adsorbed on the exposed parts of the substrate (see Fig. 15.1A).[8] According to the earlier classification, this is an indirect method. Using this technique, cowpea mosaic virus (CPMV) particles were organized on a surface using nanografting, where a gold substrate was initially coated with athiol protein resist The AFM tip was used to shave lines that were subsequently functionalized with thiol linkers. These lines determined the location for the initial virus cluster formation.[6]

15.5.2 *Dip-Pen Nanolithography*

Similar to the dip pen, which is dipped in ink and subsequently used to write on paper, in dip-pen nanolithography (DPN) an AFM probe tip is dipped into a solution containing the molecules to be printed, and subsequently contacted with the substrate in desired positions, where the molecules are deposited (see Fig. 15.1B). The molecules can then react with the surface either instantly (direct method) or after a number of additional chemical steps.[9] For example, DPN and coordination chemistry (16-mercaptohexadecanoic acid [MHA] and zinc ions) were used to immobilize individual particles of tobacco mosaic virus (TMV) in a large array. This method, which enables control over particle position and orientation, can be generalized for many different viruses with a proper protein coating.[5]

The same chemical method was used for the immobilization of antibodies and fabrication of functional antibody arrays.[10] In a different case, DPN was used to manufacture nanoscale patterns of antibodies on a gold surface. These antibody spots, spaced less than 100 nm, were able to identify human immunodeficiency virus (HIV1) p24 antigen. The detection was demonstrated by applying a solution containing gold-labeled secondary antibody. The presence of the threelayer complex (capture molecule, HIV1 p24 antigen, and secondary gold-labeled antibody) was indicated by a small height increase. This method appeared to be both more sensitive and more selective than the commonly used ELISA by more than a 1000-fold.[11]

15.5.3 *Nano–Fountain Pen*

The nano–fountain pen (NFP) uses a glass or quartz capillary tapered to a sharp tip to form a nanopipette. The tip is then filled with a solution and mounted on an AFM.[12]

Surface tension prevents the liquid from spilling until there is a contact with the surface (see Fig. 15.1C).[13] Similar to the fountain pen, the NFP allows continuous writing, without a need for repeated dipping in ink. For example, the NFP was used to fabricate nanostructures of various proteins[14] or depressions, by delivering proteolytic enzymes to a proteincoated surface.[15,16] Currently the development of a direct immunoassay with a spot radius as small as 200 nm is under development (see Fig. 15.2).[17]

The NFP can also be combined with other fabrication methods. In one example, a focused ion beam (FIB) was used to create depressions in a thin gold film that was later chemically modified with mercaptopropionic acid (MPA). The nanopipette was positioned directly over the depressions and functionalized IgG solution was then delivered. The functional antibodies were detected by confocal fluorescence microscopy.[18]

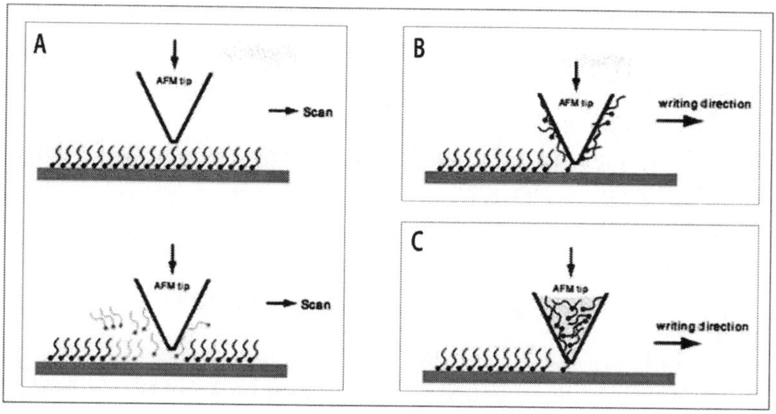

Figure 15.1 Three SPM based lithography methods (A) Nanographting: the shaving of a SAM from the sample's surface using an AFM tip (B) DPN: the AFM probe tip is dipped into a solution, and intermolecular forces cause the adhesion of the solution on the tip. (C) NFP: a glass or quartz capillary in the shape of a sharp tip forms a nanopipette.

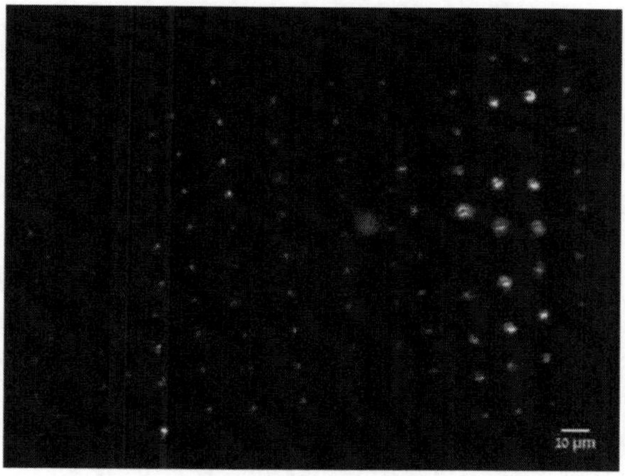

Figure 15.2 Direct immunoassay with spots as small as 200 nm.[17]

15.6 Problems Associated with Miniaturization

While miniaturizing of biochips holds many advantages, a number of problems delay their deployment and wide use.

As the spots' diameter is reduced, the amount of fluorescence emitted decreases proportionally to the area of the spot (i.e., the square of the diameter). A more detailed analysis shows that the decrease in signal emission is stronger than quadratic in the diameter, due to limited binding site density of the immobilization surface. For example, low amounts and uneven distribution of capture agents might lead to the binding of a small number of labeled target molecules. This can cause a poor fluorescence signal relative to the noise created by the surface, making detection of binding impossible. Surfaces that are appropriately designed for nanobiolithography must be developed to overcome this problem. Signal enhancement can be achieved either by increasing the surface density of binding sites or by using surfaces that produce lower fluorescence noise. Tsarfati et al introduced a model enabling the characterization of binding site density at the nanoscale and a comparison between substrates for nanoarray fabrication. This model can be useful in the development of surfaces and selection for manufacturing of nanobiochips.[17]

Another problem is the challenge of scaling up the patterning process. All of the methods above rely on an SPM for control on pattern size and location. The SPM is a slow scanning (serial) technique, which is not expected to reach fast fabrication speeds. Moreover the equipment used in nanolithography, and the need for different probes for different molecules, significantly increases the biochip's price. For mass production of nanobiochips, there is a need for a parallel writing technique, for example, "multipen" writing with "multi-ink" capabilities. A number of attempts have been reported in this direction. For example a nanoscale array of aminereactive dots was prepared using 26-pen DPN. The amine ends were bound to the protein A/G, which can subsequently bind antibodies.[19]

In another case a multi-ink parallel NFP array was created. A pattern of two different inks was delivered in the liquid phase and written using a device with two onchip reservoirs feeding the NFP array with two different inks.[20]

15.7 Conclusions

Although biochip miniaturization has not matured just yet, this approach is a promising precursor for the realization of the lab-on-a-chip concept. The continuing diversion of resources to the development of nanochips can result in various solutions to each of the challenges mentioned earlier. The ability to mass-produce multispot nanoarrays, with the subsequent development of nanofluidics, and miniaturization of the chip's transducers, will hasten the progression toward the portable biosensor vision. There is still much ground work to be done. However, as these lithography techniques become increasingly available to laboratories around the world, we will soon witness an expansion and hopefully an explosion of advancement in this field of research. Biochip fabrication will become more efficient and cost effective and will provide a widespread platform for medical diagnosis as well as viral research.

References

1. Pejcic, B., De Marco, R., Parkinson, G. The role of biosensors in the detection of emerging infectious diseases. *Analyst* 2006, **131**(10), 1079.
2. Baril, L., Need for biosensors in infectous disease epidemiology. In *Handbook of Biosensors and Biochips,* Robert S. Marks, Ed. Chichester: John Wiley, 2007, 1077.
3. Rosi, N. L., Mirkin, C. A. Nanostructures in biodiagnostics. *Chemical Reviews* 2005, **105**(4), 1547.
4. Gheber, L. A., NanoBioLithography of biochips. In *Handbook of Biosensors and Biochips,* Marks, R. S., Cullen, D. C., Karube, I., Lowe, C. R., Weetall, H. H., Eds. Wiley, 2007.
5. Vega, R. A., Maspoch, D., Salaita, K., Mirkin, C. A. Nanoarrays of single virus particles. *Angewandte Chemie (International Edition in English)* 2005, **44** (37), 6013.
6. Cheung, C. L., Chung, S. W., Chatterji, A., Lin, T., Johnson, J. E., Hok, S., Perkins, J., De Yoreo, J. J. Physical controls on directed virus assembly at nanoscale chemical templates. *Journal of the American Chemical Society* 2006, **128**(33), 10801.

7. Gheber, L. A., Nano Fountain Pen: Toward Integrated, Portable, Lab-on-Chip Devices. In *Biological Applications of Microfluidics*, Frank, A. G., Ed. John Wiley & Sons, 2008, 369.
8. Wadu-Mesthrige, K., Xu, S., Amro, N. A., Liu, G. Y. Fabrication and imaging of nanometer-sized protein patterns. *Langmuir* 1999, **15**(25), 8580.
9. Piner, R. D., Zhu, J., Xu, F., Hong, S., Mirkin, C. A. "Dip-Pen" nanolithography. *Science (New York, N.Y.)* 1999, **283**(5402), 661.
10. Vega, R. A., Maspoch, D., Shen, C. K. F., Kakkassery, J. J., Chen, B. J., Lamb, R. A., Mirkin, C. A. Functional antibody arrays through metal ion-affinity templates. *Chembiochem* 2006, **7**(11), 1653.
11. Lee, K. B., Kim, E. Y., Mirkin, C. A., Wolinsky, S. M. The use of nanoarrays for highly sensitive and selective detection of human immunodeficiency virus type 1 in plasma. *Nano Letters* 2004, **4**(10), 1869.
12. Hong, M. H., Kim, K. H., Bae, J., Jhe, W. Scanning nanolithography using a material-filled nanopipette. *Applied Physics Letters* 2000, **77**(16), 2604.
13. Lewis, A., Kheifetz, Y., Shambrodt, E., Radko, A., Khatchatryan, E., Sukenik, C. Fountain pen nanochemistry: Atomic force control of chrome etching. *Applied Physics Letters* 1999, **75**, 2689.
14. Taha, H., Marks, R. S., Gheber, L. A., Rousso, I., Newman, J., Sukenik, C., Lewis, A. Protein printing with an atomic force sensing nanofountain-pen. *Applied Physics Letters* 2003, **83**(5), 1041.
15. Ionescu, R. E., Marks, R. S., Gheber, L. A. Nanolithography using protease etching of protein surfaces. *Nano Letters* 2003, **3**(12), 1639.
16. Ionescu, R. E., Marks, R. S., Gheber, L. A. Manufacturing of nanochannels with controlled dimensions using protease nanolithography. *Nano Letters* 2005, **5**(5), 821.
17. Tsarfati, I., Gheber, L. A., Miniaturization of immunoarrays into nano arrays using nano fountain pen. In preparation.
18. Bruckbauer, A., Zhou, D., Kang, D. J., Korchev, Y. E., Abell, C., Klenerman, D. An addressable antibody nanoarray produced on a nanostructured surface. *Journal of the American Chemical Society* 2004, **126**(21), 6508.
19. Lee, S. W., Oh, B. K., Sanedrin, R. G., Salaita, K., Fujigaya, T., Mirkin, C. A. Biologically active protein nanoarrays generated using parallel dip-pen nanolithography. *Advanced Materials* 2006, **18**(9), 1133.
20. Moldovan, N., Kim, K. H., Espinosa, H. D. A multiink linear array of nanofountain probes. *Journal of Micromechanics and Microengineering* 2006, **16**(10), 1935.

Chapter 16

Optical Fiber Immunosensors and Genosensors for the Detection of Viruses

Yael Liebes[a] **and Robert S. Marks**[a,b,c]
[a] *Department of Biotechnology Engineering, Ben-Gurion University of the Negev, Beer-Sheva, Israel*
[b] *The National Institute of Biotechnology in the Negev, Ben-Gurion University of the Negev, Beer-Sheva, Israel*
[c] *The Ilse Katz Center for Meso and Nanoscale Science and Technology, Ben-Gurion University of the Negev, Beer-Sheva, Israel*
libesy@gmail.com, rsmarks@bgu.ac.il

16.1 Issues in Biothreat Detection

Biological warfare is the intentional use of infectious microorganisms and toxins to disperse disease and death in humans, livestock, and crops.[1,2] The relative ease of attaining bioweapons[1] and recent bioterrorist events[3] have emphasized the immediate need to detect and identify biothreat agents.[4]

The detection platform for a bioterrorism event has to fulfill many demands. It should be rapid, accurate, and reliable; made directly from a complex sample with no false results; and have high reproducibility. Furthermore, the instrument should be portable,

Viral Diagnostics: Advances and Applications
Edited by Robert S. Marks, Leslie Lobel, and Amadou Alpha Sall

ISBN 978-981-4364-43-0 (Hardcover), 978-981-4364-44-7 (eBook)
www.panstanford.com

user friendly, and capable of simultaneously testing multiple agents. These requirements encourage the development of novel diagnostics and set numerous challenges to conventional laboratory methods, which are often time consuming, require extensive training, and sometimes are not sensitive enough.[5] Various tests have been developed to detect and identify biothreat agents, and although many of them claim to fulfill those needs, only a few have been extensively evaluated under field conditions.[3] Rapid and sensitive detection of the viral agent will determine the success of strategies for disease zoning, control, and eradication, even before symptoms are apparent.[6]

The absence of a suitable detection platform has led to the development of biosensors in biodefense. Biosensors,[7] as defined before by the International Union of Pure and Applied Chemistry (IUPAC),[8] seem most promising, providing real-time, qualitative, quantitative, and kinetic information with high sensitivity, low cost, minimal power requirements, microfabrication, and miniaturization potential.[2] These rapid and specific methods involve biomolecular recognition (antibodies and oligonucleotides), which can be incorporated into biosensors, and, in the case of optical biosensors, provide an optoelectronic signal right after the recognition event occurs.[9]

Biosensors, which comprise a biological system intimately connected to a physicochemical transducer, can be divided into groups according to those elements (see Fig. 16.1). Developments in the fields of electronics, transducers, and chemical immobilization in biosensors are being applied increasingly for clinical applications and environmental monitoring.[10]

16.2 Optical Fibers as Optical Transducers: Why Optical Fibers to Begin With?

The transducer element of the biosensor is used to convey the biological recognition step into a measurable signal that can be coupled to a microprocessor for control and display. In the particular case of an optical transducer, the measurable signal is light emission. Fiber optic biosensors are analytical devices in which the fiber optic

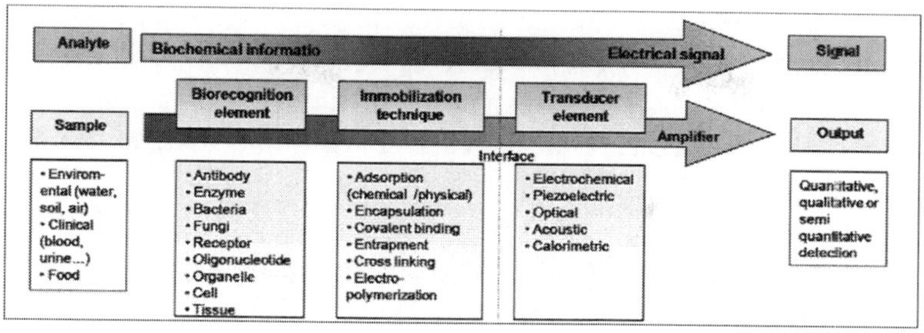

Figure 16.1 Biosensor's elements. The different elements comprising a biosensor are designed according to the aim and need. The recognition element (the bioreceptor) interacts with the analyte and responds in a manner that is detected by the transducer and followed by amplification, with the output recorded.

serves as the transduction element as well as a solid phase for the immobilization of the bioactive components at the end face, hence achieving biorecognition potential. Thus, a signal is produced in correlation to the analyte concentration, and both qualitative and quantitative information is thus obtained.[10,11]

16.2.1 *Optical Fibers: Pros and Cons*

Optical fiber sensors that exploit biomolecular interactions are ideal transducers having the following advantages: (1) geometric convenience, flexibility, and robustness with high tensile strength, enabling sensing in otherwise nonaccessible environments; (2) low production cost; (3) inertness (thus being nonhazardous, safe, and biocompatible to the human body) in in vivo applications; (4) dielectric, freedom from signal interference, and protection against atmospheric disturbance; (5) small volume (therefore with economical use of reagents and enabling of portability with miniaturization potential); (6) silica composition enabling macromolecular immobilization via silanization and solid-phase characterization to the analyte; (7) high-efficiency coupling in the blue region of light, ideal for chemiluminescence; (8) optical multiplicity; (9) amenability to mass production; (10) polyvalence as an optrode system as it is easily adaptable from various biological systems; (11) transmission of optical signal with minimal intensity loss; (12) potential for greater information transition than electrical transducers; (13) environmentally rugged with high tolerance to harsh conditions; (14) fast response—light propagation within the fiber is two-thirds that of the light speed in vacuum; and (15) ability to probe surfaces and films in a nondestructive manner.[10,12–14]

Optical fibers, however, have some drawbacks that need to be taken into account when designing the biosensor: (1) appropriate reagents, (2) background caused by ambient light interference, (3) limited dynamic range compared to electronic sensors, (4) reactions involving mass transfer slowing response time, (5) immobilization chemistry not always easy to accomplish, (6) limited stability of the immobilized receptor, and (7) fiber-to-fiber differences causing great difficulty in normalizing the signal from one to another.[15,16]

16.2.2 *The Basic Physics behind Optical Fiber Operation*

When light (usually monochromatic) is passed through a sample, several actions can transpire. The light can either be reflected back or be transmitted through the sample. The process that occurs will depend on the wavelength of the light, the angle of incidence, the sample concentration, and the sample thickness. The exploited energy from the electromagnetic spectrum can be used to provide information about the changes in the local environment surrounding the analyte. In practice, optical fibers, governed by Snell's law and total internal reflection (TIR), can be coupled with all optical techniques, thus increasing their versatility and suitability as transducers to analytical biosensors.[6,17,18] A simplified use of optical physics enables the understanding of the physical principles behind optical biosensors.[15]

16.2.2.1 Snell's law and TIR

The discussion in this chapter will be limited to multimode step index fibers, a common type in biosensors. Also, it will be characterized only by geometrical optics with no consideration to the electromagnetic theory, which is valid for single-mode optical fibers.

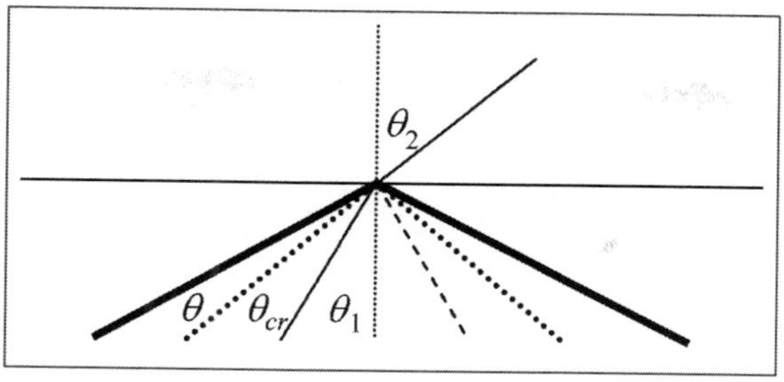

Figure 16.2 Light rays on a high to low refractive index interface.

An optical fiber consists of two concentric, circular, symmetric, and homogenous loss-free dielectrics, with the inner dielectric being the core with an index of refraction n_{core} and the outer dielectric, cladding, with a slightly lower index of refraction n_{clad} and a jacket for protecting the glass from mechanical forces.[19]

An optical fiber is designed to be a lengthwise guide light due to the light propagation via the fiber by TIR, as governed by Snell's law. When a ray of light strikes a boundary of two materials with different indexes of refraction, the ray will be refracted and reflected according to Snell's law (see Eq. 16.1, Fig. 16.2).

$$n_1 \sin\theta_1 = n_2 \sin\theta_2 \tag{16.1}$$

Minimal loss of light will occur when the whole ray will reflect back into the fiber core, that is, the incidence angle will be higher than the critical angle, θ_{cr} (see Eq. 16.2, Fig. 16.2).

$$\sin\theta_{\text{cr}} = \frac{n_2}{n_1} \tag{16.2}$$

A ray strikes at an angle θ_1, it refracts at angle θ_2 (continuous line), and part of it undergoes internal reflection at angle θ_1 (dashes). The limiting case of TIR shows the critical ray at an angle θ_{cr} (dots). TIR is where $\theta > \theta_{\text{cr}}$ (in bold).

Not every radiation that strikes the end face of the optical fiber will enter the fiber's core. Only rays at an angle lower than the acceptance angle (θ_a) will facilitate entrance to the fiber. According to Snell's law, when $\theta_1 = \theta_a$ and the refracted angle into the fiber is θ_2, it can be written (Eq. 16.3):

$$\sin\theta_{\text{a}} = \frac{1}{n_0}(n_1^2 - n_2^2)^{\frac{1}{2}} \tag{16.3}$$

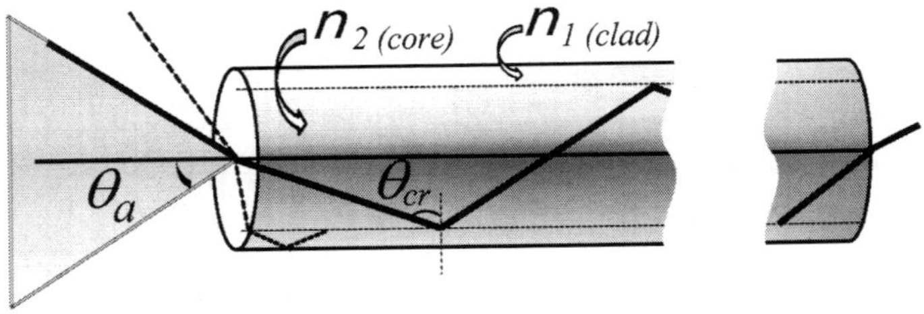

Figure 16.3 Meridional rays of light propagate in an optical fiber.

Another way to describe the acceptance angle is the numerical aperture, NA, as described in Eq. 16.4:

$$\mathrm{NA}\,\theta n_0 \sin = \theta_{\mathrm{a}}(n_1^2 - n_2^2)^{\frac{1}{2}} \tag{16.4}$$

Therefore, a high NA allows a wider angle's range to enter the fiber but has an enlarged dispersion as a consequence; thus, a compromise occurs between a high acceptance angle and refraction.

The actual ray's propagation within the optical fiber is not as simple as described before, and the following analysis will be separated into two cases, meridional rays and skew rays. Meridional rays propagate along the longitudinal axis of the optical fiber, cross the core's center, and represent the simple propagation route. Defects in the core-clad interface will cause significant losses, and thus high-quality optical fibers are desired. The above Eqs. 16.1–16.4 are valid for meridional rays and are illustrated in Fig. 16.3.

The acceptance angle (θ_{a}) and cone (marked as a gray triangle) are shown. Allowed rays, within the acceptance cone (full line) propagate until the end of the optical fiber. Otherwise, rays (dotted line) are eventually lost.

The second type of rays are skew rays, which propagate in the tangent direction, near the peripheral annular of the core, without crossing the fiber core axis (see Fig. 16.4), and thus propagate in a helical shape with a $2 \cdot \gamma$ change of direction each time the ray strikes the core-clad interface. Most light radiation entering the fiber are from these type of rays, dramatically increasing the light intensity. The analysis for the different parameters of the fiber is becoming more complicated, as described in Fig. 16.4.

When a ray strikes at the end face of the optical fiber at an angle of θ_{s}, it refracts at an angle of θ (according to Snell's law)

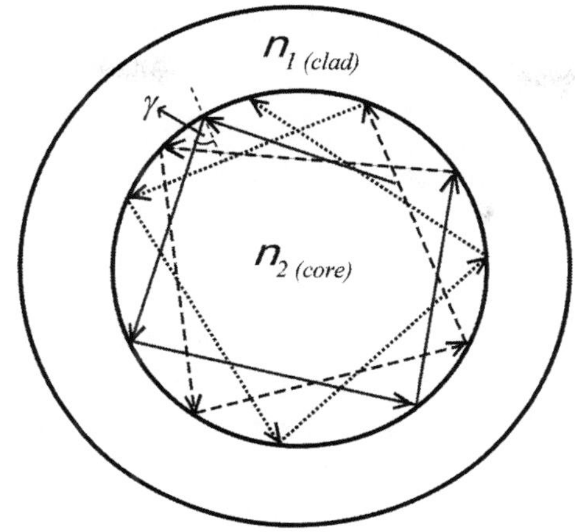

Figure 16.4 Skew rays of light propagation in an optical fiber.

and undergoes TIR at the core-clad interface at an angle of φ. The projection of the refracted ray with the core's radius at the plane of the end face gives an angle of γ. It should be noted that there are two perpendicular planes, the end-face plane and the plane between the core axis and the core-clad interface at the TIR point. Thus Eqs. 16.1–16.4 can be reset to:

$$n_0 \sin\theta_S = n_1 \sin\theta \tag{16.5a}$$

$$\Phi > E = \frac{1}{e}E_0 \tag{16.5b}$$

$$\cos\Phi = \cos\gamma \cdot \sin\theta \tag{16.6}$$

$$NA = (n_1^2 - n_2^2)^{\frac{1}{2}} = n_0 \sin\theta_a \cos\gamma \tag{16.7}$$

Skew rays path down the fiber. Top view: each type of dash represents one round of the helix. When comparing Eq. 16.4 and 16.7, it shows that skew rays are accepted at larger axial angles in a given fiber than meridional rays, since $(n_1^2 - n_2^2)^{\frac{1}{2}}$ is constant for a given fiber. Thus, θ_a is the maximum acceptance angle for meridional rays and minimal for skew rays.[20]

Hence, the main portion of the light power collected by the detector is found within the section of the solid angle defined as:

$$\arcsin\frac{n_\gamma}{n_1}\theta = \theta \arcsin\frac{n_0}{n_{core}} \tag{16.8}$$

16.2.3 *Relevance of Optical Fibers as a Waveguide to Chemiluminescence*

Luminescence is the emission of light, which occurs when a molecule in an excited state relaxes to its ground state. In chemiluminescence, the energy is produced by a chemical reaction with no outer radiation. The chemical reaction in chemiluminescence produces a sufficient amount of energy (approximately 300 kJ/mol for blue light emission and 150 kJ/mol for red light emission) to induce the transition of an electron from its ground state to an excited electronic state.[21] Chemiluminescence has great advantages for immunoassays[22] with superior sensitivity over absorption spectroscopy and over fluorimetry, between 5 and 1,000 orders of magnitude, respectively.[23]

The chemiluminescence reaction is triggered by a biocatalyst marker, horseradish peroxidase (HRP), which provides high specific activity and steady-state kinetics, while being proportional to the label's concentration. This photodynamic protein generates an enhanced chemiluminescence reaction with the appropriate substrate, for example, luminol (see Fig. 16.5). The outcome of such a chemiluminescence reaction is light emission in the visible light region ($\lambda = 430$ nm), ideal for labeling molecules in use for immunoassays coupled to an optical fiber as a waveguide.[17,19,24]

HRP catalyzes, breaking itself into oxygen radicals that oxidize luminol to yield aminophtalic acid. The excited part of the oxidized molecules of luminol emits the quanta of light.

The value of a chemiluminescence signal is estimated according to Eqs. 16.9 and 16.10:[25]

$$\text{signal}_{\text{CL}} = k_1 \cdot k_2 \cdot V \tag{16.9}$$

$$k_1 = k_{\text{chem}} \cdot k_{\text{excited}} \cdot k_{\text{f}} \tag{16.10}$$

where:

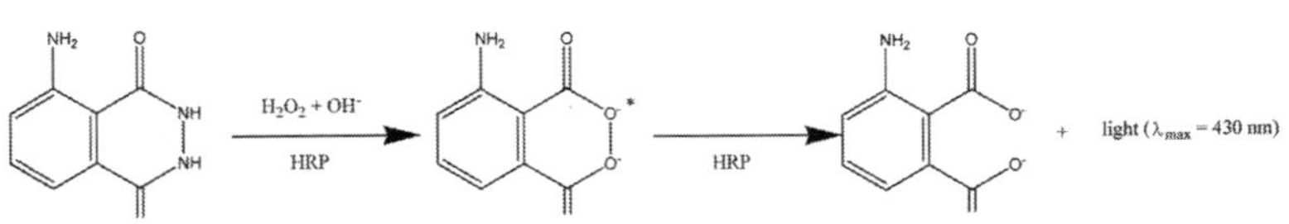

Figure 16.5 Chemiluminescent reaction catalyzed by HRP.

- k_1 is the quantum output of the chemiluminescence (ratio of the number of emitted photons and the number of reacted molecules).
- k_2 is the detection efficiency (ratio of the number of registered photons and the number of emitted photons).
- V is the number of photons registered per unit time.
- k_{chem} is the ratio of the number of luminol molecules reacted to yield light to the total number of molecules precipitated in the reaction.
- k_{excited} is the ratio of the number of molecules in the electron excited state to the number of molecules reacted to yield light.
- k_{f} is the quantum output of fluorescence or chemiluminescence.

16.2.4 *Evanescent Wave Principles Useful in Fluorescence-Based Optical Fiber Sensors*

Light can be described as oscillating radiating electric dipoles placed within a dielectric medium. This causes light propagation not only as rays, as discussed before, but also as waves (usually referred to as the duality of light). Propagation of electromagnetic waves of light, within an optical fiber via TIR, induces an evanescent electromagnetic field in the surrounding dielectric media, because a field cannot be discontinuous at the boundary between core and clad. Thus, there is an exponential decay of the electromagnetic wave power with distance from the surface as detailed in Eqs. 16.11 and 16.12:

$$E = E_0 \exp\left(\frac{-Z}{d_{\text{p}}}\right) \tag{16.11}$$

$$d_{\text{p}} = \frac{\lambda}{2\pi n_{\text{core}}(\sin^2\theta - \sin^2\theta_{\text{cr}})^{\frac{1}{2}}} \tag{16.12}$$

where E_0 is the electric field amplitude at the waveguide–media interface, d_{p} is the penetration depth (in it $E = \frac{1}{e}E_0$), and z is the distance from the surface (see Fig. 16.6B). Usually, the penetration distance is about a few hundred nanometers deep. The penetration depth of the evanescent wave is usually employed for excitation of

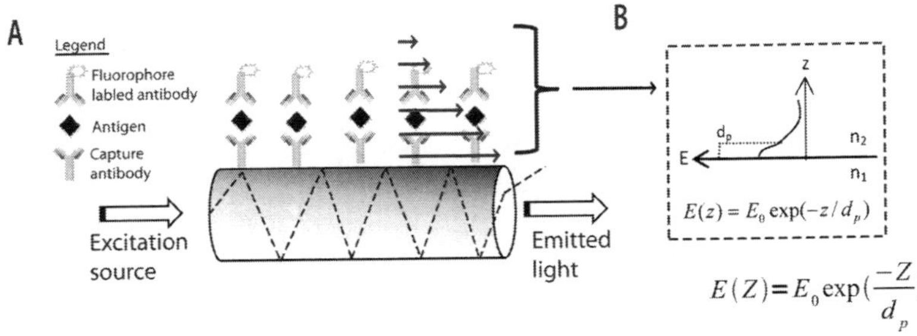

Figure 16.6 Evanescent wave–based biosensor. (A) Demonstration of the induced evanescent wave and a sandwich fluoroimmunoassay. The capture antibody, analyte, and fluorescent-labeled antibodies are immobilized on the fiber's surface as the sandwich fluoroimmunoassay. Light propagation as meridional rays is shown (dashes), and the induced evanescent wave (black line) is sketched. (B) Electrical field amplitude, E, on both sides of the core-clad interface of a waveguide. Zoom onto the core-clad interface. In the lower index medium (n_2, cladding) the electrical field amplitude of the evanescent wave decays exponentially, with a penetration distance, d_p.

fluorophore-labeled secondary antibodies attached to the analyte captured onto the waveguide surface.

In brief, a fluorescent probe bound to the surface and within the penetration distance will be excited due to an evanescent wave, and some of the released energy will be coupled back into the fiber. Nonbound fluorophores will encounter a low level of evanescent wave and thus are not effectively excited, and thus this method provides considerable protection from background levels.[5,12]

An evanescent wave biosensor is an ideal system for direct detection of pathogens in complex matrices (even in a turbid one) using a sandwich format of fluoroimmunoassay: a selective receptor immobilized on the optical fiber's surface facilitates "fishing out" the wanted analyte and further introducing of fluorophore antibodies (see Fig. 16.6A).

However, evanescent wave technology suffers from the necessity of an external excitation light source as well as the use of filters.[19]

Many immunosensors[13,26–30] were designed on the principle of evanescence wave and fluorescence, and two commercial products are available on the market, Analyte 2000 and RAPTOR.[5,31,32]

16.3 Bioreceptor Immobilization: Chemical Modification to Optical Fibers

16.3.1 *Immobilization to Solid Supports*

A key advantage of biosensors is the close proximity between the physicochemical interface and the biological components, which both theoretically and practically will increase sensitivity of the assay. Solving the problem of their integration is the challenge.[33]

For a biosensor to be successful, it is necessary to modify the surface of the transducer into a discrete (irreversible immobilization of several receptor units) or a continuous form (characteristics of the surface) to achieve special affinity and orientation with the appropriate selectivity. Immobilization techniques must satisfy the practical use of biosensors, retain substantial biological activity, remain tightly associated with a sensor surface, and allow long-term stability and durability and a high degree of specificity.[6,12,15]

The interactions with the solid phase, that is, the immobilization, can be oriented or random[34] via any one of the following methods: adsorption, entrapment, and covalent binding (see Table 16.1). And they can be in one, two, or three dimensions, which result in different spacing from the solid-phase surface.[35–37] Usually a minimum distance from the solid phase is recommended in order to minimize steric hindrance and help expose the bioreceptors to the liquid phase.[38]

Immobilizing proteins to the solid phase is much more complicated than immobilizing oligonucleotides. The protein itself is more complex than oligonucleotides, with high heterogeneity between different proteins (in means of structure, function, and affinity). There are also complex cases, such as enzymes, with an easily lost configuration and activity when immobilized or stored, because of changes in temperature, humidity, pH, ionic strength, and redox reactions. Also, proteins can lose activity due to postimmobilization storage: the active site, steric hindrance, conformational change, and denaturation. Hence, immobilization methods alter surface properties and the activity of the receptor and should be carefully chosen.[34,36]

Table 16.1 Immobilization techniques on a glass surface. Several common strategies for binding receptors on glass surfaces, including a general binding mechanism and functional groups

Immobilization method	Surface derivation	Mechanism of binding to proteins	Coupling procedure	Remarks	Ref.
Chemical and Physical adsorption	Van der Waal	Electrostatic and VDW forces, hydrophobic bonds, ionic bonds	Incubation	Random orientation, low coverage, and reversibility	39, 40
Direct immobilization	Primary amines	Reactive to primary amines and forming substituted imidocarbonate	Cyanogens bromide activation to the surface followed by incubation with protein incubation	Direct and effective binding	25, 41
Covalent, random	Succinimidyl ester (NHS), epoxy-, aldehyde-, carbodiimide-	Reactive primary amines of lysines, and arginines		Different cross linkers (activators) that bind any amine group of the protein results in non oriented binding	36
Covalent, oriented	Maleamides, disulphide derivatives and arylating agents	Mostly directed to cysteins in the antibodys hinge region	Reduction of disulfide bridges of cysteines	Orientation at the hinge region	36
Covalent, oriented	Amino groups and hydrazines	Carbohydrate residues of Fc region of the antibody	Oxidation of carbohydrates by sodium metaperiodate	Orientation at the carbohydrate moiety region	36, 42
Specific, oriented binding	Protein A, G or L	Affinity binding of amino groups in the Fc region	Incubation	Usually protein A, G or L are bound to prior bifunctional reagent	42

Introduction of specific functional groups	Suitable functional group	Strong and specific interaction between high affinity groups	Incubation	e.g.: avidin-biotin	38, 42
Covalent, partly oriented	Silanization with desired functional group	Covalent binding to functional groups of the protein (usually amine)	Silanization, cross linking and incubation	Many possible techniques	14, 19, 33, 43–48
Covalent	Benzophenone (Bz)	UV illumination causes formation of radical on Bz alkyl, recombination, and abstraction	photoimmobilization	The PMMA fiber optic is electro-polymerized twice: with poly-pyrrole (pPy) and then with pPy-Bz	49
Entrapment and specific functional group	Silanization with $-NH_2$ residual, cross linking with gluteraldehyde and streptavidin conjugation	Polymer matrix	Bacteria encapsulated within biotin-alginate microspheres	Different parameters are restricted by mass transfer and diffusion.	50

Many procedures to immobilize the receptors to the solid phase are reported in the literature, and here we will review the common methods used in the biosensors field.

16.3.2 *Immobilization via Functional Group-Terminated Silane Reagents*

Organosilane compounds with the general structure of X_3-Si-$(CH)_n$-R (X usually is methoxy or ethoxygroup and R is a functional head group).[13] Because the Si–X bond is ionic and long, compared to the Si–C bond, Si–X bond can thus be easily hydrolyzed and form siloxane bonds (Si–O–Si) with hydroxyls from the substrate's surface.[51] This hydrolysis is the driving force for the organization of the silane molecules into a self-assembly monolayer (SAM) over the silica substrate. Creating the SAM layer is essential in order to achieve the homogenous surface for the receptor binding and is very complicated because of the significant influence of humidity, temperature, solvents, substrate, and the silane itself (alkyl chain length, head group, and active group).[52,53] There are many ways to silanize the surface, and the most common procedures to silanize glass optical fibers are liquid- or gas-phase deposition, spraying, and spin coating. There are also many silane reagents with different linkers, hydrolyzable groups, and head groups, so the reagents can be chosen according to needs.[54]

Immobilization via silanization is comprised of three major steps, each of them with many different options: (1) cleaning the surface and surface hydroxyl exposure,[55] (2) silanization, and (3) cross linking of the receptor.[42,47] Table 16.2 details several common silanization procedures in the field of optical fiber biosensors.

16.3.3 *Immobilization via Electrochemical Procedures*

A novel approach of biomolecule immobilization and transduction of the biological process in the field of biosensors is based on electrogenerated polymers. This process has advantages over many conventional immobilization techniques, including adsorption, entrapment, covalent binding, and anchoring by affinity interactions of different biomolecules. Electrochemical procedures enable the

Table 16.2 Common silanization techniques. Different silanization techniques, including the silane reagent, general mechanism of silanization, cross linker, functional groups, and the immobilized receptor

Silane agent (hydrolysable group, active group)	Silanization method	Cross linker	Immobilization technique to receptor	Receptor (binding group for immobilization)	Ref.
(3-mercaptopropyl) trimethoxy silane (methoxy, thiol)	Vapor phase deposition under Ar atmosphere, in ambient atmosphere for 20 min.	6-Maleimidohexanoic acid *N*-succinimidyl ester (EMCS)	Incubation	Antibody ($-NH_2$)	48
				GIPC-1 protein (NH_2)	46
				WNV inactivated virions (NH_2)	44
Chlorodimethylsilane (chloro, hydrogen) Dichloromethylsilane (chloro, hydrogen)	Liquid phase deposition from 1mmole/L silane in toluene with $(C_2H_5)N$ as catalyst at ambient atmosphere for 8 hr.	4-allyloxybenzophe-none	Phtoimmobilization (illumination at 385 nm).	Cholera toxin β-subunit (hydrocarbon of amino acid side chain)	45
3-aminopropyltrimethoxysilane (methoxy, silane)	Liquid phase deposition from pure solution for 1hr, 900°C.	1,3,5-tri-chlorotriazine	Incubation	BSA or *S. pneumoniae* polusaccharides (hydroxyl)	14, 41
3-glycidoxytrimethoxysilane (methoxy, epoxy)	Liquid phase deposition from pure solution for 1hr, 900°C.	None. However, oxidation was done to provide aldehydes	Incubation followed incubation with sodium cyanoborohydride for covalent binding.	Cholera toxin β-subunit (amino resiude)	13, 19

(*Contd.*)

Table 16.2 *(Contd.)*

(3-mercaptopropyl) trimethoxy silane (methoxy, thiol)	Liquid phase deposition from pure solution for 1hr, 700°C.	NHS-EDC chemistry		Incubation	Protein ($-NH_2$)	13, 19
3-aminopropyltriethoxysilane (ethoxy, amino)	Liquid phase deposition from 20% toluene solution for 8 hr at RT while mixing	Carbonyldiimidazole		Incubation with agitation	Anti Newcastle disease virus IgG	29
(3-mercaptopropyl) trimethoxy silane (methoxy, thiol)	Liquid phase deposition from 2% solution in toluene for 2 hr under N_2 atmosphere	N-succinimidyl 4-maleimidobutyrate	–	Incubation with and without avidin followed incubation with receptor	Biotinylated antibody ($-NH_2$ of biotin), antibody ($-NH_2$)	56–58
3-aminopropyltriethoxysilane (ethoxy, amino)	Vapors in vacuum container for 12 hr	gluteraldehyde		Incubation	Antibody or antigen ($-NH_2$)	25
(3-mercaptopropyl) trimethoxy silane (methoxy, thiol)	Liquid phase deposition from 2% solution in toluene	N-succinimidyl 4-maleimidobutyrate	–	Incubation with and without protein A/G followed incubation with receptor	Antibody (residues in the CH_2 and CH_3 domains of Ig heavy chains)	42, 59
(3-mercaptopropyl) trimethoxy silane (methoxy, thiol)	Liquid phase deposition from 2% solution in toluene	none		Carbohydrate's oxidation with sodium metaperiodate and incubation	Antibody (oxidized carbohydrate moiety)	42

Abbreviations: NHS, N-hydroxysuccinimide; EMCS, 6-Maleimidohexanoic acid *N*-succinimidyl ester; BSA, bovine serum albumin.

production of simple and reproducible organic films with easy control over the properties of the polymeric layers, including thickness and morphology.[60]

Indium tin oxide (ITO) has gained popularity in the fields of optics and optoelectronic devices since it is transparent and highly conductive with low resistance. Many procedures offer ITO deposition methods, with sputtering as the most viable technique for high-quality ITO thin-film deposition, including on nonconductive materials such as optical fibers.[61]

After ITO sputtering, electropolymerization of pyrrole derivative monomers is achieved by using the ITO-coated fiber as the working electrode and by applying a voltage using a potentiometer.[62–65] After pyrrole is polymerized, covalent attachment is done by different strategies, depending on the pyrrole derivative and the bioreceptor's functional groups. Three common strategies are summarized in Fig. 16.7.

16.3.4 *Immobilization via an Avidin–Biotin Bridge*

Avidin–biotin interactions were discovered and have been used since 1965, when biotin was bound to cellulose in order to extract avidin, a glycoprotein from egg white.[66] Since then, many improvements and developments have been achieved regarding avidin–biotin interactions and uses.

The avidin–biotin interaction is the strongest-known noncovalent, biological interaction ($K_a = 10^{15}$ M^{-1}) between a protein and a ligand. The bond (Fig. 16.8) is formed randomly and rapidly and, once formed, is unaffected by a wide range of parameters, such as pH, temperature, organic solvents, or other denaturizing agents, and also is resistant to enzymatic proteolysis within the digestive tract. However, detergents decrease the binding strength of avidin to biotinylated molecules. Another important advantage of the biotin-avidin system is the relative ease of conjugation of biotin to a variety of biomolecules via simple and well-described conjugation techniques.[67]

The above advantages explain the high use of avidin-biotin affinity-based immobilization in the field of biosensors, indirect

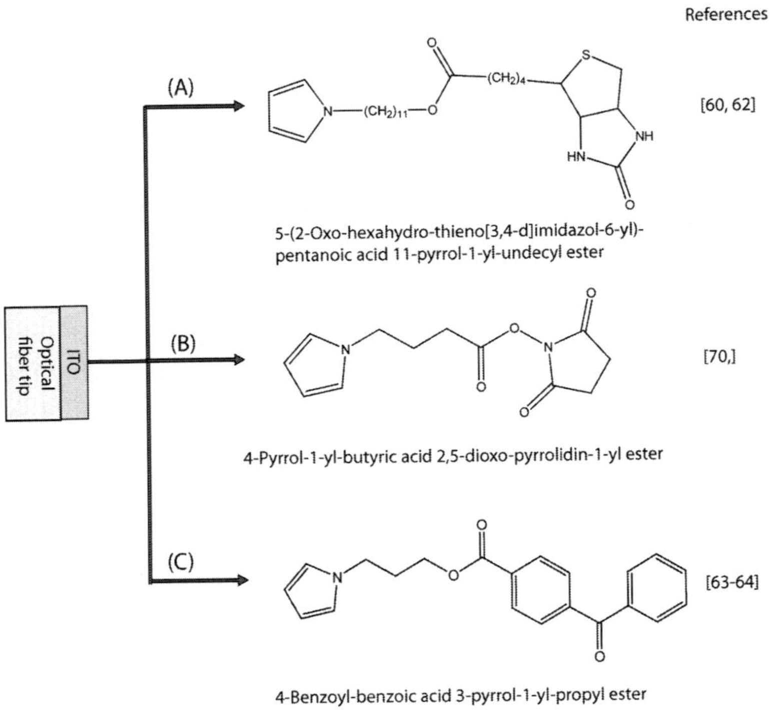

Figure 16.7 Possible strategies for biomolecule immobilization on optical fibers using electropolymerization. (A) poly(pyrrole-biotin), (B) poly(pyrrole-NHS), and (C) poly(pyrrole-benzophenone).

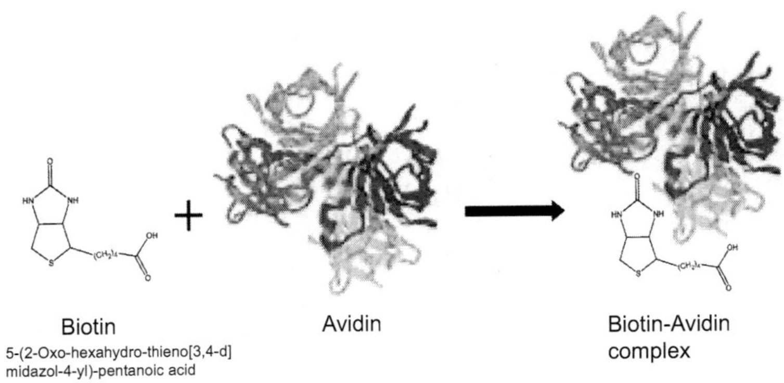

Figure 16.8 Illustration of the biotin structure and interaction with avidin. The avidin is sketched according to its tetrameric glycoprotein structure, having four sites for the high-affinity interaction with biotin.

adsorption, indirect conjugation through silanization (most common), and electropolymerization for both protein and oligonucleotides receptors.

Direct adsorption is one of the simplest methods used for antibody immobilization on avidin-precoated glass optical fibers. This method only includes the incubation of the avidin-coated optical fibers with biotinylated immuoreagents. This avidin–biotin affinity-based immobilization method was used for both biotinylated antibodies in the Analyte 2000 automated biosensor[32] and biotinylated oligonucleotide probes.[68]

Using silanization, avidin is covalently bound to the solid phase through a cross linker, having the advantages of stability and spacing between the receptor and the solid phase, thanks to the spacer arm of the cross linker. Several silane reagents and different cross linkers are used for avidin immobilization, detailed in Table 16.3.

Table 16.3 Common procedures for introducing avidin to a fiber optics surface using silane and a cross linker. General description of the immobilization procedure, including the reagent and special remarks

Silane reagent	Cross linker	Remarks	Ref.
Trimethoxysilyl-propyldiethyl-enetriamine	Sulfosuccinimidyl-6-(biotinamido) hexanoate (NHS-LC-biotin)	Immobilization of molecular beacons conjugated to biotin and fluorescent moiety	30
(3-mercaptopropyl) trimethoxy silane	N-succinimidyl – 4-maleimidobutyrate	Incubation with Immuno-PureNeutrAvidin followed by incubation with biotinylated antibodies or with streptavidin and then biotinylated antibodies	42, 57, 58
(3-mercaptopropyl) trimethoxy silane	Biotin maleamide	Coating with streptavidin and then with biotinylated antibodies	42
3-aminopropyltriethoxysilane	NHS-LC-biotin	Coating with streptavidin followed by incubation with biotinylated oligonucleotides probe	26

The third method of covalent binding of the receptor using avidin-biotin affinity is electropolymerization, as mentioned in the previous section. Basically, electropolymerization is performed at room temperature in a conventional three-electrode cell and a potentiostat. The ITO-coated fiber optic (see Fig. 16.13C, right side) is used as the working electrode with 2mM of the pyrrole monomers in 0.1M $LiClO_4$ in CH_3CN, 10mM Ag/Ag^+ in CH_3CN as the reference electrode, and a platinum wire as a counterelectrode, with controlled potential oxidation about 0.8–0.95 V.[63,64,69]

It is also important to note that biotin–avidin affinity is not only used for immobilization of the receptor but also used for attaching the labeled molecule in the final steps of the immunoassay. Several papers describe the attachment of biotinylated enzyme to an avidin-coated molecule prior to signal capture.[69–71]

16.4 Signal Measurements: State-of-the-Art Photodetectors

The detector is an essential component of the biosensor and is crucial since it influences the overall performance of the biosensor. Its function is to convert the received optical signal (usually referred to as relative light units [RLUs]) into an electrical signal, which is then amplified before further processing. It is important to define requirements for a detector such as (1) sensitivity at the operating wavelengths, (2) high reliability of the received signal, (3) large electrical response to the received optical signal, that is, high gain or quantum efficiency (the percentage of photoelectrons produced per photon, ranging from a low percentage to more than 50%), (4) short response time, (5) minimal noise or background signals, (6) performance stability in different measurement conditions, (7) detector size, and (8) low cost.[20] The current photodetectors incorporate analysis software.[24,72–74]

Several photodetectors are on the market, and some are reviewed here (Fig. 16.9).

Charge-coupled device (CCD) sensors and cameras are best described as semiconductor chips with one face being sensitive

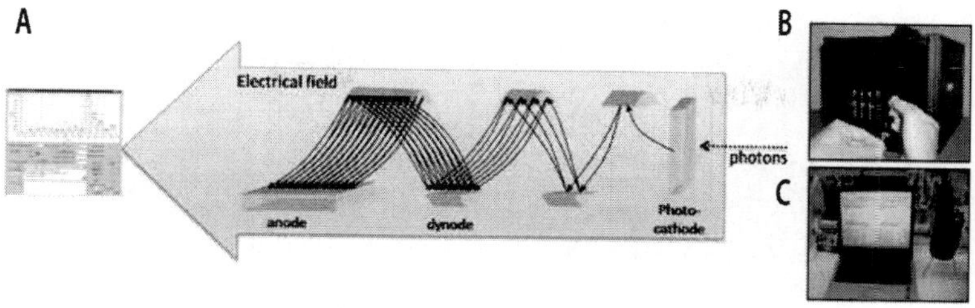

Figure 16.9 Photodetectors used in biosensors. (A) Schematic illustration of the optoelectronic process within the PMT. Photons (dotted line) strike the photocathode, transduced into electrons (black arrows) and multiplied by dynodes in an electrical field till the anode, where a current pulsed is triggered and translated into a signal (credits: Sebastian Herrman). (B) PMT integrated in a homemade black box. (Reprinted with permission)[75] (C) SPAD as designed by Sebastian Herrmann.

to light. The semiconductor chip is rectangular in shape and is subdivided into a grid of discrete rectangular areas—pixels. Upon photon arrival on a pixel, a small electrical charge is generated and stored for later readout, and it increases, cumulatively, as more photons strike the surface. With an amplifier and a fast analog-to-digital converter (ADC), it is possible to evaluate the amount of light that has fallen onto each pixel by examining the amount of charge it has stored. Special software and electronics display the image created from the charge data. The main advantages of CCDs are their high sensitivity, dynamic range, and linearity.[24]

Photomultiplier tubes (PMTs) are efficient and accurate instruments for light detection. A PMT is a glass tube with high vacuum that houses a photocathode, a series of dynodes, and an anode. When a photon strikes the photocathode with energy over a threshold, an electron will be released by the photoelectric effect. This electron has the energy of the photon and is accelerated in the magnetic field, thus having greater energy, and multiplied through the set of dynodes, each having a higher voltage than the previous one. Each dynode releases more low-energy electrons, which again are accelerated in the magnetic field. Finally, the accumulated electrons reach the anode, and a current pulse is triggered. Such devices are very attractive due to their high gain, low noise, high-frequency response, large area of collection, and variety of operatable wavelengths.[72]

Avalanche photodiodes (APDs) are a new, emerging technology. Based on silicon photodiodes with effective areas of a few square millimeters, they have proven to be inexpensive and suitable for in-field measurement due to their small size and battery operation. APDs are based on a reversed biased p-n junction. However, these diodes are not suitable for single-photon-counting purposes.[44]

Single-photon avalanche diodes (SPADs) are used for single-photon detection, consist of a reverse bias on a p-i-n junction, and operate in a nonproportional multiplication mode analogous to a Geiger–Müllerr tube.[73]

16.4.1 *Evolution of Photodetector Instrumentations*

Presently, photon counting using PMTs still seems to be the most cost-effective way to detect very low light levels, and PMTs have quite large photocathode areas. SPAD detectors will surely become attractive as time goes by, as their development proceeds further and the prices decrease. Also, CCD detectors will become relatively low-cost instruments in the future.[24]

A next generation of photodetectors will include not only the photon-counting unit but also the whole biosensor itself integrated with close proximity between the bioassay unit and the detecting unit. The first one will be inexpensive and disposable, while the second will be long lasting with the disposable part fitted to it. Such a device, referred to as the BioPen, is under development by our group and is based on a chemically modified optical fiber with a capture molecule, the receptor, immobilized onto the optical fiber as the disposable part, a PMT, and a microfluidic system containing all the reagents needed for the immunoassay (see Fig. 16.10). The BioPen will be a generic platform of a chemiluminescent optical fiber immunosensor with the integration of the above elements and low-cost mass production. The authors believe such a device can play a role in future field diagnostics.

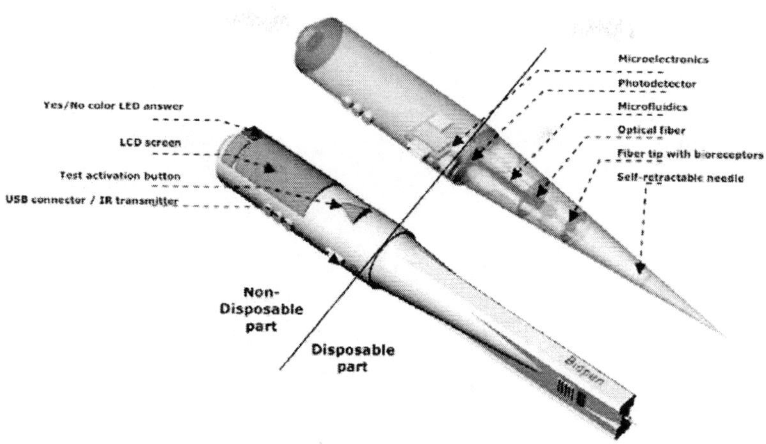

Figure 16.10 BioPen CAD prototype. Visualization of the different elements of the BioPen and major key components. CAD was designed by S. Herrmann. *Abbreviation*: CAD, computer-aided design.

16.5 Fiber Optic Immunosensors Applications for Use in Viral Infections

In the field of diagnostics, especially for biowarfare detection, it is best to identify the causative agent as soon as possible. Thus it is preferable to identify the virus itself rather than antibodies or to identify IgM antibodies, which occur prior to IgG antibodies. However, most of the work has been done on IgGs as analytes and little has been done in the actual detection of the viruses because in many cases, viremia disappears once symptoms kick in. Yet, in cases such as human immunodeficiency virus (HIV) it is always detectable. Next, one will find some examples in the monitoring of human sera for antiviral immunoglobulins.

16.5.1 *Biosensors under Research or Development: Antibody Detection*

16.5.1.1 Detection of anti-West Nile virus IgG antibodies

A chemiluminescence optical fiber immunosorbent (CLOFI) assay for the detection of anti-West Nile virus (WNV) antibodies was

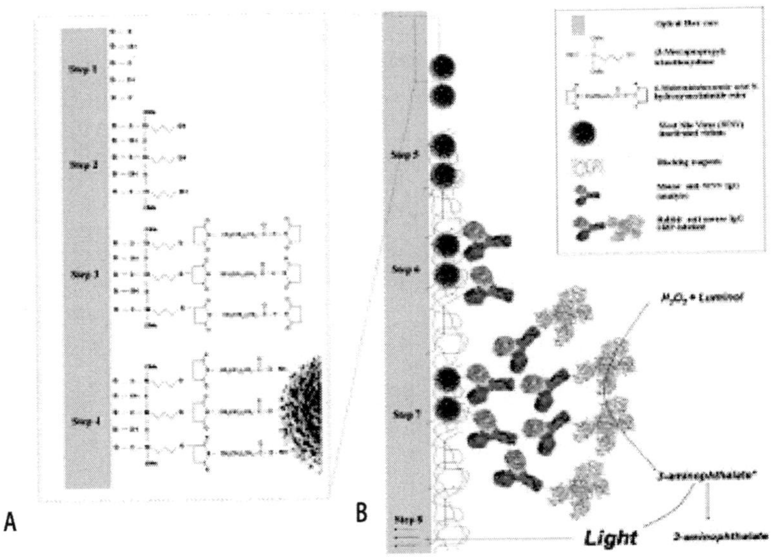

Figure 16.11 CLOFI modification and immunoassay rational. (A) Modification of the optical fiber core: the hydroxyl groups exposed on the fiber core by exposure to sulfuric acid (step 1) allow the covalent attachment of bifunctional silane molecules (step 2). The thiol group of the silane is then allowed to react with the maleimide function of the bifunctional cross linker (step 3). Finally the ester moiety of the cross linker allows the covalent linking of an amino group (step 4). (B) Immunoassay steps: after adding the blocking reagents to the surface to minimize nonspecific binding (step 5), the serum sample is tested for anti-WNV IgG presence (step 6), followed by the chemiluminescent detection of mouse IgGs (step 7). The luminous signal is finally captured by the fiber core itself (step 8), measured, and analyzed. (Reprinted with permission.)[44]

developed by Herrmann et al.[44] The assay, as described in Fig. 16.11, includes three main stages:

(1) Chemical modification to the optical fiber (steps 1–3), as described previously in Table 16.2 (entrance 1).
(2) Biological assay based on the conventional enzyme-linked immunosorbent assay (ELISA) technique (steps 4–7).
(3) Detection of the produced chemiluminescence signal (steps 7–8). The chemiluminescence signal was coupled to the fiber and propagated until collected and measured in a PMT.

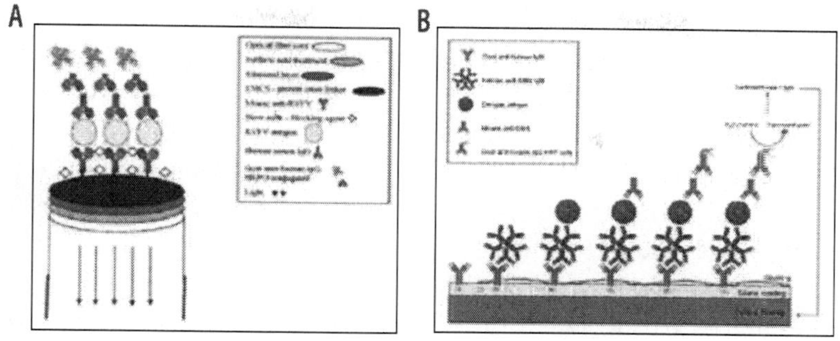

Figure 16.12 CLOFI schematic representation. (A) CLOFI modification for detection of anti-RVF antibodies (reprinted with permission).[48] (B) OFIS modification for detection of anti-DEN IgM antibodies (reprinted with permission).[43] *Abbreviation*: OFIS, optofluidic intracavity spectroscopy.

This procedure using the CLOFI methodology was done on the same rational basis as the conventional colorimetric and chemiluminescence ELISA and shown to be more sensitive, with a lower limit of detection and wider linear range. The additional advantages were shorter assay times and a lower volume of required reagents.

Detection of anti-Rift Valley fever (RVF) IgG antibodies (see Fig. 16.12A)[48] and antidengue virus (DEN) IgM antibodies (see Fig. 16.12B)[43] was also developed.

In the anti-RVF IgG case, after a spiking the antibodies in 2% (w/v) skim milk, 0.1% (v/v) phosphate buffer saline (PBS)-tween buffer and showing a theoretical detection limit of 1:1,638,400 using CLOFI (over 30 times more sensitive when compared to chemiluminescence ELISA), 242 field-collected sera samples were blind-tested and compared to results obtained by standard colorimetric ELISA. CLOFI diagnostics showed a sensitivity of 97.22% and a specificity of 98.86%. These results demonstrate that CLOFI technology is highly accurate when compared to the conventional and gold standard techniques in the case of RVF. Adding ease of performance in the OFIS technique, the potential to use CLOFI in a portable format is clear.

A technology transfer of a routinely used colorimetric ELISA for detection of anti-DEN IgM antibodies[76] was adapted to the CLOFI platform at the CNR-Arbovirus Laboratory, Institute Pasteur, French

Guiana. In this work, calibration was achieved using a seven-point panel comprised of several dilutions of a sera pool containing anti-DEN IgM antibodies. Once the CLOFI system was optimized and calibrated, comparisons between CLOFI and chemiluminescence ELISA, colorimetric ELISA, and the commercial Panbio kit were performed. Comparison of results showed a lower detection limit using the chemiluminescence techniques. Seventy-nine human sera samples were then screened, and results showed an assay sensitivity of 98.1% with a specificity of 86.4%. This work is unique and one of the few assays toward IgM antibodies.

16.5.1.2 Detection of viral antibodies using an "electroptode"

After the establishment of the electropolymerization of polypyrrole derivates,[61–63,69] an "electroptode" for the detection of anti-hepatitis C virus (HCV) IgG antibodies was done[64] and later for detection of anti-Ebola virus IgG antibodies.[65] As detailed in section 16.3.3 and illustrated in Fig. 16.13, optical fibers were coated with ITO to achieve a transparent and conductive solid phase named an "electroptode." After an electrode was layered onto an optical fiber transducer, it was further electropolymerized with pyrrole-benzophenone to form a thin film of poly(pyrrole-benzophenone) (surface coverage, $\text{Å} = 1.95 \cdot 10^{-9}$ mol $\cdot$ cm^{-2}) and the benzophenone was used for binding HCV-E2 proteins after illumination with UV light ($\lambda = 345$ nm). After finalizing the rest of the ELISA procedure, the chemiluminescence signal was measured using a PMT. It was shown that CLOFI enhanced significantly HCV serological standard testing, compared with conventional methods, which showed more false-positive results.

In another work,[65] the poly(pyrrole-benzophenone)-functionalized fiber optic was the solid phase for the covalent immobilization of inactivated Ebola virus antigen. Again, after finalizing conventional ELISA steps, the chemiluminescence signal was measured using a PMT. This method is used for detecting IgG antibodies to the Ebola virus from two strains, Zaire and Sudan, at low titers of 1:960,000 and 1:1,000,000, respectively, and with low cross reactivity between the two strains.

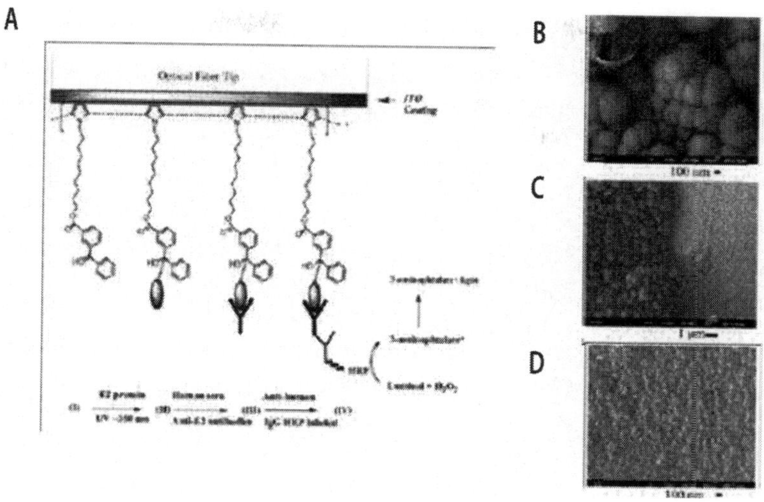

Figure 16.13 Anti-HCV IgG antibodies detection. (A) Biosensor scheme describing the various steps involved in the immunoassay using ITO-poly(pyrrole-benzophenone)-coated optical fibers for the detection of anti-HCV in human sera samples. (B) SEM micrograph of poly(pyrrole-benzophenone) film formed on an ITO-coated fiber optic. (C) The border between the polymerized pyrrole-benzophenone on ITO-coated fiber (left) and bare ITO fiber (right). (D) ITO-coated fiber. (Reprinted with permission.)[64]

16.5.2 *Virus Detection*

16.5.2.1 Newcastle disease virus

Detection of the virus itself was directed against Newcastle disease virus (NDV).[29] An evanescent wave fiber optic biosensor with a sandwich immunoassay and fluorescent marker was developed. The polyclonal antibody against NDV was covalently bound to the optical fiber via 3-aminopropyltriethoxy silane and carbonyldiimi dazole as a cross linker, described in Table 16.2, entrance 6. The detecting antibody was an anti-NDV antibody fluorescein labeled, and the signal was revealed using a fiber optic waveguide with a 4 W quartz halogen lamp for excitation and a photodetector for measuring. The assay time was approximately 15 minutes, and the limit of detection was 10 ng/mL of purified virus.

16.5.2.2 MS2 bacteriophage

An array of biosensors for the simultaneous identification of bacterial, viral, and protein analytes was done using an MS2 bacteriophage.[57,58] The sensor utilizes a sandwich assay format, using antigen-specific capture antibodies and fluorescent-labeled antibodies. Immobilization of capture antibodies in a patterned array on a surface planar waveguide was achieved by immobilization with (3-mercaptopropyl) trimethoxy silane and *N*-succinimidyl-4-maleimidobutyrate as a cross linker (see Table 16.2, entrance 7). NeutrAvidin was then covalently bound, and biotinylated anti-MS2 antibodies were immobilized. The immobilization of capture antibodies in known points gave addressability to the identification made by a CCD-based optical readout. This method facilitated the detection of 10^7 pfu/mL of the MS2 bacteriophage in a multianalyte format.

16.5.3 *Detection of Oligionucleotides*

Genosensors utilize nucleic acid–based detection, which uses the specificity of base pair matching,[3] to diagnose genetic diseases in early stages[11] when not only phenotypic identification is desired but also genetic identification.[77] Gene probe assays are known to be one of the most specific assays in the molecular laboratory, and polymerase chain reaction (PCR) or other amplification further enhances sensitivity[68] with advantages over immunologically based detection of specificity and sensitivity.[77]

A very common strategy for nucleic acid detection is to immobilize to a solid support probe, a capture target, and a second probe labeled with a reporter (directly with a radioisotope or an enzyme or indirectly with a mediator such as digoxigenin or biotin and antibody or streptavidin conjugated to the label), which hybridizes to another portion of the target. If the target nucleic acid is in the form of a duplex, a denaturizing step should be added.[77]

An automated, reusable, optical, fiber optic evanescent wave biosensor has been developed for detection of 16-mer oligonucleotides in a DNA hybridization assay.[26]

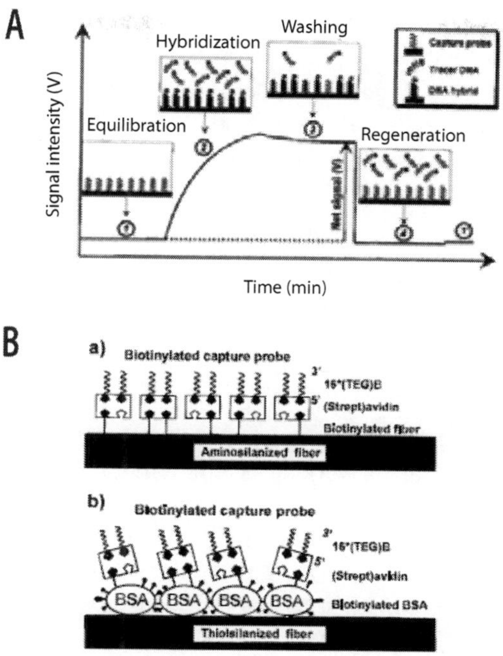

Figure 16.14 Schematic representation of a model or a genosensor. (A) Schematic diagrams of the possible arrangement of immobilized capture probes via (a) amino-silanized fiber and biotinylated with NHS-LC-biotin or (b) biotinylated BSA was bound hydrophobically to a thiolsilanized quartz fiber. Streptavidin and biotinylated oligonucleotides were bound to the surface in two consecutive incubation steps. (B) Idealized signal development during the different steps of a hybridization assay cycle. (Reprinted with permission.)[26]

Abel et al. have shown a proof of concept of hybridization detection with a fluorescent label via binding to a silanized optical fiber and avidin-biotin affinity system. Silanization provided amino- or thio-active groups onto the glass surface, and then NHS-LC biotin was covalently bound to the amine group or adsorbed hydrophobically with biotinylated BSA (see Fig. 16.14A). Streptavidin was then bound as well as the biotinylated capture oligonucleotides. Hybridization with a fluorescein probe revealed a signal, while chemical or thermal denaturation allowed reuse of the probe (see Fig. 16.14B). No significant differences in sensor stability and performance were found by immobilizing the biotin either covalently or hydrophobically. The sensitivity achieved allowed detection of 132 fM oligonucleotides, which is insufficient for in vitro diagnostic tests.

A fiber optic evanescence wave biosensor with immobilized 13-mer-oligonucleotide DNA targets, either by direct coupling via avidin adsorption and an avidin–biotin bridge or covalently via carbodiimidazole and 1-ethyl-3-(3-dimethylaminopropyl)-carbodiimide chemistry, was developed.[68] The complementary probe strand was fluorescent labeled. The shelf life of the optical fiber with the oligonucleotides' probe was three months at -18°C. The optical fiber with the immobilized capture probe was set in a flow cell in order to introduce DNA samples. The detection limit was 30 fm, and single-base mismatches were detected.

The feasibility of oligonucleotide detection has been shown but not for virus detection. Sintov et al.[78] have used a model covalent-bound amino-terminated cDNA probe to a thio-functionalized fiber optic via silanization (see Table 16.2, entrance 1) and EMCS as a cross linker (see Fig. 16.15A). Hybridization to a complementary synthetic analyte strand was followed by a complementary hybridization of a second biotinylated probe and then treated with avidin-HRP to reveal the chemiluminescence signal (see Fig. 16.15B). The assay was specific, showing no cross reactivity with other checked possible analytes and a limit of detection of 0.715 pM (100 times more sensitive than conventional Southern dot blot).

Further work is needed in this field in order to improve the sensitivity of the assay as well as to develop assays for more diverse viral agents.

16.6 Commercial Products

The maturation of the fiber optic–related technologies provides new optical platforms for diagnostics, offering the advantages of high sensitivity, remote near-real-time detection, and surface-sensitive detection for analysis in complex matrices. The Naval Research Laboratory (NRL) has developed such a sandwich fluoroimmunoassay biosensor for biowarfare with an immobilized

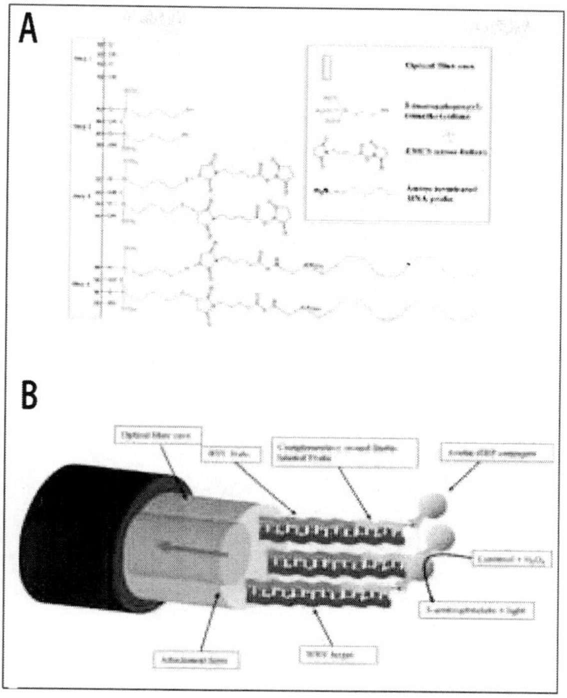

Figure 16.15 Chemiluminescence fiber optic biosensor for the detection of WNV cDNA. (A) Molecular detection model upon optical fiber core's surface. (B) The optical fiber and ELISA assembly scheme of the assay.[78]

capture antibody. On this basis there was a biosensor evolution and three commercials biosensors were created. The first prototype, the Analyte 2000 (see Fig. 16.16A), suffered from portability and power requirement problems; thus a second generation, the MATNIS (Man-portable Analyte Identification System) was developed. The second generation had a pneumatic fluidic system, where the problem of long endurance and reliable pneumatic materials was not found. Thus, the third generation, the RAPTOR,[31] was developed with peristaltic pumps for different needed reagents. Analyte detection at levels below 1 ppb from a few hundred microliter sample was enabled by the integration between optics, electronics, microfluidics, and software for monitoring. The fluorescent label is excited with a 635 nm laser diode, and a quartz optical fiber

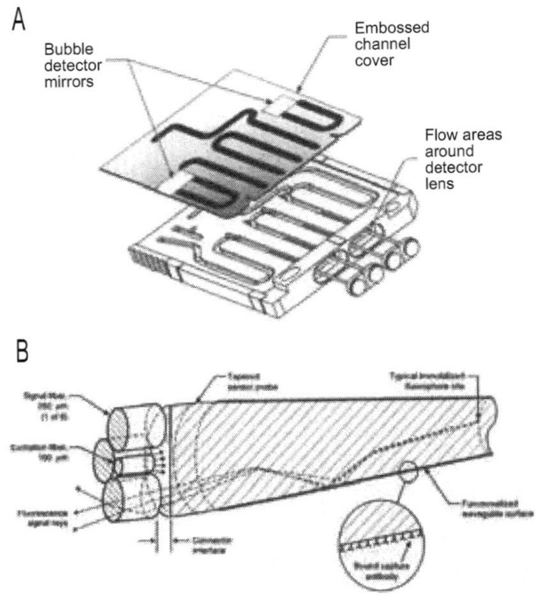

Figure 16.16 From Analyte 2000 to RAPTOR. (A) Fiber bundle jumper interface with fiber optic probe. A tapered multimode waveguide, the basis for the Analyte 2000 instrument (reprinted with permission).[28] (B) Schematic of assay cartridge. Representing orientation of optical waveguides, adhesive polyethylene channel cover, and bubble detector mirrors (reprinted with permission).[79]

is the waveguide till the reaction and measurements' points. The end product is a small portable multichannel fluorometer, user friendly and adaptable to a wide variety of sensor types in detecting different bacteria, toxicants, and viral agents.[31,32,79] In Analyte 2000, biotinylated anti-Vaccinia antibodies were immobilized with NHS-LC-biotin to an optical fiber passively coated with streptavidin, and Cy5 (emission at 670–710 nm)-labeled antibodies were used for the detection of Variola virus in spiked or throat samples with a limit of detection of $2.5 \cdot 10^5$ pfu/mL of Vaccinia virus. Time to obtain results for a single sample was 20 minutes when preparation of the waveguides and background levels were performed in advance.[27]

RAPTOR was used for the detection of MS2 coliphage (see Fig. 16.16B) with a detection limit of 10^9 pfu/mL, with no false-positive results, and an assay time of 10 minutes with no sample pretreatment.

In its most current configuration, RAPTOR fiber optic biosensor demonstrates parallel testing of bacterial, viral, and protein analytes. With a repertoire of successful immunoassays and the capability for continuous monitoring, RAPTOR is an instrument for field detection. Lightweight, portable, and fully self-contained, RAPTOR fulfills the need for a sensitive, user-friendly device for the rapid identification of biological hazards.[31,32]

16.7 Issues in Developing Fiber Optic–Based Immunosensors

Fiber optic immunosensors have many advantages, as reviewed in this chapter. Indeed, many immunosensors are based on this platform for the diagnoses of viruses, antibodies, toxins, and bacteria, only a few of which are mentioned here. However, in practice, fiber optic immunosensors have some problems and issues, which need to be taken into consideration. Those issues usually appear when constructing biosensors, and some of them, if not all of them, appear only while performing experimentation.

A major issue to overcome is fiber-to-fiber reproducibility. Fibers, after the same treatment, must give the same signal with accepted standard deviation in order to be a reliable diagnostic platform. First and foremost, from a mechanical point view, fiber cleaving must provide a clear-cut end face. Manual cleaving with a diamond pen is very cheap but requires technical cutting skills, and even with the most skilled personnel, the end face will not be perfect. To overcome it, specialized devices for fiber cleaving or polishing need to be used. However these devices are very expensive. Second, cleaning the fiber is also known as a key point. There are several cleaning methods all utilizing several steps with acid solutions, but to achieve a perfectly clean surface, the assistance of special instrumentation with anion and cation exchangers can be used. Those instruments are also quite expensive and occupy a large area. The third aspect, involving reproducibility, is chemical modification and immobilization onto the fiber. These steps must provide a modified surface with homogeneity between all the fibers. To achieve that, a controlled reproducible procedure should be developed. Furthermore, a quality control assay should be developed to assess the reproducibility after each step, including light microscopy, atomic force microscopy (AFM), scanning electron microscopy (SEM), energy-dispersive X-ray spectroscopy (EDS)-SEM, X-ray diffraction, and conventional chemical methods to confirm both fiber surface resemblance and the chemical groups involved.

Another important issue when developing self-contained biosensors is the need to incorporate in the biosensor an automated microfluidic system to introduce all the required reagents to the area of reaction and thus facilitate ease of use to the nonprofessional users. All of these aspects demand serious attention when, in particular, developing biosensors and optical fiber immunoassays in order to achieve a reliable diagnostic platform. The development is multidisciplinary and will have to combine specialized personnel such as those from the fields of immunology, optics, surface chemistry, and microfluidics to achieve a successful product.

16.8 The Future Role of Optical Fiber Biosensors

Optical fiber biosensors offer several advantages over other biosensor platforms. Their small size, flexibility, remote working capability, and noise immunity make them ideal in clinical applications, environmental monitoring, process control with easy operation, and further miniaturization potential. Optical fibers have a variety of immobilization methods intimate between the solid fiber phase and the biological elements and a relatively facile adaptation ability for the detection of other analytes.

Biosensors have the potential to play a key role in the development and improvement of public health by finding applications in areas where remote and rapid detection, high sensitivity, and specificity are important, that is, in bioterrorism scenarios. For a biosensor to be in use there is a need not only to find the appropriate application niche but also to be successfully commercialized. Although there are numerous papers published in the field, there is little commercialization progress. For complete commercialization and further use of biosensor technologies, there is a need to consider cost, integration to other systems, stability and sensitivity issues, quality assurance, and competitive technologies.[80]

Hence, industrial collaboration is needed to speed up progress from the research bench to the field stage and finally commercialization.[10]

The next-generation battlefield detection devices will most likely be stand-alone, multianalyte, remote sensing, and fully automated with integrated sample preparation and biosensing elements. The authors strongly believe that optical fiber–based biosensors should play a key role in such diagnostics, especially in bioterrorism scenarios.

Acknowledgments

This work was supported by Yeshaya Horowitz Association, project BioPen #84534101, and the EC6th framework under the DEN-FRAME program #517711. We would also like to thank Gina Hughes for her expert editorial help. This Research is conducted by NTU-HUJ-BGU Nanomaterials for Energy and Water management Programme under the Campus for Research Excellence and Technological Enterprise (CREATE), that is supported by the National Research Foundation, Prime Minister's Office, Singapore.

References

1. DaSilva, E.J., Biological warfare, bioterrorism, biodefence and the biological and toxin weapons convention. *EJB Electronic Journal of Biotechnology*, 1999. **2**(3), 109–139.
2. Sadik, O.M., W.H. Land, and J. Wang, Targeting chemical and biological warfare agents at the molecular level. *Electroanalysis*, 2003. **15**(14), 1149–1159.
3. Lim, D.V., J.M. Simpson, E.A. Kearns, and M.F. Kramer, Current and developing technologies for monitoring agents of bioterrorism and biowarfare. *Clinical Microbiology Reviews*, 2005. **18**(4), 583–607.
4. Paddle, B.M., Biosensors for chemical and biological agents of defence interest. *Biosensors and Bioelectronics*, 1996. **11**(11), 1079–1113.

5. Lim, D.V., Detection of microorganisms and toxins with evanescent wave fiber-optic biosensors. *Proceedings of the IEEE*, 2003. **91**(6), 902–907.
6. Pejcic, B., R. De Marco, and G. Parkinson, The role of biosensors in the detection of emerging infectious diseases. *Analyst*, 2006. **131**(10), 1079–1090.
7. IUPAC defines a biosensor as *"a self-contained integrated device," w.i.c.o.p.s.q.o.s.-q.a.i.u.a.b.r.e.b.r.w.*
8. Thevenot, D.R., K. Toth, R.A. Durst, and G.S. Wilson, Electrochemical biosensors: recommended definitions and classification. *Biosensensors and Bioelectronics*, 2001. **16**(1–2), 121–131.
9. Ligler, F.S., and D. Phil, Biosensors for detection of Bioterrorist threats, in *Optical Chemical Sensors*, ed. A.N.C. F. Baldini, J. Homola, and S. Martellucci, 2006, Springer Netherlands, **224**, 437–455.
10. Mehrvar, M., C. Bis, J.M. Scharer, M. Moo-Young, and J.H. Luong, Fiber-optic biosensors: trends and advances. *Analytical Sciences*, 2000. **16**(7), 677–692.
11. Marazuela, D., and M.C. Moreno-Bondi, Fiber-optic biosensors: an overview. *Analytical and Bioanalytical Chemistry*, 2002. **372**(5–6), 664–682.
12. Collings, A.F. and F. Caruso, Biosensors: recent advances. *Reports on Progress in Physics*, 1997. **60,** 1397–1445.
13. Hale, Z.M., F.P. Payne, R.S. Marks, C.R. Lowe, and M.M. Levine, The single mode tepered optical fibre loop immunosensor. *Biosensors and Bioelectronics*, 1996. **11**(1/2), 137–148.
14. Marks, R.S., A. Margalit, A. Bychenko, E. Bassis, N. Porat, and R. Dagan, Development of a chemiluminescent optical fiber immunosensor to detect Streptococcus pneumoniae antipolysaccharide antibodies. *Applied Biochemistry and Biotechnology*, 2000. **89**(2–3), 117–126.
15. Ramsden, J.J., Optical biosensors. *Journal of Molecular Recognition*, 1997. **10**(3), 109–120.
16. Vo-Dinh, T., and B. Cullum, Biosensors and biochips: advances in biological and medical diagnostics. *Fresenius' Journal of Analytical Chemistry*, 2000. **366**(6–7), 540–551.
17. Bosch, M.E., A.J.R. Sanchez, F.S. Rojas, and C.B. Ojeda, Recent development in optical fiber biosensors. *Sensors*, 2007. **7**, 797–859.
18. Wolfbeis, O.S., Fiber-optic chemical sensors and biosensors. *Analytical Chemistry*, 2004. **76**(12), 3269–3283.

19. Marks, R.S., E. Bassis, A. Bychenko, and M. Levine, Chemiluminescent optical fiber immunosensor for detecting cholera antitoxin. *Optical Engineering*, 1997. **36**(12), 3258–3264.

20. Senior, J.M., Optical fiber communications, in *Optoelectronics*, 2nd ed., ed. J. Midwinter, A. Snyder, and B. Weiss, 1992, Prentice Hall, 922.

21. Dodeigne, C., L. Thunus, and R. Leheune, Chemiluminescnce as diagnostic tool. A review. *Talanta*, 2000. **51,** 415–439.

22. Rongen, H.A., R.M. Hoetelmans, A. Bult, and W.P. van Bennekom, Chemiluminescence and immunoassays. *Journal of Pharmaceutical and Biomedical Analysis*, 1994. **12**(4), 433–462.

23. Salama, K., H. Eltoukhy, A. Hassibi, and A. El-Gamal, Modeling and simulation of luminescence detection platforms. *Biosensors and Bioelectronics*, 2004. **19**(11), 1377–1386.

24. Herrmann, S., and R.S. Marks, Chemiluminescnt optical fiber immunosensor, in *Biosensors and Biochips*, ed. R.S. Marks, et al., 2007, Wiley.

25. Starodub, N.F., P.Y. Arenkov, N.A. Starodub, and V.A. Berezin, Construction and biomedical application of immunosensors based on fiber optics and enhanced chemiluminescence. *Optical Engeeniring*, 1994. **33**(9), 2958–2963.

26. Abel, A.P., M.G. Weller, G.L. Duveneck, M. Ehrat, and H.M. Widmer, Fiber-optic evanescent wave biosensor for the detection of oligonucleotides. *Analytical Chemistry*, 1996. **68**(17), 2905–2912.

27. Donaldson, K.A., M.F. Kramer, and D.V. Lim, A rapid detection method for Vaccinia virus, the surrogate for smallpox virus. *Biosensors and Bioelectronics*, 2004. **20**(2), 322–327.

28. Golden, J.P., E.W. Saaski, L.C. Shriver-Lake, G.P. Anderson, and F.S. Ligler, Portable multichannel fiber optic biosensor for field detection. *Optical Engineering*, 1997. **36**(4), 1008–1013.

29. Lee, E.W., and H.G. Thompson, Detection of Newcastle disease virus using an evanescent wave immuno-based biosensor. *Canadian Journal of Chemistry*, 1996. **74**(5), 707–712.

30. Liu, X., and W. Tan, A fiber-optic evanescent wave DNA biosensor based on novel molecular beacons. *Analytical Chemistry*, 1999. **71**(22), 5054–5059.

31. Anderson, G.P., and D.A. McCrae, RAPTOR: development of a fiber-optic biosensor. Biosensors and biochips, ed. R.S. Marks, et al., 2007, Wiley, 1500.

32. King, K.D., J.M. Vanniere, J.L. Leblanc, K.E. Bulloc, and G.P. Anderson, Automated fiber optic biosensor for multiplexed immunoassays. *Envi-romental Science Technology*, 2000. **34**, 2845–2850.

33. Starodub, N.F., I.M. Samodumova, and V.N. Starodub, Usage of organosilanes for integration of enzymes and immunocomponents with electrochemical and optical transducers. *Sensors and Actuators B*, 1995. **24**(1–3), 173–176.

34. Wilchek, M., and T. Miron, Oriented versus random protein immobilization. *Journal of Biochemistry and Biophysics Methods*, 2003. **55**(1), 67–70.

35. Kusnezow, W., and J.D. Hoheisel, Antibody microarrays: promises and problems. *Biotechniques*, 2002. Suppl, 14–23.

36. Kusnezow, W., and J.D. Hoheisel, Solid supports for microarray immunoassays. *Journal of Molecular Recognition*, 2003. **16**(4), 165–176.

37. Nisnevitch, M., and M.A. Firer, The solid phase in affinity chromatography: strategies for antibody attachment. *Journal of Biochemical and Biophysical Methods*, 2001. **49**(1–3), 467–480.

38. Peluso, P., D.S. Wilson, D. Do, H. Tran, M. Venkatasubbaiah, D. Quincy, B. Heidecker, K. Poindexter, N. Tolani, M. Phelan, K. Witte, L.S. Jung, P. Wagner, and S. Nock, Optimizing antibody immobilization strategies for the construction of protein microarrays. *Analytical Biochemistry*, 2003. **312**(2), 113–124.

39. Lin, J.N., J. Herron, J.D. Andrade, and M. Brizgys, Characterization of immobilized antibodies on silica surfaces. *IEEE Trans Biomedical Engineering*, 1988. **35**(6), 466–471.

40. Lutanie, E., J.C. Voegel, P. Schaaf, M. Freund, J.P. Cazenave, and A. Schmitt, Competitive adsorption of human immunoglobulin G and albumin: consequences for structure and reactivity of the adsorbed layer. *Proceeding of the National Academy of Sciences of the U S A*, 1992. **89**(20), 9890–9894.

41. Strashnikova, N., V. Papper, P. Parkhomyuk, G.I. Likhtenshtein, B. Ratner, and R.S. Marks, Local medium effects in the photochemical behavior of substituted stilbenes immobilized on quartz surfaces. *Journal of Photochemistry and Photobiology A*, 1999. **122**(2), 133–142.

42. Vijayendran, R.A., and D.E. Leckband, A quantitative assessment of heterogeneity for surface-immobilized proteins. *Analytical Chemistry*, 2001. **73**(3), 471–480.

43. Atias, D., Y. Liebes, V. Chalifa-Caspi, L. Bremand, L. Lobel, R.S. Marks, and P. Dussart, Chemiluminescent optical fiber immunosensor for the

detection of IgM antibody to dengue virus in humans. *Sensors and Actuators B*, 2009. **140**(1), 206–215.

44. Herrmann, S., B. Leshem, S. Landes, B. Rager-Zisman, and R.S. Marks, Chemiluminescent optical fiber immunosensor for the detection of anti-West Nile virus IgG. *Talanta*, 2005. **66,** 6–14.
45. Leshem, B., G. Sarfati, A. Novoa, I. Breslav, and R.S. Marks, Photochemical attachment of biomolecules onto fibre-optics for construction of a chemiluminescent immunosensor. *Luminescence*, 2004. **19**(2), 69–77.
46. Salama, O., S. Herrmann, A. Tziknovsky, B. Piura, M. Meirovich, I. Trakht, B. Reed, L.I. Lobel, and R.S. Marks, Chemiluminescent optical fiber immunosensor for detection of autoantibodies to ovarian and breast cancer-associated antigens. *Biosensors and Bioelectronics*, 2007. **22**(7), 1508–1516.
47. Shriver-Lake, L.C., B. Donner, R. Edelstein, K. Breslin, S.K. Bhatia, and F.S. Ligler, Antibody immobilization using heterobifunctional crosslinkers. *Biosensors and Bioelectronics*, 1997. **12**(11), 1101–1106.
48. Sobarzo, A., J.T. Paweska, S. Herrmann, T. Amir, R.S. Marks, and L Lobel, Optical fiber immunosensor for the detection of IgG antibody to Rift Valley fever virus in humans. *Journal of Virological Methods*, 2007. **146**(1–2), 327–334.
49. Atias, D., K. Abu-Rabeah, S. Herrmann, J. Frenkel, D. Tavor, S. Cosnier, and R.S. Marks, Poly(methyl metacrylate) conductive fiber optic transducers as dual biosensor platform. *Biosensors and Bioelectronics*, 2009. **24**(12), 3683–3687.
50. Polyak, B., S. Geresh, and R.S. Marks, Synthesis and characterization of a biotin-alginate conjugate and its application in a biosensor construction. *Biomacromolecules*, 2004. **5**(2), 389–396.
51. Thames, S.F., and K.G. Panjnani, Organosilane polymer chemistry: a review. *Journal of Inorganic and Organometallic Polymers*, 1996. **6**(2), 59–94.
52. Ulman, A., Formation and structure of self-assembled monolayers. *Chemical Reviews*, 1996. **96**(4), 1533–1554.
53. Wang, M., K.M. Liechti, Q. Wang, and J.M. White, Self-assembled silane monolayers: fabrication with nanoscale uniformity. *Langmuir*, 2005. **21**(5), 1848–1857.
54. Wink, T., S.J. Van Zuilen, A. Bult, and W.P. van Bennekom, Self-assembled monolayers for biosensors. *The Analyst*, 1997. **122,** 43R–50R.
55. Cras, J.J., C.A. Rowe-Taitt, D.A. Nivens, and F.S. Ligler, Comparison of chemical cleaning methods of glass in preparation for silanization. *Biosensors and Bioelectronics*, 1999. **14,** 683–688.

56. Bakaltcheva, I.B., F.S. Ligler, C.H. Patterson, and L.C. Shriver-Lake, Multi-analyte explosive detection using a fiber optic biosensor. *Analytica Chimica Acta*, 1999. **399,** 12–20.
57. Rowe, C.A., S.B. Scruggs, M.J. Feldstein, J.P. Golden, and F.S. Ligler, An array immunosensor for simultaneous detection of clinical analytes. *Analytical Chemistry*, 1999. **71**(2), 433–439.
58. Rowe, C.A., L.M. Tender, M.J. Feldstein, J.P. Golden, S.B. Scruggs, B.D. MacCraith, J.J. Cras, and F.S. Ligler, Array biosensor for simultaneous identification of bacterial, viral, and protein analytes. *Analytical Chemistry*, 1999. **71**(17), 3846–3852.
59. Anderson, G.P., M.A. Jacoby, F.S. Ligler, and K.D. King, Effectiveness of protein A for antibody immobilization for a fiber optic biosensor. *Biosensors and Bioelectronics*, 1997. **12**(4), 329–336.
60. Cosnier, S., Recent advances in biological sensors based on electrogenerated polymers: a review. *Analytical Letters*, 2007. **40,** 1260–1279.
61. Konry, T., and R.S. Marks, Physico-chemical studies of indium tin oxide-coated fober optic biosensors. *Thin Solid Films*, 2005. **492,** 313–321.
62. Cosnier, S., Biosensors based on immobilization of biomolecules by electrogenerated polymer films. New perspectives. *Applied Biochemistry and Biotechnology*, 2000. **89**(2–3), 127–138.
63. Konry, T., A. Novoa, S. Cosnier, and R.S. Marks, Development of an "electroptode" immunosensor: indium tin oxide-coated optical fiber tips conjugated with an electropolymerized thin film with conjugated cholera toxin B subunit. *Analytical Chemistry*, 2003. **75**(11), 2633–2639.
64. Konry, T., A. Novoa, Y. Shemer-Avni, N. Hanuka, S. Cosnier, A. Lepellec, and R.S. Marks, Optical fiber immunosensor based on a poly(pyrrole-benzophenone) film for the detection of antibodies to viral antigen. *Analytical Chemistry*, 2005. **77**(6), 1771–1779.
65. Petrosova, A., T. Konry, S. Cosnier, I. Trakht, J. Lutwama, E. Rwaguma, A. Chepurnov, E. Muhlberger, L. Lobel, and R.S. Marks, Development of a highly sensitive, field operable biosensor for serological studies of Ebola virus in central Africa. *Sensors and Actuators B*, 2007. **122**(2), 578–586.
66. Scouten, W.H., Affinity chromatography. Chemical analysis, ed. P.J. Elving, J.D. Winefordner, and I.M. Kolthoff. Vol. 1. 1992: Wiley.
67. Savage, M.D., G. Mattson, S. Desai, G.W. Nielander, S. Morgensen, and E.J. Conklin, Avidin-Biotin chemistry: a handbook. 1992: Pierce. 467.
68. Kleinjung, F., F.F. Bier, A. Warsinke, and F.W. Scheller, Fibre-optic genosensor for specific determination of femtomolar DNA oligomers. *Analytica Chimica Acta*, 1997. **350,** 51–58.

69. Marks, R.S., A. Novoa, T. Konry, R. Krais, and S. Cosnier, Indium tin oxide-coated optical fiber tips for affinity electro-polymerization. *Materials Science and Engineering C*, 2002. **21**(1–2), 189–194.

70. Ionescu, R.E., C. Gondran, S. Cosnier, L.A. Gheber, and R.S. Marks, Comparison between the performances of amperometric immunosensors for chol era antitoxin based on three enzyme markers. *Talanta*, 2005. **66,** 15–20.

71. Ionescu, R.E., S. Herrmann, S. Cosnier, and R.S. Marks, A polypyrrole cDNA electrode for the amperometric detection of the West Nile virus. *Electrochemistry Communications*, 2006. **8,** 1741–1748.

72. Buzhan, P., B. Dolgoshein, L. Filatov, A. Ilyin, V. Kantzerov, V. Kaplin, A. Karakash, F. Kayumov, S. Klemin, E. Popova, and S. Smirnov, Silicon photomultiplier and its possible applications. *Nuclear Instruments and Methods in Physics Research A*, 2003. **504**, 48–52.

73. Cova, S., M. Ghioni, A. Loitto, I. Rech, and F. Zappa, Evolution and prospects for single-photon avalanche diodes and quenching circuits. *Journal of Modern Optics*, 2004. **51**(9–10), 1267–1288.

74. Michalet, X., O.H.W. Siegund, J.V. Vallerga, p. Jelinsky, J.E. Millaud, and S. Weiss, Detectors for single-molecule fluorescence imaging and spectroscopy. *Journal of Modern Optics*, 2006. **54**(2–3), 239–281.

75. Ivask, A., T. Green, B. Polyak, A. Mor, A. Kahru, M. Virta, and R. Marks, Fibre-optic bacterial biosensors and their application for the analysis of bioavailable Hg and As in soils and sediments from Aznalcollar mining area in Spain. *Biosensors and Bioelectronics*, 2007. **22**(7), 1396–1402.

76. Kuno, G., I. Gomez, and D.J. Gubler, Detecting artificial anti-dengue IgM immune complexes using an enzyme-linked immunosorbent assay. *The Americal Journal of Tropical Medicine and Hygiene*, 1987. **36**(1), 153–159.

77. Iqbal, S.S., M.W. Mayo, J.G. Bruno, B.V. Bronk, C.A. Batt, and J.P. Chambers, A review of molecular recognition technologies for detection of biological threat agents. *Biosensors and Bioelectronics*, 2000. **15**(11–12), 549–578.

78. Sintov, E., S. Herrmann, and R.S. Marks, unpublished data.

79. King, K.D., G.P. Anderson, K.E. Bullock, M.J. Regina, E.W. Saaski, and F.S. Ligler, Detecting staphylococcal enterotoxin B using an automated fiber optic biosensor. *Biosensors and Bioelectronics*, 1999. **14**(2), 163–170.

80. Weetall, H.H., Chemical sensors and biosensors, update, what, where, when and how. *Biosensors and Bioelectronics*, 1999. **14,** 247–242.

Chapter 17

Aptamers, a New Class of Binders, with Particular Focus on Diagnostics and Bioactivity in the Field of Virology

Andreas Kage[a] and Leslie Lobel[b]

[a] *IAptaRes AG, Berlin, Germany*

[b] *Department of Virology, Faculty of Health Sciences, Ben-Gurion University of the Negev, Beer-Sheva, Israel*

a.kage@aptares.de, llobel@bgu.ac.il

17.1 General Facts about Aptamers

Since the first reports of the isolation of protein- and nucleotide-binding oligonucleotides isolated from a combinatory library[1,2] of molecules, oligonucleotide aptamers have become a widely established class of binders being an alternative to protein-based antibodies. Oligonucleotide aptamers are short sequences of single-stranded DNA and RNA molecules. The spectrum of targets is large, ranging from small molecules, such as caffeine and peptides over proteins, to complex molecular structures, such as viruses and bacteria. This wide variety of molecules can be addressed with high specificity and affinity. So far reports exist

Viral Diagnostics: Advances and Applications

Edited by Robert S. Marks, Leslie Lobel, and Amadou Alpha Sall

ISBN 978-981-4364-43-0 (Hardcover), 978-981-4364-44-7 (eBook)

www.panstanford.com

about hundreds of different targets that have successfully been used for aptamer selection.[2a] In addition, the animal-free in vitro selection of specific aptamers from a library of about 10^{12} to 10^{16} randomized oligonucleotide sequences enables the use of haptens and toxic substances as targets.[3] The chemical synthesis of the identified binders enables a site-directed introduction of a large variety of modifications. Bulk-scale synthesis guaranties fast and relatively uncomplicated upscaling and high batch stability. Moreover, oligonucleotide aptamers possess proteinase and heat stability.[3] The heat treatment permits sterilization processes, which makes aptamers attractive binder alternatives for antibodies but also for specific affinity materials in downstream processes or plasmaphresis. DNA aptamers show an appropriate stability toward DNases, permitting their use in biological fluids such as blood and serum. Moreover, known from the antisense technology, small chemical variations of the molecules will improve their stability.[3]

These properties show the advantages of oligonucleotide binders, especially DNA binders over protein antibodies. They make these molecules excellent candidates for a variety of purposes in analytics[3a] and pharmaceutics.[3,3b]

17.2 Selection Procedures

The classic strategy for the selection of aptamers called SELEX requires several rounds of selection and amplification in order to exponentially enrich high-affinity binders from the large variety of oligonucleotide molecules.[4]

In contrast to this protocol the MonoLex aptamer selection protocol (patents held by AptaRes AG, Germany[5]) uses a straightforward one-step affinity chromatographic protocol. Low-affinity binders are removed from the column by extensive washing. High-affinity binders are released from the affinity resin by physical segmentation of the column and subsequent amplification of the oligonucleotide binders by polymerase chain reaction (PCR). The affinity of the aptamers differs and shows a decreasing affinity gradient downstream along the column.

The very high efficiency in the selection of available binders in the combinatory library enables the use of significantly less random positions in the template, as was originally proposed for the SELEX aptamer library.

MonoLex has been proven to be an appropriate method for the selection of specific and highly affine binders against viruses such as the vaccinia virus[5a] and the influenza virus.[5b,c]

17.3 Aptamers for Analytical Purposes (Detection and Quantification)

The detection and quantification of molecules play an essential role in basic research as well as in environmental and clinical practice. An improvement of the existing technology by the introduction of new binding molecules showing different qualities compared to antibodies might open up new possibilities for analytes, which were so far not appropriate for the classical antibody technology. This is, for example, true for highly toxic substances or haptens. Oligonucleotide aptamers are a class of such new binder molecules, suitable for applications based on molecular target recognition with high levels of specificity and affinity. They are therefore an excellent alternative for antibodies to be included into analytical and therapeutical methods, partly even replacing antibodies.

With respect to molecule identification, aptamers have already shown their great potential in both chemical analytics and clinical diagnostics.[5d] An aptamer-based test detecting *Francisella tularensis* has proven its superiority over an existing enzyme-linked immunosorbent assay (ELISA).[6] Aptamers have been exploited for dipstick development.[7] Further tests are under development.[8,9] The latest development is a quantitative test for cocaine detection[10,11] and affinity support for SERS in influenza.[32a,b]

17.4 Polyvalent, Polyspecific Aptamer Constructs

One major advantage of aptamers with respect to their use in analytics and therapy is the easy and uncomplicated design of constructs carrying several binders showing affinities to the same (polyvalent) or different (polyspecific) target domains.[5c,12] This offers a great opportunity to use the same molecule constructs for capturing and detecting analytes.[13a,b] Besides analytical advantages, such as the reduction of the detection limit being able to detect sterical isoforms of an analyte in one sample (comparable to the capability of polyclonal antibodies), fabrication vantages are also very obvious.[5c]

17.5 Aptamers for Therapeutic Purposes

The molecular nature of aptamers and their qualities make these molecules desirable candidates as therapeutic agents. So far one protein-binding aptamer has already obtained approval from the Food and Drug Administration (FDA) and Europe, the Middle East, and Africa (EMEA) for therapeutic purposes. This aptamer, which targets the VEGF-molecule, serves for the treatment of neurovascular wet age-related macular degeneration.[13]

Further aptamer candidates are in preclinical [14a,b] and advanced stages of clinical testing.[3,15]

17.6 Aptamers in Virology

17.6.1 *General Facts about Aptamers in Virology*

All of the above-described general facts about the therapeutical and analytical application of aptamers are also applicable to the field of virology. In fact, one of the first aptamers selected was directed against a viral protein, reverse transcriptase of thean immunodeficiency virus (HIV).[16]

In the field of virology aptamers are thought to be exploitable for analytics, diagnostics, and therapy. The latter has to take into account virus-specific characteristics, which are based on the fact that a virus population does not grow via cell division but via the joint use of the host cell replication machinery. The sustainment of a virus population is therefore based on the following general processes: **attachment penetration uncoating replication** and **assembly**, in addition to a variety of additional family- and subtype-specific processes.

With the availability of highly specific and affine binders from each single step of a viral infection, possible targets for a point of action could be deduced. First of all the virus infection itself, which means the **attachment** and **entry**, into the organism or the single cell might be an excellent target for an intervention. In this case the surface proteins or glycoproteins of the virus should be addressed by an aptamer, interfering with the virus attachment. This has been successfully described neutralizing the surface unit (SU) glycoprotein in HIV-1 by Khati et al. in 2003.[17]

Targeting the influenza virus is another successful example that shows the applicability of this principle.[18] The authors were able to demonstrate that the antihemagglutinin DNA aptamer was not only able to significantly reduce the in vitro infectivity of the influenza virus but also able to reduce the virus burden in treated animals. These are only two crucial and impressive examples out of a pool of recent publications dealing with this specific matter.

It is generally assumed today that aptamers are not able to penetrate cell membranes[19] because of physicochemical features such as high hydrophilicity. To overcome these limitations, the virus host cells can be treated either by using aptamer-carrying vectors[20] or by applying expression systems/cassettes.[21] Being in the virus host cell, the aptamers can effectively interfere with virus replication, addressing different possible targets. Here, as mentioned above, the **genome replication** of the virus has been in the focus of the research since the 80th.[16] This technique has been applied, improved, and further investigated over the years by the same research group[22] and by others.[23,24] The whole replication machinery offers, however, more possible points for attack, for example, virus RNA-binding proteins such as the[25] Aptamers against Rev have been selected, described,[26] an successfully tested.[20] Butthis is only one example of many unmentioned targets and possibilities at this stage of virus replication.

This holds true for virus **assembly** as well. The regulation of encapsidation goes partly back to a control machinery, which is also entrapped in different growth stages. The fast development of resistance toward existing antiviral therapeutics is the reason for an urgent search for alternative therapeutical strategies. The encapsidation process is a new target for antiviral therapy in general,[27] in addition to the field of aptamers.[28]

As for any other stage of population sustainment, the process of **virus maturation** is very complex, strongly depending on virus family and subtype characteristics. The participation of core proteins in maturation processes is accepted for a variety of viruses, such as Gag for retroviruses,[29] while other authors also stress the importance of peripheral and integral membrane proteins for budding processes.[30] For the hepatitis C virus, the nonstructural protein 3, a serine protease that essentially contributes in the maturation process, has been the target for aptamer selection.[31] These aptamers were able to inhibit the protease activity with high efficiency, exploiting corresponding in vitro assays.

MonoLex-selected aptamers against complete Vaccinia virus particles[5a] showed inhibitory virus-spreading activity, not only for the selection target, but also for a wide spectrum of orthopox-like Cowpox viruses.[5a]

17.6.2 *Next Steps into the Future of Aptamers in Virology*

First of all we would like to stress that the present article is not a complete review, systematically analyzing the existing literature in the field. It is just a brief introduction into the world of oligonucleotide aptamers, with a special focus on their applicability in virology. Although the therapeutical aspect of antiviral aptamers is clearly pronounced, their analytical usefulness for target finding and other basic research should not be underestimated.

An unmet potential of aptamers is the short time for development and scaling up the chemical synthesis. This will enable the identification of leads for diagnostics and therapeutic intervention within a few weeks if new virus strains are showing up, either by natural or by purposely origin.

Despite being only in the beginning of the era of aptamer technology, with respect to virology, aptamers have already shown their future potential, and we are looking forward to see more exciting applications of aptamers in virology in the near future.

17.7 AptaRes AG: MonoLex Aptamers

AptaRes AGhas exclusive international rights of the MonoLex patent[5] for the selection of aptamers, which enables a one-step isolation of oligonucleotide aptamers by using a microaffinity support. The MonoLex-selected aptamers have already been proven to be appropriate for the selection of specific, highly affine, and functionally active aptamers against virus particles using orthopoxviruses and influenza viruses as targets.[5a,32a,b]

References

1. Joyce, G. F. 1989. In *Molecular Biology of RNA* (Cech, T. R., ed), Alan Liss, New York, pp. 361–371
2. Yarus, M. 1989. In *Molecular Biology of RNA* (Cech, T. R., ed), Alan Liss, New York, pp. 373–374.

2a. Pendergrast, P. S., H. N. Marsh, D. Grate, J. M. Healy, and M. Stanton. Nucleic acid aptamers for target validation and therapeutic applications. 2005. *J. Biomol. Techniques*, **16**, 224–234.

3. Rimmele, M. Nucleic acid aptamers as tools and drugs: recent developments. 2003. *ChemBioChem*, **4**, 963–971.

3a. Mairal, T., V. C. Özalp, P. L. Sánchez, M. Mir, I. Katakis, and C. K. O'Sullivan. Aptamers: molecular tools for analytical applications. 2008. *Anal. Bioanal. Chem.*, **390**, 989–1007.

3b. Blank, M., and M. Blind. Aptamers as tools for target validation. 2005. *Curr. Opin. Chem. Biol.*, **9**, 336–342.

4. Tuerk, C., and L. Gold. Systematic evolution of ligands by exponential enrichment: RNA ligands to bacteriophage T4 DNA polymerase. 1990. *Science*, **249**, 505–510.

5. Kage, A. Method of selection, by two-dimensional separation, of nucleic acids that bind to a target with high affinity. 2002. WO 002002029094 A3.

5a. Nitsche, A., A. Kurth, A. Dunkhorst, O. Pänke, H. Sielaff, W. Junge, D. Muth, F. Scheller, W. Stöcklein, C. Dahmen, G. Pauli, and A. Kage. One-step selection of vaccina virus binding DNA-aptamers by MonoLEX. 2007. *BMC Biotechnol.*, **7**, 48.

5b. Dahmen, C., W. Stöcklein, H. Kost, A. Haberland, and A. Kage. 2007. Virus specific aptamers: new analytical tools for virology. 5th annual conference, Washington DC.

5c. Dahmen, C., A. Haberland, and A. Kage. Tetrameric aptamer-costructs for a sandwich binding assay (aptamers selected using the MonoLex technology). 2007. *Clin. Chem. Lab. Med.*, **45**, A111 (abstract *Congress of Clinical Chemistry and Laboratory Medicine*, Vienna).

5d. Yan, A. Aptamers as potential diagnostic reagents for diabetes. 2002. *Diabetes Technol. Ther.*, **4**, 339–346.

6. Vivekananda, J., and J. L. Kiel. Anti-Francisella tularensis DNA aptamers detect tularemia antigen from different subspecies by aptamer-linked immobilized sorbent assay. 2006. *Lab. Invest.*, **86**, 610–618.

7. Liu, J., D. Mazumdar, and Y. Lu. A simple and sensitive "dipstick" test in serum based on lateral flow separation of aptamer-linked nanostructures. 2006. *Angew. Chem., Int. Ed.*, **45**, 1–5.

8. Drolet, D. W., L. Moon-McDermott, and T. S. Romig. An enzyme-linked oligonucleotide assay. 1996. *Nat. Biotechnol.*, **14**, 1021–1025.

9. Yan, X. R., X. W. Gao, L. H. Yao, and Z. Q. Zhang. Novel methods to detect cytokines by enzyme-linked oligonucleotide assay. 2004. *Sheng Wu Gong Cheng Xue Bao*, **20**, 679–682 (abstract available in English).

10. Baker, B. R., R. Y. Lai, M. S. Wood, E. H. Doctor, A. J. Heeger, and K. W. Plaxco. An electronic, aptamer based small-molecule sensor for the rapid, label-free detection of cocaine in adulterated samples and biological fluids. 2006. *J. Am. Chem. Soc.*, **128**, 3138–3139.

11. Eisenstein, M. Coke and chips: a winning combination? 2006. *Nat. Methods*, **3**, 244.

12. Di Giusto, D. A., and G. C. King. Construction, stability, and activity of multivalent circular anticoagulant aptamers. 2004. *J. Biol. Chem.*, **279**, 46483–46489.

13. Chakravarthy, U., A. P. Adamis, E. T. Cunningham, Jr., M. Goldbaum, D. R. Guyer, B. Katz and M. Patel (clinical trial group). VEGF inhibition study in ocular neovascularization (V.I.S.I.O.N.); year 2 efficacy results of 2 randomized controlled clinical trials of pegeptanib for neovascular age-related macular degeneration. 2006. *Ophthalmology*, **113**, 1508.

14a. Haberland, A., G. Wallukat, C. Dahmen, A. Kage and I. Schimke. Aptamer neutralization of ß1-adrenoceptor autoantibodies isolated from patients with cardiomyopathies. 2011. *Circ. Res.*, **109**, 986–992.

14b. Wallukat, G., A. Haberland, S. Berg, A. Schulz, E. J. Freyse, C. Dahmen, A. Kage, M. Dandel, R. Vetter, E. Salzsieder, R. Kreutz and I. Schimke.

The first aptamer-apheresis column specifically for clearing blood of ß1-receptor autoantibodies. 2012. *Circ. J.*, **76**, 2449–2455.

15. Cerchia, L., and V. de Franciscis. Nucleic acid-based aptamers as promising therapeutics in neoplastic diseases. 2007. *Methods Mol. Biol.*, **361**, 187–200.
16. Tuerk, C., S. MacDougal, and L. Gold. RNA pseudoknots that inhibit human immunodeficiency virus type 1 reverse transcriptase. 1992. *Proc. Natl. Acad. Sci. USA*, **89**, 6988–6992.
17. Khati, M., M. Schüman, J. Ibrahim, Q. Sattentau, S. Gordon, and W. James. Neutralization of infectivity of diverse R5 cliical isolates of human immunodeficiency virus type 1 by gp120-binding 2'F-RNA aptamer. 2003. *J. Virol.*, **77**, 12692–12698.
18. Jeon, S. H., B. Kayhan, T. Ben-Yedidia, and R. Arnon. A DNA aptamer prevents influenza infection by blocking the receptor binding region of the viral hemagglutinin. 2004. *J. Biol. Chem.*, **279**, 48410–48419.
19. Valigra, L. 2005. First aptamer-based drugs on tough diseases. *Drug Discovery Dev.* (http://www.dddmag.com/articles/2007/09/first-aptamer-based-drugs-take-tough-diseases).
20. Düzgünes, N., E. Pretzer, S. Simões, V. Slepushkin, K. Konopka, D. Flasher, and M. C. de Lima. Liposome-mediated delivery of antiviral agents to human immunodeficiency virus-infected cells. 1999. *Mol. Membr. Biol.*, **16**, 111–118.
21. Kolb, G., S. Reigadas, D. Castanotto, A. Faure, M. Ventura, J. J. Rossi, and J. J. Toulmé. Endogenous expression of an anti-TAR aptamer reduces HIV-1 replication. 2006. *RNA Biol.*, **3**, 150–156.
22. Chen, H., D. G. McBroom, Y. Q. Zhu, L. Gold, and T. W. North. Inhibitory RNA ligand to reverse transcriptase from feline immundeficiency virus. 1996. *Biochemistry*, **35**, 6923–6930.
23. Kensch, O., B. A. Connolly, H. J. Steinhoff, A. McGregor, R. S. Goody, and T. Restle. HIV-1 reverse transcriptase-pseudoknot RNA aptamer interaction has a binding affinity in the low picomolar range with high specificity. 2000. *J. Biol. Chem.*, **275**, 18271–18278.
24. Andreola, M. L., F. Pileur, C. Calmels, M. Ventura, L. Tarrago-Litvak, J. J. Toulmé, and S. Litvak. DNA aptamers selected against the HIV-1 Rnase H display in vitro antiviral activity. 2001. *Biochemistry*, **40**, 010087–010094.
25. Malim, M. H., J. Hauber, S. Y. Le, J. V. Maizel, and B. R. Cullen. The HIV-1 rev trans-activator acts through a structured target sequence to

activate nuclear export of unspliced viral mRNA. 1989. *Nature*, 338, 254–257.

26. Ye, X., A. Gorin, A. D. Ellington, and D. J. Patel. Deep penetration of an alpha-helix into a widened RNA major groove in the HIV-1 rev peptide-RNA aptamer complex. 1996. *Nat. Struct. Biol.*, **3**, 1026–1033.
27. Visalli, R. J., and M. van Zeijl. DNA encapsidation as a target for anti-herpesvirus drug therapy. 2003. *Antiviral Res.*, **59**, 73–87.
28. Kim, S. J., M. Y. Kim, J. H. Lee, J. C. You, and S. Jeong. Selection and stabilization of the RNA aptamers against the human immunodeficiency virus type-1 nucleocapsid protein. 2002. *Biochem. Biophys. Res. Commun.*, **291**, 925–931.
29. Salzwedel, K., D. E. Martin, and M. Sakalian. Maturation inhibitors: a new therapeutic class targets the virus structure. 2007. *AIDS Rev.*, **9**, 162–172.
30. Garoff, H., R. Hewson, and D. J. Opstelten. Virus maturation by budding. 1998. *Microbiol. Mol. Biol. Rev.*, **62**, 1171–1190.
31. Fukuda, K., D. Vishnuvardhan, S. Sekiya, J. Hwang, N. Kakiuchi, K. Taira, K. Shimotohno, P. K. Kumar, and S. Nishikawa. Isolation and characterization of RNA aptamers specific for the hepatitis C virus nonstructural protein 3 protease. 2000. *Eur. J. Biochem.*, **267**, 3685–3694.

32a. Negri, P., A. Kage, A. Nitsche, D. Naumann and R. A. Dluhy. Detection of viral nucleoprotein binding to anti-influenza aptamers via SERS. 2011. *Chem. Commun. (Camb).*, **47**, 8635–8637.

32b. Negri, P., G. Chen, A. Kage, A. Nitsche, D. Naumann, B. Xu and R. A. Dluhy. Direct optical detection of viral nucleoprotein binding to an anti-influenza aptamer. 2012. *Anal. Chem.*, **84**, 5501–5508.

Chapter 18

Pseudotyped Viruses: A New Sero-Diagnostic Tool

Jean-Michel Garcia
HKU-Pasteur Research Centre, Hong Kong, SAR China
jmgarcia@hku.hk

18.1 Brief Historical Review of Pseudotyped Viruses

Looking at the literature, pseudotyped viruses can be found as early as in the seventies. In 1978, Zavada et al.[1] described the use of pseudotype particles as a sensitive assay for the detection of neutralizing antibodies (N-Abs), in particular for antigens from noncytopathogenic or slow-growing viruses.[2] At that time, the so-called pseudoparticles, or phenotypically mixed particles, referred to replicative virions from doubly infected cells that incorporated an envelope of one virus on the surface of the second one. According to these authors, it was possible the rhabdovirus vesicular stomatitis virus (VSV) could acquire surface glycoproteins (GPs) of all retroviruses. This approach successfully produced pseudotype particles from a wild range of viruses extensively used for the detection of N-Abs. Nevertheless, there were some disadvantages, in particular the tedious purification and characterization process,

Viral Diagnostics: Advances and Applications
Edited by Robert S. Marks, Leslie Lobel, and Amadou Alpha Sall

ISBN 978-981-4364-43-0 (Hardcover), 978-981-4364-44-7 (eBook)
www.panstanford.com

the absence of a reporter signal for the direct observation of the infection, and the biosafety concern related to the use of live viruses. Research pioneered by Zavada demonstrated that retroviruses, like human immunodeficiency virus (HIV), can rather easily incorporate heterologous GPs from other viruses, including other viral families, through phenotype mixing. These early observations were the cornerstone to the subsequent design of HIV-1-based lentiviral vector strategies.

With the progress of molecular biology in 1990, Page et al. were the first to engineer the HIV-1 vector system.[3] A milestone was reached with this work because the pseudotype particles produced could be handled more safely as they were rendered replication deficient. Furthermore, the promoter region was modified to be inducible, which could thereby report on the infection. These two very important features, replication deficiency and reporter gene, will be conserved in all the successive generations of pseudotype viral particles. As a supplementary benefit of this strategy, only the envelope that has been brought as an additional expression plasmid, encoding for the heterologous GP, is incorporated on the surface of the virions, leading to a better characterized system.

Three components can be distinguished in these pseudoparticles: 1) the surface GP, 2) the vector or core of the virion, and 3) the transgene. In the nineties, boosted by blooming research on HIV and gene therapy, these tools evolved quickly by application-driven modifications of each of the three constituents. Several reviews cover these topics and it is suggested they be read for a better understanding of the subject.[4,5,6]

18.2 Present Lentiviral Production Technologies and Their Limitations

This section focuses on well-characterized production by reverse genetics. Although different alternative backbones can be used (e.g., from adenoviruses, baculoviruses, retroviruses, or vesiculoviruses), the retroviral system developed here is that of the most used vector. Because the retrovirus genome is relatively small and well characterized, it is possible to engineer vector-packaging systems that produce vectors that only encode transgenes and do not replicate competent viruses. The most widely used retrovirus vectors to date are based on murine leukemia viruses (MLVs) or HIV-1. A detailed comparison of these three different systems can be found in the review by Baum et al.[4] In this chapter, discussion will be primarily about the development of HIV systems, although many principles can be applied to other retroviral systems. The main advantage of lentiviruses is their ability to transduce a large range of dividing or nondividing cells, the tropism essentially being driven by the coating envelope pseudotyped.[5,6]

Although relatively small, the HIV genome codes for a group of proteins: three structural polyproteins, *gag* that forms the structure of the virion, *env* (the surface GP that will primarily interact with the host cell in the first steps of infection), and *pol* (the viral replication enzymes), as well as several additional genes with regulatory (*tat*, *rev*) or accessory functions (*vif*, *vpr*, *vpu*, *nef*) flanked by long terminal repeats (LTRs) essential for integration and replication processes.

The first generation of the lentiviral vector was made of the nearly intact HIV genome, with the envelope gene disrupted or partially deleted (to prevent the generation of wild-type HIV replicative virion). The surface GP to be incorporated was trans-complemented with a second plasmid coding for the corresponding gene.[7,8] Subsequent developments led to the design of a second and then a third generation.[9] This was possible thanks to a deeper understanding of the molecular biology of HIV that demonstrated most genes were not essential for the production of lentiviral particles and could be inactivated, trans-complemented, or even removed (accessory genes *vif*, *vpr*, *vpu*, *nef*, or even the regulatory gene *tat*), leaving only the *cis*-acting sequence necessary for the assembly of a viral particle and for a single round of replication. These modifications considerably improved the biosafety issues faced by the first generation.[10–13] Finally, a responsive element to *rev* gene (RRE), introduced in trans as a fourth plasmid, was added to the packaging vector in 1998 by Dull et al.[14] Apart from this plasmid coding for *rev*, the system contains a first plasmid coding for the heterologous envelope, a second for the packaging proteins, and a third coding for a reporter gene (or a gene to be vectorized). It is important to note that only the third one encodes the encapsidation signal ψ that will be recognized by the *gag* protein. Overall, this four-plasmid system design is the safest vector available, as it is less prone to interplasmid recombination and transfers neither the packaging nor the envelope sequences to transduced cells.

Despite all these successive improvements, envelopes from some viruses presently cannot be pseudotyped with the lentiviral system nor with little efficacy. Several reasons can explain this drawback. Using lentiviruses, the budding of the particles, and therefore the incorporation of the envelope proteins, takes place at the plasma membrane most probably in lipid raft.[15–17] The incorporation of most of the surface GP by lentiviruses follows a passive process requiring their colocalization at the budding site.[18–20] The passivity of the process may be revised; as recent data on HIV indicates there may be specific elements in the envelope required for envelope incorporation on the virion surface.[21] Furthermore, some envelopes, like the surface GPs of dengue, autoassemble in the endoplasmic reticulum (ER), forming virus-like particles that mature in transport vesicles until they are released at the cell surface. Therefore, the chance to be incorporated in lentiviral particles at the cell surface is very limited, if not null. To improve this, Kowolik and Yee showed that replacing the cytoplasmic tail of Sendai virus F protein with the one from an HIV-1 envelope increases its incorporation into the pseudotyped particles.[22] This strategy does not always work, and it has been shown that some proteins are still excluded despite their colocalization.[18,23,24]

Once the compatible sequences have been found, the production can be done either by transient transfections or by establishment of a stable cell line. These packaging cell lines, if successful, allow direct or inducible production on a larger scale and with more reproducibility. They are also safer as they can be characterized before use and tested for replication-competent viruses.[25–28] In all cases, the transfection conditions for the production of lentiviral particles are crucial to reach the maximum yield.[29–32] Among those parameters should be mentioned the following: cell type, cell density, growth medium composition, transfection reagent, time of harvest, concentration, and purification of particles from the supernatant. The two different processes used to introduce the nucleic acids into the producing cells are based on physical (mainly electroporation) or chemical methods based on mixing of negatively charged DNA by cationic reagents such as salt (calcium phosphate), polymer (diethylaminoethyl [DEAE]-dextran), or lipids (lipofectants).[33,34] Finally, depending on the nature of the surface GP, some particles may need additional treatments for efficient release during production. For example, lentiviral particles pseudotyped with hemagglutinin of influenza require neuraminidase for their release from the surface of producing cells followed, for so-called "low pathogenic" influenza, by trypsin activation before infection.[35–37]

It should also be mentioned that the sequence itself can drastically improve production yield. In particular, it is now possible to synthesize a sequence with additional features: codon optimization for expression in mammalian (or other) cell types, introduction of the Kozak motif for proper initiation,[38] addition of a tag for better characterization (care must be taken as this may affect the formation of particles) of the second stop codon for efficient termination or of subcloning sites, and depletion of cryptic splices' sites. The choice of the expression cell type should take into consideration the differences in the glycosylation pattern that may affect the final activity of the protein and, by extension, of the pseudoparticles (Fig. 18.1 upper part).

18.3 Pseudoparticles Characterization and Titration

Once the pseudoviruses have been produced, several layers of characterization and titration are required. Biochemical analysis will demonstrate that the envelope is properly incorporated on viral particles (e.g., by analysis of successive fraction of the sucrose gradient using western blotting to codetect the envelope and the core protein *p24* and by polymerase chain reaction [PCR] for the encapsidated reporter gene). Ideally, this should be completed by electromicroscopy to assess the shape and size of the particles, with, if available, a gold-labeled antibody against the envelope. For most applications other than gene vectorization, the pseudoparticles will have to faithfully reproduce the surface of the original live virus

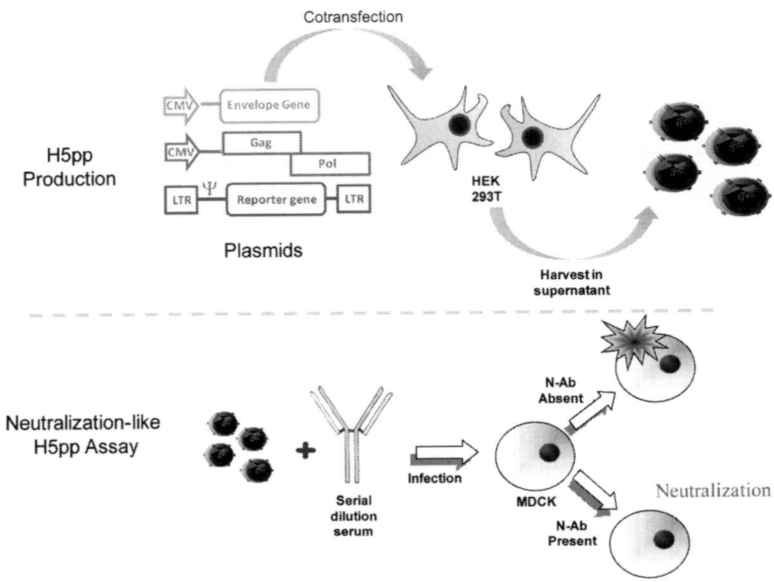

Figure 18.1 (Upper part) Pseudotyped lentiviral particles are produced in HEK 293 cells by cotransfection of plasmids coding, respectively, for the envelope of the virus of interest, for the *gag/pol* polyprotein that will make the core of the virion, and for the reporter gene. Only the RNA from the last plasmid containing the encapsidation signal (ψ) will be incorporated in the particle. The pseudoparticles are harvested in the supernatant of the producing cells, clarified and concentrated by ultracentrifugation, and titered. (Lower part) This is the neutralization assay. First a standardized amount of pseudoparticles are added to serial dilution of sera and incubated for 1 to 2 hours at 37°C before being adding to the monolayer of MDCK cells. The transduced cells are incubated for 48 hours at 37°C before the readout. Either the sera tested contains N-Abs that prevent the infection and no signal from the reporter gene can be measured, or there is no N-Ab and there is the same expression of the reporter gene as the "control" condition of infection in the absence of sera. *Abbreviation*: MDCK, Madin–Darby canine kidney.

(receptor binding, detection by antibodies, infection). Therefore, a functional characterization will be needed and should demonstrate the proper folding, oligomerization, and surface density of the envelope.

Regarding the titration, several methods are currently used. These methods include the assessment of vector RNA in the supernatant, DNA in transduced cells, and vector expression in transduced cells (using a reporter gene). A comparison of these methods can be found in the works published by Sastry et al.[39] and Geraerts et al.[40] Nevertheless, this is insufficient when the envelope is the main focus of the application, as these methods do not correlate with the quantification of the amount of envelope incorporated at the surface of the particles (the methods' postinfection does it indirectly). Furthermore, when the particles are used in a neutralization assay instead of a replicative virus, it is important to differentiate between the functional particles that have a proper surface and an encapsidated reporter gene from the particles that do not incorporate the reporter gene (but that will still be recognized by antibodies if they have the proper surface) or from particles that have the reporter gene but did not (or at lower level) take the envelope. All three kinds of particles can be found in the supernatant of producing cells and may introduce batch-to-batch variations. Therefore, both biochemical and functional characterization are required for each batch to ensure normalization and robustness of the assay using the pseudotyped particles (Fig. 18.1 lower part).

18.4 Applications to a Neutralization-Based Sero-Diagnostic Assay

As mentioned earlier, the range of applications of pseudoviruses is only limited by imagination and the knowledge of the degree of similarity between the recombinant particles and the original viruses mimicked by the pseudoparticle. Pseudotyped viral particles have been used successfully in neutralization tests (Table 18.1) from earlier works at the end of the seventies (using replicative chimera from coinfected cells for extended tropism).[2,41–43] At this time, nonreplicated recombinant particles are preferred as a substitute for live virus for sera-diagnostic assays.[35,44–49] Compared to the live virus, pseudoparticles present several advantages: are safer (replication deficient, no need to grow the virus, particularly important with viruses that require biosafety level 3 or 4 containment),[47,50–52] are more reproducible once fully characterized and optimized (e.g., does not generate adaptive mutation by too frequent passage),

Table 18.1 Illustration of a range of envelope GPs from BSL-4 to BSL-2 pseudotyped viruses used in sero-neutralization

Surface GP	Core	Report gene	References
NiV-F/G	VSV	Luc	50
EboV-G	VSV	GFP	47
RabV-G	HIV/MLV	GFP/Luc	52
EBLV-G	HIV/MLV	GFP/Luc	52
HtnV-G	VSV	GFP	46
SeoV-G	VSV	GFP	46
SARS-CoV-S	VSV	GFP	49, 51
SARS-CoV-S	MLV	GFP	48
HPAI-H5	HIV	Luc	35–37, 53
Seasonal influenza HA	HIV	Luc	37
HCV-E1/E2	MLV	GFP	45
HIV-gp140	MLV	β-Gal	44
HTLV	VSV	None	42, 43
BLV	VSV	None	1, 41
CMV	VSV	None	2

Abbreviations: NiV, Nipah virus; EbV, Ebola virus; RabV, rabies virus; EBV, European bat virus; HtnV, Haanta virus; SeoV, Seol virus; SARS-CoV, severe acute respiratory syndrome coronavirus; H5N1, HPAI, highly pathogenic avian influenza; HCV, hepatitis C virus; HTLV, human T-cells leukemia; BLV, bovine leukemia virus; CMV, cytomegalovirus; VSV, vesicular stomatitis virus; MLV, murine leukemia virus; Luc, luciferase; GFP, green fluorescent protein; β-Gal, β-galactosidase.

could be produced with a high yield (codon optimization), and are flexible (e.g., easy to introduce point mutation for molecular study). An illustration of production and characterization of avian influenza H5 pseudotyped lentiviral particles can be found in Ref. 35. These particles have been successfully used in sero-diagnosis/sero-epidemiology, and the thorough validation of this assay has been described in Ref. 53.

18.5 Perspectives for the Use of Pseudoparticles in Serology and Other Applications

Despite these apparent successes, neutralization-like pseudo-particle-based assays present some limitations apart from those common to other serological assays (e.g., cross-reactivity, window of detection, need of continual adaptation to circulating strain). The most important, as mentioned before, is that not all the envelopes can pseudotype with lentiviral vectors, in particular those like the flaviviruses, which bud intracellularly (e.g., ER).[54] Tomorrow's challenge will be to address that question with a new kind of vector derived from one of these viruses. This will require a better understanding of viral budding at a molecular, if not atomic, level and the identification of encapsidation signals. A second challenge will be to have better standardization in both biochemical and functional characterizations as well as in the titration of such particles. In particular, analysis will have to demonstrate the degree of similitude of the surface between the pseudoparticle and the original virus; modern techniques of electromicroscopy and/or imaging definitively will play a crucial role. The function of a protein at the surface of the pseudoparticle may differ from one assay to the other (e.g., receptor binding as in a hemagglutination assay, conformational antigen exposure to antibodies as in a neutralization assay); therefore, the functional characterization also may be different. Finally, serological multiplexing assays are currently in development, following what was done with molecular diagnostics (multiplex PCR).[55-57] The multiplex approach, although more complex to validate, is important in sample sparing strategies, for example, when the volume of the original samples is limited (when obtaining samples from small animals like bats or from young children) or when a large number of analyses are expected to be performed on a given sample.

References

1. Zavada, J., L. Cerny, A. D. Altstein, and Z. Zavadova. Pseudotype particles of vesicular stomatitis virus with surface antigens of bovine leukaemia virus—VSV (BLV)—as a sensitive probe for detecting antibodies in the sera of spontaneously infected cattle. 1978. *Acta Virol.*, **22**, 91–96.
2. Vonka, V., H. Zavadova, and D. Polakova. A vesicular stomatitis virus (cytomegalovirus) pseudotype and its use in neutralization tests. 1979. *Intervirology*, **12**, 32–38.

3. Page, K. A., N. R. Landau, and D. R. Littman. Construction and use of a human immunodeficiency virus vector for analysis of virus infectivity. 1990. *J. Virol.*, **64**, 5270–5276.
4. Baum, C., A. Schambach, J. Bohne, and M. Galla. Retrovirus vectors: toward the plentivirus? 2006. *Mol. Ther.*, **13**(6), 1050–1062.
5. Stevenson, M. Portals of entry: uncovering HIV nuclear transport pathways. 1996. *Trend Cell. Biol.*, **6**, 9–15.
6. Mochizuki, H., J. P. Schwartz, K. Tanaka, R. Brady, and J. Reiser. High-titer human immunodeficiency virus type 1-based vector systems for gene delivery into nondividing cells. 1998. *J. Virol.*, **72**(11), 8873–8883.
7. Helseth, E., M. Kowalski, D. Gabuzda, U. Olshevsky, W. Haseltine, and J. Sodroski. Rapid complementation assays measuring replicative potential of human immunodeficiency virus type 1 envelope glycoprotein mutants. 1990. *J. Virol.*, **64**, 2416–2420.
8. Page, K. A., N. R. Landau, and D. R. Littman. Construction and use of a human immunodeficiency virus vector for analysis of virus infectivity. 1990. *J. Virol.*, **64**, 5270–5276.
9. Buchschacher, G. L., and F. Wong-Staal. Development of lentiviral vectors for gene therapy for human diseases. 2000. *Blood*, **95**(8), 2499–2504.
10. Reiser, J., G. Harmison, S. Kluepfel-Stahl, R. O. Brady, S. Karlsson, and M. Schubert. Transduction of nondividing cells pseudotyped defective high-titer HIV type 1 particles. 1996. *Proc. Natl. Acad. Sci. U S A*, **93**, 15266–15271.
11. Zufferey, R., D. Nagy, R. J Mandel., L. Naldini, and D. Trono. Multiply attenuated lentiviral vector achieves efficient gone delivery in vivo. 1997. *Nat. Biotechnol.*, **15**, 871–875.
12. Kim, V. N., K. Mitrophanous, S. M. Kingsman, and A. J. Kingsman. Minimal requirement for a lentivirus vector based on human immunodeficiency virus type 1. 1998. *J. Virol.*, **72**, 811–816.
13. Srinivasakumar, N., N. Chazal, C. Helga-Maria, S. Prasad, M. Hammarskjold, and D. Rekosh. The effect of viral regulatory protein expression on gene delivery by human immunodeficiency virus type 1 vectors produced in stable packaging cell lines. 1997. *J. Virol.*, **71**, 5841–5848.
14. Dull, T., R. Zufferey, M. Kelly, R. J Mandel., M. Nguyen, D. Trono, and L. A. Naldini. Third-generation lentivirus vector with a conditional packaging system. 1998. *J. Virol.*, **72**, 8463–8471.
15. Pickl, W. F., F. X. Pimentel-Muinos, and B. Seed. Lipid rafts and pseudotyping. 2001. *J. Virol.*, **75**(15), 7175–7183.

16. Sandrin, V., and F. L. Cosset. Intracellular versus cell surface assembly of retroviral pseudotypes is determined by the cellular localization of the viral glycoprotein, its capacity to interact with Gag, and the expression of the Nef protein. 2006. *J. Biol. Chem.*, **281**(1), 528–542.
17. Sandrin, V., D. Muriaux, J. L. Darlix, and F. L. Cosset. Intracellular trafficking of Gag and Env proteins and their interactions modulate pseudotyping of retroviruses. 2004. *J. Virol.*, **78**(13), 7153–7164.
18. Cantin, R., S. Methot, and M. J. Tremblay. Plunder and stowaways: incorporation of cellular proteins by enveloped viruses. 2005. *J. Virol.*, **79**(11), 6577–6587.
19. Hammarstedt, M., and H. Garoff. Passive and active inclusion of host proteins in human immunodeficiency virus type 1 gag particles during budding at the plasma membrane. 2004. *J. Virol.*, **785**(11), 686–697.
20. Kolegraff, K., P. Bostik, and A. A. Ansari. Characterization and role of lentivirus-associated host proteins. 2006. *Exp. Biol. Med.*, **231**(3), 252–263.
21. Yue, L., L. Shang, and E. Hunter. Truncation of the membrane-spanning domain of human immunodeficiency virus type 1 envelope glycoprotein defines elements required for fusion, incorporation and infectivity. 2009. *J. Virol.*, **83**(22), 11588–11598.
22. Kowolik, C. M., and J. K. Yee. Preferential transduction of human hepatocytes with lentiviral vectors pseudotyped by Sendai virus F protein. 2002. *Mol. Ther.*, **5**(6), 762–769.
23. Christodoulopoulos, I., and P. M. Cannon. Sequences in the cytoplasmic tail of the gibbon ape leukemia virus envelope protein that prevent its incorporation into lentivirus vectors. 2001. *J. Virol.*, **75**(9), 4129–4138.
24. Hammarstedt, M., and H. Garoff. Passive and active inclusion of host proteins in human immunodeficiency virus type 1 gag particles during budding at the plasma membrane. 2004. *J. Virol.*, **785**(11), 686–697.
25. Kuate, S., R. Wagner, and K. Uberla. Development and characterization of a minimal inducible packaging cell line for simian immunodeficiency virus-based lentiviral vectors. 2002. *J. Gene Med.*, **4**(4), 347–355.
26. Ni, Y., S. Sun, I. Oparaocha, L. Humeau, B. Davis, R. Cohen, G. Binder, Y. N. Chang, V. Slepushkin, and B. Dropulic. Generation of a packaging cell line for prolonged large-scale production of high-titer HIV-1-based lentiviral vector. 2005. *J. Gene Med.*, **7**(6), 818–834.
27. Cockrell, A. S., H. Ma, K. Fu, T. J. McCown, and T. Kafri. A trans-lentiviral packaging cell line for high-titer conditional self-inactivating HIV-1 vectors. 2006. *Mol. Ther.*, **14**(2), 276–284.

28. Ghani, K., S. Cottin, A. Kamen, and M. Caruso. Generation of a high-titer packaging cell line for the production of retroviral vectors in suspension and serum-free media. 2007. *Gene Ther.*, **14**(24), 1705–1711.

29. Karolewski, B. A., D. J. Watson, M. K. Parente, and J. H. Wolfe. Comparison of transfection conditions for a lentivirus vector produced in large volumes. 2003. *Human Gene Ther.*, **14**(14), 1287–1296.

30. Devitt, G., M. Thomas, A. M. Klibanov, T. Pfeiffer, and V. Bosch. Optimized protocol for the large scale production of HIV pseudovirions by transient transfection of HEK293T cells with linear fully deacylated polyethylenimine. 2003. *J. Virol. Methods*, **146**(1–2), 298–304.

31. Mitta, B., M. Rimann, and M. Fussenegger. Detailed design and comparative analysis of protocols for optimized production of high-performance HIV-1-derived lentiviral particles. 2005. *Metab. Eng.*, **7**(5–6), 426–436.

32. Sena-Esteves, M., J. C. Tebbets, S. Steffens, T. Crombleholme, and A. W. Flake. Optimized large-scale production of high titer lentivirus vector pseudotypes. 2004. *J. Virol. Methods*, **122**(2), 131–139.

33. Graham, F. L., and A. J. van der Eb. A new technique for the assay of infectivity of human adenovirus 5 DNA. 1973. *Virology*, **52**, 456–467.

34. Lee, J. T., and R. Jaenisch. A method for high efficiency YAC lipofection into murine embryonic stem cells. 1996. *Nucleic Acids Res.*, **24**(24), 5054–5055.

35. Nefkens, I., J. M. Garcia, C. S. Ling, N. Lagarde, J. Nicholls, D. J. Tang, M. Peiris, P. Buchy, and R. Altmeyer. Hemagglutinin pseudotyped lentiviral particles: characterization of a new tool for avian influenza H5N1 influenza serodiagnostic. 2007. *J. Clin. Virol.*, **39**(1), 27–33.

36. Temperton, N. J., K. Hoschler, D. Major, C. Nicolson, R. Manvell, V. M. Hien, Q. Ha do, M. de Jong, M. Zambon, Y. Takeuchi, and R. A. Weiss. A sensitive retroviral pseudotype assay for influenza H5N1-neutralizing antibodies. 2007. *Influenza Other Respir. Viruses*, **1**(3), 105–112.

37. Wang, W., E. N. Butler, V. Veguilla, R. Vassell, J. T. Thomas, M. Moos, Jr., Z. Ye, K. Hancock, and C. D. Weiss. Establishment of retroviral pseudotypes with influenza hemagglutinins from H1, H3, and H5 subtypes for sensitive and specific detection of neutralizing antibodies. 2008. *J. Virol. Methods*, **153**(2), 111–119.

38. Kozak, M. An analysis of 5'-noncoding sequences from 699 vertebrate messenger RNAs. *Nucleic Acids Res.*, **15**, 8125–8148; M. Kozak (1990) Downstream secondary structure facilitates recognition of initiator codons by eukaryotic ribosomes. 1987. *Proc. Natl. Acad. Sci. U S A*,

87, 8301–8305; M. Kozak An analysis of vertebrate mRNA sequences: intimations of translational control. 1991. *J. Cell Biol.*, **115**, 887–903.

39. Sastry, L., T. Johnson, M. J. Hobson, B. Smucker, and K. Cornetta. Titering lentiviral vectors: comparison of DNA, RNA and marker expression methods. 2002. *Gene Ther.*, **9**, 1155–1162.
40. Geraerts, M., S. Willems, V. Baekelandt, Z. Debyser, and R. Gijsbers. Comparison of lentiviral vector titration methods. 2006. *BMC Biotechnol.*, **6**(July 12), 34.
41. Zavada, J., L. Cerny, Z. Zavadova, J. Bozonova, and D. Alstein. A rapid neutralization test for antibodies to bovine leukemia virus, with the use of rhabdovirus pseudotypes. 1979. *J. Natl. Cancer Inst.*, **62**(1), 95–101.
42. Clapham, P., K. Nagy, and R. A. Weiss. Pseudotypes of human T-cell leukemia virus types 1 and 2: neutralization by patients' sera. 1984. *Proc. Natl. Acad. Sci. U S A*, **81**, 2886–2889.
43. Hoshino, H., P. R. Clapham, R. A. Weiss, I. Miyoshi, M. Yoshida, and M. Miwa. Human T-cell leukemia virus type I: pseudotype neutralization of Japanese and American isolates with human and rabbit sera. 1985. *Int. J. Cancer*, **36**(6), 671–675.
44. Kim, Y. B., M. K. Lee, D. P. Han, and M. W. Cho. Development of a safe and rapid neutralization assay using murine leukemia virus pseudotyped with HIV type 1 envelope glycoprotein lacking the cytoplasmic domain. 2001. *AIDS Res. Hum. Retroviruses*, **17**(18), 1715–1724.
45. Bartosch, B., J. Bukh, J. C. Meunier, C. Granier, R. E. Engle, W. C. Blackwelder, S. U. Emerson, F. L. Cosset, and R. H. Purcell. In vitro assay for neutralizing antibody to hepatitis C virus: evidence for broadly conserved neutralization epitopes. 2003. *Proc. Natl. Acad. Sci. U S A*, **100**(24), 14199–14204.
46. Ogino, M., H. Ebihara, B. H. Lee, K. Araki, A. Lundkvist, Y. Kawaoka, K. Yoshimatsu, and J. Arikawa. Use of vesicular stomatitis virus pseudotypes bearing hantaan or seoul virus envelope proteins in a rapid and safe neutralization test. 2003. *Clin. Diagn. Lab. Immunol.*, **10**(1), 154–160.
47. Ito, H., S. Watanabe, A. Takada, and Y. Kawaoka. Ebola virus glycoprotein: proteolytic processing, acylation, cell tropism, and detection of neutralizing antibodies. 2001. *J. Virol.*, **75**(3), 1576–1580.
48. Temperton, N. J., P. K. Chan, G. Simmons, M. C. Zambon, R. S. Tedder, Y. Takeuchi, and R. A. Weiss. Longitudinally profiling neutralizing antibody response to SARS coronavirus with pseudotypes. 2005. *Emerg. Infect. Dis.*, **11**(3), 411–416.

49. Fukushi, S., T. Mizutani, M. Saijo, I. Kurane, F. Taguchi, M. Tashiro, and S. Morikawa. Evaluation of a novel vesicular stomatitis virus pseudotype-based assay for detection of neutralizing antibody responses to SARS-CoV. 2006. *J. Med. Virol.*, **78**(12), 1509–1512.

50. Tamin, A., B. H. Harcourt, M. K. Lo, J. A. Roth, M. C. Wolf, B. Lee, H. Weingartl, J. C. Audonnet, W. J. Bellini, and P. A. Rota. Development of a neutralization assay for Nipah virus using pseudotype particles. 2009. *J. Virol. Methods*, **160**(1–2), 1–6.

51. Ge, J., Z. Wen, X. Wang, S. Hu, Y. Liu, X. Kong, H. Chen, and Z. Bu. Generating vesicular stomatitis virus pseudotype bearing the severe acute respiratory syndrome coronavirus spike envelope glycoprotein for rapid and safe neutralization test or cell-entry assay. 2006. *Annu. N. Y. Acad. Sci.*, **1081**, 246–248.

52. Wright, E., N. J. Temperton, D. A. Marston, L. M. McElhinney, A. R. Fooks, and R. A. Weiss. Investigating antibody neutralization of lyssaviruses using lentiviral pseudotypes: a cross-species comparison. 2008. *J. Gen. Virol.*, **89**(Pt 9), 2204–2213.

53. Garcia, J. M., N. Lagarde, E. S. Ma, M. D. de Jong, and J. S. Peiris. Optimization and evaluation of an influenza A (H5) pseudotyped lentiviral particle-based serological assay. 2010. *J. Clin. Virol.*, **47**(1), 29–33.

54. Welsch, S., S. Miller, I. Romero-Brey, A. Merz, C. K. Bleck, P. Walther, S. D. Fuller, C. Antony, J. Krijnse-Locker, and R. Bartenschlager. Composition and three-dimensional architecture of the dengue virus replication and assembly sites. 2009. *Cell Host Microbe*, **5**(4), 365–375.

55. Markoulatos, P., N. Siafakas, and M. Moncany. Multiplex polymerase chain reaction: a practical approach. 2002. *J. Clin. Lab. Anal.*, **16**, 47–51.

56. Henegariu, O., N. A. Heerema, S. R. Dlouhy, G. H. Vance, and P. H. Vogt. Multiplex PCR: critical parameters and step-by-step protocol. 1997. *BioTechniques*, **23**, 504–511.

57. Drummond, J. E., E. E. Shaw, J. M. Antonello, T. Green, G. J. Page, C. O. Motley, K. A. Wilson, A. C. Finnefrock, X. Liang, and D. R. Casimiro. Design and optimization of a multiplex anti-influenza peptide immunoassay. 2008. *J. Immunol. Methods*, **334**, 11–20.

Chapter 19

Nucleic Acid Isothermal Amplification Technologies and Point-of-Care Diagnostics

Tanya M. Ferguson and Angelika Niemz
Keck Graduate Institute of Applied Life Sciences, Claremont, CA, USA
tanya_ferguson@kgi.edu

Over the past two decades, the development of nucleic acid amplification technologies (NAATs) and molecular diagnostics has enhanced the detection, characterization, and monitoring of infectious agents, which includes monitoring the emergence of drug resistance worldwide. In developed regions such as Europe and North America, molecular diagnostics is by now the standard of care for many applications but generally involves complex instrumentation operated by highly skilled personnel in a centralized laboratory environment. In developing countries, molecular diagnostics has proven to be equally critical to providing appropriate health care, yet a great need exists for technologies that enable molecular diagnostics to be performed at the point of care (POC) in low-resource settings.[1] Traditional pathogen identification and disease monitoring in low-resource settings often involves simple microscopic tests (e.g., smear tests for malaria and tuberculosis [TB]), but these tests suffer

Viral Diagnostics: Advances and Applications
Edited by Robert S. Marks, Leslie Lobel, and Amadou Alpha Sall

ISBN 978-981-4364-43-0 (Hardcover), 978-981-4364-44-7 (eBook)
www.panstanford.com

from limited sensitivity and specificity. Culture-based methods can increase sensitivity and specificity but are time consuming, highly specialized, and expensive, thereby being often unsuitable for low-resource settings. Rapid diagnostic tests based on lateral flow immunoassays address the needs of low-resource settings and are used increasingly. However, these immunoassays again suffer from limited sensitivity and are often unable to distinguish new from persistent infections. Thus, lateral flow–based diagnostic tests are of limited utility in disease monitoring and in aiding treatment decisions (e.g., for human immunodeficiency virus [HIV]). With the advent of NAATs, the detection of pathogens, including those that are unculturable and/or highly contagious, may become possible in a timely, sensitive, and inexpensive manner.[2] Moving molecular diagnostics to the POC is advantageous for rapid pathogen diagnosis and disease surveillance and to facilitate effective treatment and containment, particularly concerning life-threatening disease or disease outbreaks in both developed and developing countries.

The polymerase chain reaction (PCR), particularly in real time, has revolutionized the sensitive, specific, and rapid detection of infectious agents present in clinical samples as compared to culture-based techniques. Real-time PCR-based systems have been successfully implemented for the detection of infectious agents at the POC, including GeneXpert® from Cepheid (Sunnyvale, CA), PCR-Light® from Enigma Diagnostics (Wiltshire, UK), Liat™ (laboratory-in-a-tube) analyzer from IQ^{uum} Inc. (Allston, MA), and HyBeacon® -OptiGENE technologies (LGC, Middlesex, UK; Osmetech, GA).[3] These four POC systems are all self-contained real-time thermocycling benchtop instruments capable of amplifying DNA or RNA targets in 20 minutes to 2 hours. Despite the obvious advantages of PCR-based systems, there are some definite disadvantages, particularly when focusing on their application in low-resource settings. In particular, the need for rapid thermocycling increases the expense and complexity of the systems involved. To properly address the need of low-resource settings a systems approach is required, which should take into account instrumentation requirements, including cost and complexity; reagent and consumables costs, reagent storage, and supply chain issues; technician training and time; and interpretation of the results for proper patient treatment.

Isothermal NAATs require a single reaction temperature for nucleic acid amplification, which translates into less complex and less expensive machinery, a significant advantage compared to PCR for the diagnosis of infectious diseases at the POC, particularly in low-resource settings. Some of the isothermal amplification schemes being used for pathogen identification (see Table 19.1) include loop-mediated amplification (LAMP),[4] strand displacement amplification (SDA),[5] exponential amplification reaction (EXPAR),[6] helicase-dependent amplification (HDA),[7] rolling-circle amplification (RCA)[8] and ramification amplification reaction (RAM),[9] nucleic acid sequence-based amplification (NASBA)[10] and transcription-mediated amplification (TMA),[11] strand-mediated amplification of RNA technology (SMART),[12] recombinase polymerase reaction (RPA),[13] nicking enzyme amplification reaction (NEAR),[14] and single-primer isothermal amplification (SPIA).[15]

19.1 Isothermal Amplification Technologies

Nucleic acid isothermal amplification technologies can be classified into three major groups according to the method of target detection: (1) methods based on RNA transcription, which include NASBA, TMA, and SMART; (2) methods based on DNA replication, which include HDA and RPA; and (3) methods based on strand displacement utilizing a single enzyme, which include RCA, RAM, and LAMP, or multiple enzymes, as in SDA and EXPAR.

19.1.1 *Target Detection via RNA Transcription*

NASBA and TMA are two closely related target amplification technologies based on in vitro transcription. In both methods, the initial target and amplicon consist of RNA rather than DNA. To initiate NASBA, first-stranded cDNA is synthesized on the RNA target through the action of avian myeloblastosis virus reverse transcriptase (AMV-RT) using a primer containing a T7 RNA polymerase promoter region at the 5′ end and a target-specific region at the 3′ end. The RNA strand of the resulting RNA-DNA heteroduplex is then removed through digestion with RNAse H. A

Table 19.1 Isothermal amplification reactions and reported targets

Amplification method	Manufacturer	Reaction time	Reaction temp.	Reported target(s)	References
EXPAR	Keck Graduate Institute (Claremont, CA)	10–20 min	55°C	HSV-1 and HSV-2	6, 17, 18
HDA	BioHelix Corp. (Beverly, MA)	30 min–2 h	37°C	Ebola (armored RNA), MRSA, *H. pylori*, *B. anthracis*, *C. difficile*, *N. gonorrhoeae*, *Treponema*	7, 19–23
LAMP	Eiken Chemical Co. (Tochigi, Japan)	1 h	65°C	Dengue virus, HBV, SARS, influenza (HA1/HA3), *M. tuberculosis*, West Nile, *V. cholerae*, *Salmonella*, *Legionella*, *Listeria*, *E. coli*, *Campylobacter*	24–27
NASBA	Organon-Teknika, Corp. (Durham, NC); bioMérieux, Inc. (Marcy l'Etoile, FR)	90 min	41°C	HCV, HIV-1, *Pneumoniae sp.*, *V. cholerae*, *T. brucei*, *C. trachomatis*, *N. gonorrhoeae*, norovirus, bocavirus	28–33
RPA	TwistDx (Cambridge, UK)	10–30 min	24°C –45°C (37°C is optimal)	MRSA	13, 34
RCA/RAM	Molecular Staging, Inc. (New Haven, CT)/New York University (NY)	1–2 h	63°C–65°C	HBV, SARS, HIV1, HPV, Borrelia, EBV, *C. trachomatis*, *E. coli* O157:H7	9, 35–39
SDA	Becton-Dickinson (Sparks, MD)	15 min–1 h	50°C–60°C	HCV, HIV, *C. trachomatis*, *N. gonorrhoeae*, *M. tuberculosis*	5, 40, 41
SMART, cytAMP	Cytocell, Ltd. (Banbury, UK); British Biocell Int. (Cardiff, UK)	2.5–3 h	41°C	HBV, *M. tuberculosis*, MRSA, *E. coli*	12, 42, 43
TMA	Gen-Probe (San Diego, CA)	15–30 min	42°C	HCV, HIV, HBV, West Nile, dengue virus, *C. trachomatis*, *N. gonorrhoeae*, *M. tuberculosis*, *Streptococcus sp.*	2, 11, 44–47

Abbreviations: HSV, herpes simplex virus; HCV, hepatitis C virus; HBV, hepatitis B virus; MRSA, methicillin-resistant *Staphylococcus aureus*; SARS, severe acute respiratory syndrome; HPV, human papillomavirus; EBV, Epstein–Barr virus; *C. trachomatis*, *Chlamydia trachomatis*; *H. pylori*, *Helicobacter pylori*; *B. anthracis*, *Bacillus anthracis*; *C. difficile*, *Clostridium difficile*; *N. gonorrhoeae*, *Neisseria gonorrhoeae*; *M. tuberculosis*, *Mycobacterium tuberculosis*; *V. cholera*, *Vibrio cholerae*; *E. coli*, *Escherichia coli*; *T. brucei*, *Trypanosoma brucei*.

second sequence-specific primer anneals to the now single-stranded cDNA, and through the DNA polymerase activity of AMV-RT a double-stranded DNA template is produced. T7 RNA polymerase then recognizes the promoter region at the upstream end of this double-stranded DNA template, and through RNA transcription approximately 100–1000 copies of antisense single-stranded RNA transcripts are produced per DNA template. Newly transcribed RNA reenters the amplification cycle, with DNA primers annealing in the reverse order, thereby producing 10^{10}-fold exponential amplification.[16]

NASBA has been licensed to BioMérieux (Marcy l'Etoile, FR), which has commercialized the technology internationally for the detection of many different infectious agents, including HIV, methicillin-resistant *Staphylococcus aureus* (MRSA), human papillomavirus (HPV); pathogens associated with lower respiratory tract infections (LRTIs), including respiratory syncitial virus (RSV), Human metapneumovirus (hMPV), influenza, and severe acute respiratory syndrome (SARS); and pathogens associated with central nervous system (CNS) infections, including herpes simplex virus (HSV), enterovirus, and cytomegalovirus (CMV). Other laboratories have developed NASBA-based assays for the detection of bocavirus (respiratory infections),[28] norovirus (viral gastroenteritis),[30] HIV-1,[48] and hepatitis C virus (HCV). BioMérieux initially coupled NASBA with an electrochemiluminescence-based readout but has recently switched to detection using molecular beacons. In research settings, the reaction has been optimized with varying detection schemes, including real-time fluorescence and the colorimetric end-point assay enzyme-linked oligonucleotide capture (EOC). Instrumentation requirements for the real-time detection technologies are too expensive for most low-resource settings, and therefore EOC may be more suitable due to its simple readout.

TMA is similar to NASBA but uses Moloney murine leukemia virus reverse transcriptase (MMLV-RT) instead of AMV-RT. MMLV-RT contains endogenous RNAse H activity; therefore TMA requires two enzymes (MMLV-RT and RNA polymerase), whereas NASBA requires three (MMLV-RT, RNAse H, and RNA polymerase). Gen-Probe, Inc. (San Diego, CA) has focused its efforts on commercializing TMA. For many reactions involving TMA, optical detection of the RNA amplified from an infectious agent is achieved using a hybridization protection assay (HPA). HPA utilizes chemiluminescent acridinium ester-labeled probes and a two-step process to differentiate between a probe bound to a target versus an unbound single-stranded probe in the reaction. Differentiation is based on sterics: for an unhybridized, single-stranded labeled probe, the acridinium ester is destroyed by alkaline hydrolysis. The acridinium ester in hybridized probes is protected and not subject to hydrolysis. In the first step, any free probe is thus destroyed; in the second step, the hybridized probe is dissociated and hydrolyzed and the chemiluminescence can be detected by a luminometer.[45] TMA-HPA has been used in the detection of HCV, HIV,[11,45] *C. trachomatis, N. gonorrhoeae, M. tuberculosis,* and West Nile virus.[44]

A third isothermal nucleic acid amplification scheme that utilizes RNA transcription for target amplification is SMART. SMART relies on the formation of a three-way junction structure, involving two single-stranded oligonucleotide probes and a DNA or RNA target site. The two probes hybridize to adjacent sites on the target and also to each other, forming a "T" structure, with the assistance of facilitator probes. A DNA polymerase extends one of the hybridized areas to form an active double-stranded RNA polymerase promoter region within the junction, thereby producing large amounts of RNA transcript.[12] The RNA signal generated from SMART is generally detectable via an end-point assay like enzyme-linked oligosorbent assay (ELOSA). The detection of *E. coli*, HBV, *M. tuberculosis*,[12] and MRSA[42] has been demonstrated using the ELOSA end-point assay, though SMART appears to offer some flexibility in terms of end-point detection more suitable for the POC in low-resource settings. An added benefit of SMART is that it is based upon the amplification of the RNA signal generated and not upon nucleic acid target amplification, thereby minimizing the occurrence of carry-over contamination of the target amplicon that may generate false positive results.

19.1.2 *Target Detection via DNA Replication*

A second approach to RNA or DNA detection and amplification is one that is performed via a reaction scheme similar to PCR. In lieu of high melting temperatures to facilitate duplex melting and primer binding, as in PCR, enzymes are used that operate at a single temperature. One isothermal nucleic acid amplification method that performs this way is HDA, which utilizes a DNA helicase to separate double-stranded DNA, as is done naturally in living organisms. Once the target strands are separated by the helicase, single-stranded DNA–binding proteins coat the DNA, allowing sequence-specific primers to hybridize to the single-stranded template, followed by extension via a DNA polymerase. The resulting double-stranded DNA then acts as a substrate for DNA helicase cycling into the next round of amplification reactions.[7] A reverse transcriptase (RT) can also be added to the reaction, resulting in the exponential amplification of cDNA for RNA targets. Overall, 10^{10}-fold exponential amplification of DNA or RNA targets can be accomplished at 37°C in less than one hour. Using HDA, diagnosis of Ebola virus,[19] MRSA,[23] and *C. difficile*[20] has been demonstrated. HDA-amplified products can be detected using real-time fluorescence monitoring, via enzyme-linked immunosorbent assay (ELISA), lateral flow, and microarray technologies. HDA offers much promise in its ability to be utilized in the POC due to low heating requirements and adaptability to different detection platforms.

Another isothermal amplification technology that relies on enzymatically driven DNA replication is RPA. In RPA, primer binding to a double-stranded target is facilitated at low temperature in an adenosine triphosphate (ATP)-driven process through the use of prokaryotic enzymes, recombinases, which guide synthetic primers to complementary target sites of the genome. Under appropriate conditions, the recombinase disassembles spontaneously upon sequence recognition and is replaced by a single-stranded DNA–binding protein, gp32. In the resulting complex, a DNA polymerase with strand displacement activity recognizes and extends the 3′ end of the primer hybridized to the template, resulting in the formation of a complete copy of the amplicon. As in PCR, exponential amplification occurs through cyclical repetition of this process. The detection of MRSA has been reported down to a single copy in 20 minutes, with other pathogen diagnostic applications being investigated. RPA operates optimally at 37°C, with a working temperature range of ambient to 45°C with little variation on assay performance as temperature fluctuates.[34] This offers the added benefit of low-powered heating sources for assay implementation. RPA is also adaptable for rapid end-point detection, like nucleic acid lateral flow (NALF) or immunoassays.[34]

19.1.3 *Target Detection via Strand Displacement*

The most common isothermal amplification method involving target detection via strand displacement using a single enzyme is LAMP. LAMP is at the forefront of isothermal amplification technologies at the POC. LAMP employs four primers (two inner and two outer) that all are specific to six regions of the DNA target. The 5′ and 3′ regions of each inner primer are complementary to two adjacent regions of the target, but the 5′ recognition sequence is located downstream of the 3′ recognition sequence; thus only the 3′ region sequence of the inner primer is able to participate in the initial hybridization event. The sequence targeted by the outer primer is located upstream of the inner primer. After initial heat denaturation, the 3′ region of the forward inner primer binds to the target and is extended by a polymerase with strand displacement activity. The forward outer primer, which has a lower Tm than the 3′ region of the forward inner primer, binds more slowly to its target region, is extended and displaces the first amplification product. After this strand displacement event, the backward inner primer binds to its recognition sequence on the first amplicon and is extended, followed by binding and extension of the backward outer primer, leading to a second strand displacement event. This second displaced strand contains the sense and antisense sequences of the backward and forward inner primers, respectively. The previously unhybridized regions of these primers are now able to fold back onto the amplicon, generating a dumbbell-shaped DNA strand. Since the forward inner primer sequence is present in the antisense direction, this folding-back generates a recessed 3′ priming site, causing a self-primed extension, which converts the dumbbell to a stem-loop structure. A new forward inner primer is able to hybridize to the loop, and the cycle of strand displacement elongation begins, forming large cauliflower-like structures that continue to feed back on the initial reaction, producing 10^9–

10^{10} copies of the target in less than one hour, with inverted repeats located within the stem-loop DNA products.[4] Loop primers have since been added to ensure that the loop structure of the amplicon remains annealed, thereby enhancing assay sensitivity.[24] The largest challenge for expanding the utility of LAMP is its complex primer design. The utility of LAMP has also been enhanced with the addition of an RT for the amplification and detection of RNA targets. RT-LAMP is commercialized by Sysmex Corporation (Kobe, Japan) under the name one-step nucleic acid amplification (OSNA). LAMP has been coupled with a simple and cost-effective turbidimetric end-point detection format. During LAMP (as well as any other NAAT), nucleotide incorporation releases an excess of pyrophosphate ions into the buffer, which forms a white precipitate in the presence of appropriate divalent metal cations. The onset of amplification can thus be detected through real-time turbidity measurements. Through the incorporation of manganese ions and calcein, fluorescence can be used as a readout for LAMP. Calcein's fluorescence is naturally quenched by binding of manganese ions. Pyrophosphate production as a by-product of LAMP removes manganese ions from the buffer through precipitation, and the increased turbidity coupled with restored calcein fluorescence enables an easy visual readout upon excitation with a handheld UV lamp. This detection format is quite easily applicable to low-resource settings in developing countries.

Another isothermal amplification method involving target detection via strand displacement using a single enzyme is RCA, which was adapted from the rolling circle replication used by many plasmids and viruses in nature. RCA requires a circular DNA template and involves two concurrent processes for generating multiple copies thereof concatenated in a linear fashion—(1) DNA polymerase synthesis of sequences complementary to the circular template and (2) unwinding of the circular duplex to allow the polymerase to advance and proceed with replication.[8] RCA's early iterations involved the use of a single circular primer, linear RCA, resulting in the "rolling out" of long, repeated sequences of DNA varying in length, with 10^5-fold amplification being observed in one to two hours. While the exact mechanism has not been completely elucidated, it is believed that a lasso-type complex is formed, allowing for spontaneous fluctuations and rotations of DNA minicircles ($\leq$100 bp in length).[49]

More recent versions of RCA have evolved that utilize multiple primers. One primer is complementary to the circular DNA template, as in linear RCA, and the other primer targets the repeated sequence that is generated by extending the first primer, thereby generating high-molecular-weight branched products. A specific example of this scheme is RAM. RAM utilizes a linear nucleotide probe (C-probe) that contains two flanking 3′ and 5′ regions complementary to the target with a sequence unrelated to the target and an interposed generic primer-binding region. The C-probe is designed such that its 5′ flanking region hybridizes to the target immediately downstream of the 3′ flanking region, thus generating a mini circle, which is then covalently closed through action of a DNA ligase.[50] Following ligation, the circular probe is helically twisted around the target. Upon addition of a DNA polymerase, forward and reverse primers, multiple primer RCA is initiated, leading to a high-molecular-weight branched product, as discussed before. These single-stranded DNA strands serve as templates for further hybridization, extension, and displacement downstream, generating a large "ramified" complex, which makes a 10^9-fold amplification achievable isothermally in one to two hours.[9] As a result of the increased specificity and sensitivity, multiprimed RCA/RAM is more amenable to molecular diagnostics at the POC. The diagnosis of pathogens like HBV,[37] coronavirus (SARS-causing virus),[38] *E. coli* O157:H7,[36] EBV,[50] and *Chlamydia trachomatis*[39] has been demonstrated when monitored in real time. Recently, RCA technologies have been coupled with microarray and immunodetection formats, making them POC friendly.[51]

SDA is one of the earliest isothermal NAATs to be developed and, as its name implies, utilizes strand displacement for target amplification. SDA relies on the combination of activities of two enzymes, the nicking activity of a restriction endonuclease and strand displacement activity of a polymerase, resulting in a cascade of exponential DNA target-generating events with a magnitude of 10^{10}-fold amplification in less than one hour. In SDA, two primers bind to a double-stranded DNA target at binding regions, marking the beginning and end of the sequence to be amplified (amplicon size generally 50–120 bp). Along with the required target recognition sequence at the 3′ end, the primers contain a 5′ overhang not complementary to the target of interest, which contains a single-stranded restriction site. Upon extension of the bottom strand by the polymerase, a double-stranded recognition site is formed, which can be acted upon by the restriction enzyme. However, one of the four deoxyribonucleotide triphosphates (dNTPs) in the reaction mixture contains an α-phosphorothioate modification; therefore the newly synthesized strand opposite the primer is resistant to cleavage. Only the unmodified recognition sequence contained within the primer is cut, generating a recessed 3′-hydroxyl group that is recognized and extended by the DNA polymerase with strand displacement activity. The strand complementary to the primer becomes the template for continued strand-displacing DNA synthesis followed by nicking of the top strand, in the form of linear amplification. Amplicons generated by the forward primer become templates for the reverse primer and vice versa; therefore the reaction becomes exponential. SDA has been adapted for the detection of RNA, where an RT is incorporated into the reaction, resulting in the successful detection of HIV-1 RNA.[52] A variety of other pathogens have been detected using SDA, including *M. tuberculosis*,[40] HCV,[41] *C. trachomatis*, and *N. gonorrhoeae*,[52] with real-time analysis using fluorescence resonance energy transfer (FRET)-based or microchip platforms.

Similar to SDA, EXPAR amplifies short oligonucleotides at 55°C 10^6- to 10^9-fold in minutes by utilizing a combination of strand extension via a thermostable polymerase and single-strand nicking via a nicking endonuclease.[6,17,18,53] The reaction amplifies a short oligonucleotide (trigger X), which can be generated from a genomic target sequence using the fingerprinting reaction.[18] EXPAR is initiated when trigger X transiently binds to the 3′ end of an amplification template consisting of two times the complementary sequence X′. Extension of the trigger sequence via DNA polymerase leads to the formation of the double-stranded nicking enzyme recognition site. The nicking enzyme nicks the top strand, generating another trigger oligonucleotide that is released from the amplification template, which is primarily due to a thermodynamic switch that occurs at 55°C, but may also be attributed to strand displacement by the polymerase. This activated, partially double-stranded, trigger-producing amplification template reenters the linear amplification cycle of extension, nicking and trigger release, leading to continued, linear amplification of the trigger sequence X. Newly generated triggers are also available to prime additional amplification templates present in excess, thus enabling true exponential amplification. HSV-1 and HSV-2 have been detected with EXPAR via real-time fluorescence monitoring, via visual end-point detection using DNA gold-nanosphere aggregation,[17,18] and NALF (unpublished data), making it adaptable and suitable for POC settings.

19.2 NAAT-Compatible End-Point Detection Platforms Suitable for Point-of-Care in Low-Resource Settings

Many of the isothermal NAATs differ in reaction format and detection schemes yet offer many advantages to molecular diagnostics and the POC. Isothermal methods can be more readily incorporated into simple, handheld portable devices for use in the field due to low power requirements as compared to thermal cyclers required for PCR. There are two main end-point detection platforms that are compatible with NAATs and handheld systems and are being explored—(1) microfluidic lab-on-a-chip pathogen-sensing systems, including microarrays, and (2) chromatographic assays, including NALF and antibody-antigen/hapten-based detection (lateral flow or flow through).

Using microfluidics to miniaturize detection systems has enabled the production of next-generation devices for POC diagnostics in low-resource settings. The ability to manipulate small sample volumes in microfluidic channels during extraction and amplification of nucleic acids has been instrumental in establishing lab-on-a-chip and microarray pathogenic sensing systems as a means of integration into one complete unit. The biggest challenge thus far has been to keep the price and complexity within the range amenable to developing countries. However, some breakthroughs have been reported in coupling NAATs and microfluidics. For example, HDA has successfully been demonstrated on-chip for the single and duplexed detection of *N. gonorrhoaea* and *Staphylococcus aureus* (*S. aureus*). Here, HDA was carried out on the surface of a glass slide, while labeled primers result in a measurable fluorescence signal upon amplification that can be detected via laser scanning.[19] The integrated microfluidic electrochemical DNA (IMED) sensor integrates amplification via PCR (though isothermal amplification systems may also work), enzymatic conversion of double-stranded amplicons to single-stranded DNA, and sequence-specific electrochemical detection via a redox reaction.[54] The IMED chips are single-use to limit contamination problems and demonstrate feasibility of NAATs in the POC.

Complementary to microfluidic systems, chromatographic assays provide a simple detection format suitable for POC diagnostic applications in low-resource settings. Lateral flow assays have long been at the forefront of rapid diagnostic technologies developed for low-resource settings and are commonly used (e.g., in the diagnosis of HIV and malaria.) These assays, with home pregnancy tests being the most recognizable, are inexpensive to manufacture, are stable for at least up to a year due (reagent stability on nitrocellulose-based membranes), are relatively fast (results within minutes), and are simple and user friendly in terms of running the test and interpreting the results. However, most current chromatographic assays are not quantitative and often times lack appropriate sensitivity and specificity adequate for diagnostic purposes.[1]

Chromatographic assays or flow-through assays do offer great flexibility in terms of the test format and detection system. Generally, colorimetric conjugates or probes (e.g., gold nanoparticles, dyed latex particles, or europium particles) are conjugated to oligonucleotides, antibodies, antigens, or haptens. Capture reagents (e.g., antibody, oligonucleotide, or antigen) are immobilized onto a nitrocellulose or nylon membrane. Depending upon the analyte being detected in the sample, a readable capture area is generated from a positive reaction when the colored conjugate first binds to the analyte present in the sample and then to the immobilized capture reagent on the membrane. Many research groups have demonstrated improvements in using lateral flow in POC settings for the detection of nucleic acids. NALF has been used to detect PCR-amplified products from infectious agents such as *Plasmodium* species, including malaria,[55] HBV, and HCV.[56] There is still a need for NAATs coupled to this technology in all areas of pathogen diagnosis in low-resource settings.

NAATs have great potential in pathogen diagnosis in developed and developing countries. Isothermal NAATs can facilitate the development of POC molecular diagnostics systems that do not rely on complex, expensive instrumentation and will not require highly trained personnel, as is the case with current microbiological methods and PCR. There are still many hurdles that need to be overcome, including assay durability and stability, systems integration, and overall cost. To be applicable for POC testing in low-resource settings, a system needs to address the entire process from sample in to answer out. Any amplification technology therefore needs to be coupled with upstream sample preparation, which may be achieved through microfluidic platforms, and with simple detection methods downstream, like lateral flow or flow-through chromatographic assays.

References

1. Yager, P., G. J. Domingo, and J. Gerdes. Point-of-care diagnostics for global health. 2008. *Annu. Rev. Biomed. Eng.*, **10**, 107–144.
2. Li, H., and Y. Tang. 2006. In vitro nucleic acid amplification: an introduction, in *Advanced Techniques in Diagnostic* (Tang, Y.-W., and Stratton, C. W., eds), Springer, pp. 158–165.
3. Holland, C. A., and F. L. Kiechle. Point-of-care molecular diagnostic systems: past, present and future. 2005. *Curr. Opin. Microbiol.*, **8**(5), 504–509.
4. Notomi, T., and H. Okayama. Loop-mediated isothermal amplification of DNA. 2000. *Nucleic Acids Res.*, **28**(12), e63.
5. Nadeau, J. G., J. B. Pitner, C. P. Pitner, J. L. Pitner, C. H. Pitner, and C. M. Pitner. Real-time, sequence-specific detection of nucleic acids during strand displacement amplification. 1999. *Anal. Biochem.*, **276**(2), 177–187.
6. Van Ness, J., L. K. Van Ness, and D. J. Van Ness. Isothermal reactions for the amplification of oligonucleotides. 2003. *Proc. Natl. Acad. Sci. U S A*, **100**(8), 4504–4509.
7. Vincent, M., Y. Xu, H., and M. Xu. Helicase-dependent isothermal DNA amplification. 2004. *EMBO Rep.*, **5**(8), 795–800.
8. Fire, A., and S. Q. Xu. Rolling replication of short DNA circles. 1995. *Proc. Natl. Acad. Sci. U S A*, **92**(10), 4641–4645.

9. Zhang, D. Y., W. Zhang, X. Li, and Y. Li. Detection of rare DNA targets by isothermal ramification amplification. 2001. *Gene*, **274**(1–2), 209–216.
10. Compton, J. Nucleic-acid sequence-based amplification. 1991. *Nature*, **350**(6313), 91–92.
11. McDonough, S. H., C. Giachetti, Y. Giachetti, D. P. Kolk, E. Billyard, and L. Mimms. High throughput assay for the simultaneous or separate detection of human immunodeficiency virus (HIV) and hepatitis type C virus (HCV). 1998. *Infusionsther. Transfusionsmed.*, **25**(2–3), 164–169.
12. Wharam, S. D., P. Marsh, J. S. Lloyd, T. D. Ray, G. A. Mock, R. Assenberg, J. E. McPhee, P. Brown, A. Weston, and D. L. N. Cardy. Specific detection of DNA and RNA targets using a novel isothermal nucleic acid amplification assay based on the formation of a three-way junction structure. 2001. *Nucleic Acids Res.*, **29**(11), E54.
13. Piepenburg, O., et al. DNA detection using recombination proteins. 2006. *PLoS Biol.*, **4**(7), 1115–1121.
14. Kong, H., T. Ranalli, and B. Lemieux. New isothermal molecular diagnostic platforms: an economical rapid amplification technology widens the applications and settings suitable for molecular assays. 2007. *In Vitro Diagn. Technol.*, **2007**, 35–40.
15. Kurn, N., et al. Novel isothermal, linear nucleic acid amplification systems for highly multiplexed applications. 2005. *Clin. Chem.*, **51**(10), 1973–1981.
16. Gracias, K. S., and J. L. McKillip. Nucleic acid sequence-based amplification (NASBA) in molecular bacteriology: a procedural guide. 2007. *J. Rapid Methods Autom. Microbiol.*, **15**(3), 295–309.
17. Tan, E., et al. Isothermal DNA amplification coupled with DNA nanosphere-based colorimetric detection. 2005. *Anal. Chem.*, **77**(24), 7984–7992.
18. Tan, E., et al. Isothermal DNA amplification with gold nanosphere-based visual colorimetric readout for herpes simplex 2 virus detection. 2007. *Clin. Chem.*, **53**(11), 2017–2020.
19. Andresen, D., M. von Nickisch-Rosenegk, and F. F. Bier. Helicase dependent onchip-amplification and its use in multiplex pathogen detection. 2009. *Clin. Chim. Acta*, **403**(1–2), 244–248.
20. Chow W. H. A., et al. Application of isothermal helicase-dependent amplification with a disposable detection device in a simple sensitive stool test for toxigenic clostridium difficile. 2008. *J. Mol. Diagn.*, **10**(5), 452–458.

21. Gill, P., et al. Colorimetric detection of Helicobacter pylori DNA using isothermal helicase-dependent amplification and gold nanoparticle probes. 2008. *Diagn. Microbiol. Infect. Dis.*, **62**(2), 119–124.
22. Goldmeyer, J., H. M. Kong, and W. Tang. Development of a novel one-tube isothermal reverse transcription thermophilic helicase-dependent amplification platform for rapid RNA detection. 2007. *J. Mol. Diagn.*, **9**(5), 639–644.
23. Goldmeyer, J., et al. Identification of Staphylococcus aureus and determination of methicillin resistance directly from positive blood cultures by isothermal amplification and a disposable detection device. 2008. *J. Clin. Microbiol.*, **46**(4), 1534–1536.
24. Mori, Y., and T. Notomi. Loop-mediated isothermal amplification (LAMP): a rapid, accurate, and cost-effective diagnostic method for infectious diseases. 2009. *J. Infect. Chemother.*, **15**(2), 62–69.
25. Parida, M., et al. Real-time reverse transcription loop-mediated isothermal amplification for rapid detection of West Nile virus. 2004. *J. Clin. Microbiol.*, **42**(1), 257–263.
26. Parida, M., et al. Rapid detection and differentiation of dengue virus serotypes by a real-time reverse transcription-loop-mediated isothermal amplification Assay. 2005. *J. Clin. Microbiol.*, **43**(6), 2895–2903.
27. Zhu, R. Y., et al. Use of visual loop-mediated isotheral amplification of RimM sequence for rapid detection of Mycobacterium tuberculosis and Mycobacterium bovis. 2009. *J. Microbiol. Methods*, **78**(3), 339–343.
28. Bohmer, A., et al. Novel application for isothermal nucleic acid sequence-based amplification (NASBA). 2009. *J. Virol. Methods*, **158**(1–2), 199–201.
29. Fykse, E. M., et al. Detection of Vibrio cholerae by real-time nucleic acid sequence-based amplification. 2007. *Appl. Environ. Microbiol.*, **73**(5), 1457–1466.
30. Lamhoujeb, S., et al. Evaluation of the persistence of infectious human noroviruses on food surfaces by using real-time nucleic acid sequence-based amplification. 2008. *Appl. Environ. Microbiol.*, **74**(11), 3349–3355.
31. Loens, K., et al. Development of real-time multiplex nucleic acid sequence-based amplification for detection of Mycoplasma pneumoniae, Chlamydophila pneumoniae, and Legionella spp. in respiratory specimens. 2008. *J. Clin. Microbiol.*, **46**(1), 185–191.
32. Loens, K., et al. Evaluation of different nucleic acid amplification techniques for the detection of M-pneumoniae, C-pneumoniae and Le-

gionella spp. in respiratory specimens from patients with community-acquired pneumonia. 2008. *J. Microbiol. Methods*, **73**(3), 257–262.

33. Mugasa, C. M., et al. Nucleic acid sequence-based amplification with oligochromatography for detection of Trypanosoma brucei in clinical samples. 2009. *J. Clin. Microbiol.*, **47**(3), 630–635.
34. Piepenburg, O., N. Armes. Biochemical solutions for portable nucleic acid testing. 2007. *BioWorld Eur.*, 2007, 38–41.
35. Johne, R., et al. Rolling-circle amplification of viral DNA genomes using Phi29 polymerase. 2009. *Trends Microbiol.*, **17**(5), 205–211.
36. Li, F., et al. Use of ramification amplification assay for detection of Escherichia coli O157 : H7 and other E-coli shiga toxin-producing strains. 2005. *J. Clin. Microbiol.*, **43**(12), 6086–6090.
37. Margeridon, S., et al. Rolling circle amplification, a powerful tool for genetic and functional studies of complete hepatitis B virus genomes from low-level infections and for directly probing covalently closed circular DNA. 2008. *Antimicrob. Agents Chemother.*, **52**(9), 3068–3073.
38. Wang, B., et al. Rapid and sensitive detection of severe acute respiratory syndrome coronavirus by rolling circle amplification. 2005. *J. Clin. Microbiol.*, **43**(5), 2339–2344.
39. Zhang, W. D., et al. Detection of Chlamydia trachomatis by isothermal ramification amplification method: a feasibility study. 2002. *J. Clin. Microbiol.*, **40**(1), 128–132.
40. Mchugh, T. D., et al. Prospective evaluation of BDProbeTec strand displacement amplification (SDA) system for diagnosis of tuberculosis in non-respiratory and respiratory samples. 2004. *J. Med. Microbiol.*, **53**(12), 1215–1219.
41. Nuovo, G. J. In situ strand displacement amplification: an improved technique for the detection of low copy nucleic acids. 2000. *Diagn. Mol. Pathol.*, **9**(4), 195–202.
42. Levi, K., et al. Evaluation of an isothermal signal amplification method for rapid detection of methicillin-resistant Staphylococcus aureus from patient-screening swabs. 2003. *J. Clin. Microbiol.*, **41**(7), 3187–3191.
43. Wharam, S. D., M. J. Hall, and W. H. Wilson. Detection of virus mRNA within infected host cells using an isothermal nucleic acid amplification assay: marine cyanophage gene expression within Synechococcus sp. 2007. *Virol. J.*, **4**, 52.
44. de Oliveira, A. M., et al. West Nile virus blood transfusion-related infection despite nucleic acid testing. 2004. *Transfusion*, **44**(12), 1695–1699.

45. Giachetti, C., et al. Highly sensitive multiplex assay for detection of human immunodeficiency virus type 1 and hepatitis C virus RNA. 2002. *J. Clin. Microbiol.*, **40**(7), 2408–2419.
46. Hofmann, W. P., et al. Comparison of transcription mediated amplification (TMA) and reverse transcription polymerase chain reaction (RT-PCR) for detection of hepatitis C virus RNA in liver tissue. 2005. *J. Clin. Virol.*, **32**(4), 289–293.
47. Munoz-Jordan, J. L., et al. Highly sensitive detection of dengue virus nucleic acid in samples from clinically ill patients. 2009. *J. Clin. Microbiol.*, **47**(4), 927–931.
48. Kievits, T., et al. NASBA isothermal enzymatic invitro nucleic-acid amplification optimized for the diagnosis of HIV-1 infection. 1991. *J. Virol. Methods*, **35**(3), 273–286.
49. Demidov, V. V. Rolling-circle amplification in DNA diagnostics: the power of simplicity. 2002. *Expert Rev. Mol. Diagn.*, **2**(6), 542–548.
50. Zhang, D. Y., et al. Amplification of target-specific, ligation-dependent circular probe. 1998. *Gene*, **211**(2), 277–285.
51. Zhang, D. et al. Amplification of circularizable probes for the detection of target nucleic acids and proteins. 2006. *Clin. Chim. Acta*, **363**(1–2), 61–70.
52. Hellyer, T. J., and J. G. Nadeau. Strand displacement amplification: a versatile tool for molecular diagnostics. 2004. *Expert Rev. Mol. Diagn.*, **4**(2), 251–261.
53. Tan, E., et al. Specific versus nonspecific isothermal DNA amplification through thermophilic polymerase and nicking enzyme activities. 2008. *Biochemistry*, **47**(38), 9987–9999.
54. Ferguson, B. S., et al. Integrated microfluidic electrochemical DNA sensor. 2009. *Anal. Chem.*, **81**(15), 6503–6508.
55. Mens, P. F., et al. Molecular diagnosis of malaria in the field: development of a novel 1-step nucleic acid lateral flow immunoassay for the detection of all 4 human Plasmodium spp. and its evaluation in Mbita, Kenya. 2008. *Diagn. Microbiol. Infect. Dis.*, **61**(4), 421–427.
56. Dineva, M. A., J. P. Allain, and H. Lee. Visual detection of multiple viral amplicons by dipstick assay: its application in screening of blood donors a welcome tool for the limited resource settings: authors' reply. 2005. *J. Clin. Microbiol.*, **43**(12), 6218–6219.

Chapter 20

Recent Ebola and Marburg Viral Hemorrhagic Fever Outbreaks in Uganda: The Need for Quick, Reliable Diagnostic Tests

Julius Julian Lutwama
Department of Arborvirology, Emerging and Re-emerging Infection, Uganda Virus Research Institute, PO Box 49, Entebbe, Uganda
jjlutwama03@yahoo.com

20.1 Introduction

Situated at the geographical heart of the African continent, Uganda lies in a region where many viral disease epidemics, leading to many deaths, have emerged in the recent past, such as Ebola hemorrhagic fever (EHF), Marburg hemorrhagic fever (MHF), yellow fever, Congo Crimean hemorrhagic fever, and Rift Valley fever, all of which have been recorded in Uganda and in the neighboring countries of Kenya, Tanzania, Sudan, and the Democratic Republic of Congo (DRC),[1–7] the most feared being the viral hemorrhagic fevers (VHFs). Over the years, outbreaks of VHFs, particularly Ebola and Marburg, have

Viral Diagnostics: Advances and Applications
Edited by Robert S. Marks, Leslie Lobel, and Amadou Alpha Sall

ISBN 978-981-4364-43-0 (Hardcover), 978-981-4364-44-7 (eBook)
www.panstanford.com

constituted a major public health problem in sub-Saharan Africa.[8–10] Since 1976 several member states in our region have been affected: Angola (Marburg 2005), DRC (Ebola 1976, 1995, 2007; Marburg 1999), Gabon (Ebola 1996–1997, 2001–2002), Republic of Congo (Ebola 2002, 2003, 2004), and Uganda (Ebola 2000, 2007–2008; Marburg 2007). There have also been outbreaks of Rift Valley fever virus in Kenya (1998, 2007) and Tanzania (2007).[2,8,10–13]

Ebola and Marburg, belonging to the family of filoviruses, are very deadly viruses that keep resurfacing as outbreaks in Africa.[10] The Ebola virus, probably the most scary and deadliest virus anywhere on earth, has a 50%–90% death rate and kills usually within two weeks of infection.[9,14,15] The lethal nature of the virus varies with the particular strains.[8,9] The first Ebola outbreaks recorded were in the DRC and Sudan in 1976,[4,16] while the first outbreak of the second member of the *Filoviridae* family, the Marburg virus (MARV), was detected in Germany and Yugoslavia among workers in laboratories who had been handling African green monkeys (imported from Uganda) in 1968.[17,18] The virulence of Marburg is also high ranging from 23%–90%.[19–21] The viruses are believed to be initially transmitted to a human via contact with an infected animal host.[13] From the first human infected, the viruses can then be transmitted by human contact with infected blood and body fluids of a diseased person and with contaminated medical equipment, such as needles.[13] Three species of fruit bat were identified as carrying the Ebola virus and did not exhibit symptoms and are now believed to be the natural host species, or reservoir, of the Ebola and Marburg viruses.[22,23]

Since 2000, when Uganda experienced the first outbreak of the Ebola VHF,[24,25] another four outbreaks of highly pathogenic viral infections in the past seven years have been experienced, three of them in the last 12 months.[2,26–28] Their reservoirs and modes of transmission to human hosts are not well understood. Unfortunately four other epidemics due to different infectious agents were ongoing at the same time. Experience during the outbreaks of Ebola and Marburg has been detailed in the following section. The need for quick differential diagnoses became more apparent in the past several months when delays in diagnostics caused losses of lives. As the magnitude of international trade and travel is continuously increasing, there is a significant risk that the hemorrhagic fever viruses could be introduced to virus-free countries from areas where they are endemic. Therefore, the development of laboratory diagnostic systems for EHF and MHF has become a vital subject even in countries without VHFs.

20.2 Outbreak Experience

20.2.1 *The 2000–2001 Ebola Outbreaks: Gulu, Masindi, and Mbarara Districts*

An outbreak of Ebola disease was reported from Uganda in October 2000 to February 2001.[15,24,25,29,30] The outbreak was characterized by fever and hemorrhagic manifestations and affected health workers and the general population of the Gulu district in Northern Uganda. Later, the outbreak spread to other parts of the country, including the Mbarara and Masindi districts (Fig. 20.1).

On October 8, 2000, an outbreak of an unusual febrile illness, with occasional hemorrhage and significant mortality, was reported to the Ministry of Health (MoH) in Kampala by the superintendent of St. Mary's Hospital in Lacor and the district director of health services (DDHS) in the Gulu district. The acting district director of health services (Ag. DDHS) of the Gulu district had received two concurrent reports concerning an unusual illness and deaths in the community and at the nongovernmental Lacor Hospital. The report, originating from the community, attributed the illness and death to a poisoning at a funeral in a remote village, Rwot Obilo, far north of Gulu. The second report, concurrently conveyed to both the Ag. DDHS and to the MoH, came from the medical superintendent (MS) of Lacor Hospital. He reported a clustering of cases and deaths, which included two dead student nurses and three critically ill. Most of the cases in the hospital reported a history of deaths with similar manifestations in their households. He suspected a possible outbreak of VHF. On October 9, a team was dispatched from the MoH to support the district team in an outbreak investigation and confirmation. A preliminary assessment conducted by the MoH team found additional cases in the Gulu

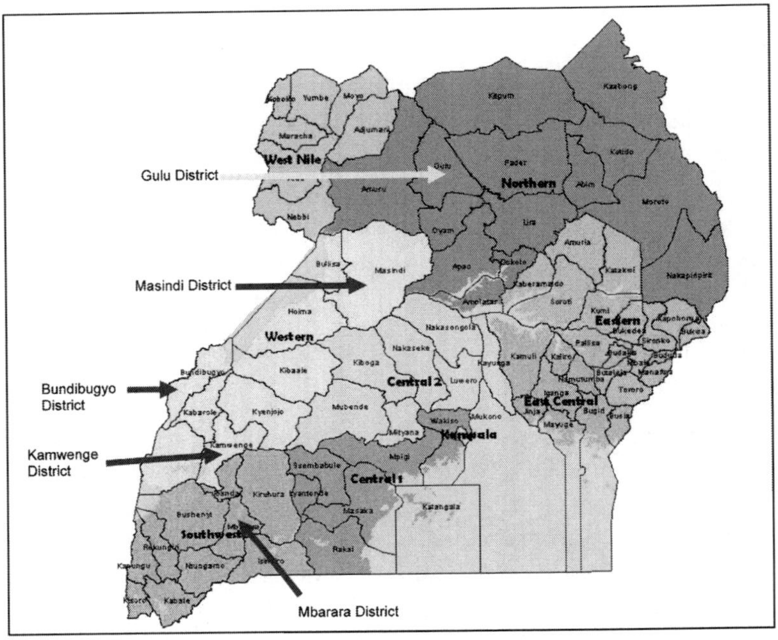

Figure 20.1 Map of Uganda, showing areas where the Ebola and Marburg outbreaks occurred.

district and in Gulu Hospital, the regional referral hospital. The team reviewed clinical notes of patients, examined patients still admitted, and collected clinical specimens from eight suspicious cases and seven contacts for confirmation. Investigations in the surrounding villages revealed many other cases and deaths in the community. Each of the patients had a history of having attended a burial in the previous few days before onset of fever. Some of the patients had lost one or more family members with similar symptoms within a short duration.

On October 12, the clinical samples from cases and contacts were forwarded through the World Health Organization (WHO) country office to a WHO collaborating laboratory in Johannesburg, South Africa—the National Institute of Virology (NIV). Suspicion of EHF was confirmed on October 15 when the NIV, South Africa, identified Ebola virus infection among specimens from patients, including health care workers at St. Mary's Hospital.

Control activities were organized around surveillance and epidemiology, clinical case management, social education and mobilization, and coordination and logistic support. An active EHF surveillance system was initiated to determine the extent and magnitude of the outbreak, identify foci of disease activity, and detect cases early. Ill persons were encouraged to be assessed at a hospital and, if indicated, to be hospitalized to reduce further community transmission. Targeted prevention activities included follow-up of contacts of identified cases for 21 days, establishment of trained burial teams for all potential and confirmed EHF deaths, community education, cessation of traditional healing and burial practices, cessation of large public gatherings, and updates of hospital infection control measures, including isolation wards.

A temporary field-screening laboratory was set up at Lacor Hospital by the Centers for Disease Control and Prevention (CDC) team on October 21, 2000. The aim was to provide on-site laboratory screening and confirmation of clinical and suspicious cases. Additional testing was carried out at the CDC, Atlanta, USA, and at the NIV, South Africa. Blood samples from alerts, suspects, and probable cases reported from different parts of the country were screened by this laboratory and confirmed cases only from the three districts of Gulu, Masindi, and Mbarara. Laboratory tests included virus antigen detection and antibody enzyme-linked immunosorbent assay (ELISA) tests and reverse transcriptase polymerase chain reaction (RT-PCR).[15] Test results were made available within 24 hours and were used to guide public health decisions and actions. Laboratory-confirmed case patients were defined as patients who met the surveillance case definitions and were either positive for Ebola virus antigen or Ebola IgG antibody. Sequence analysis revealed that the virus associated with this outbreak was Ebola-Sudan and differed at the nucleotide sequence level from earlier Ebola-Sudan isolates by 3.3% and 4.2% in the polymerase (362 nucleotides sequenced) and nucleocapsid (146 nucleotides sequenced) protein-encoding genes, respectively.[15,31]

During October 5–November 27, among 62 persons with laboratory-confirmed EHF admitted to Gulu Hospital, symptoms included diarrhea (66%), asthenia (64%), anorexia (61%), headache (63%), nausea and vomiting (60%), abdominal pain (55%), and chest pain (48%). Patients presented for care a mean of 8 days (range 2–20 days) after symptom onset. Bleeding occurred in 12 (20%) patients and primarily involved the gastrointestinal tract. Among the 62 confirmed case patients, 36 (58%) died; among patients aged 15 years, 4 of 5 died (case fatality 80%). Spontaneous abortions were reported among pregnant women infected with EHF. Patients who died usually exhibited a rapid progression of shock, increasing coagulopathy, and loss of consciousness.[15]

As of January 23, 2001, there were 425 presumptive case patients (both the laboratory-confirmed cases and those that met the clinical case definition of EHF and were epidemiologically linked to cases), with 224 deaths attributed to EHF. These were recorded from three districts in Uganda: 393 (93%) from Gulu in Northern Uganda, 27 (6%) from Masindi in Western Uganda, and 5 (1%) from Mbarara in Southwestern Uganda. The case fatality rate was 53%. The attack rate (AR) was highest in women. The average AR for the Gulu district was 12.6 cases/10,000 inhabitants when the contacts of all cases were considered and was 4.5 cases/10,000 if limited only to contacts of laboratory-confirmed cases. The secondary AR was 2.5% when nearly 5,000 contacts were followed up for 21 days.

Although the cluster of cases in early October triggered identification of the outbreak and response measures, investigations (i.e., case record review and interviews with surviving patients or their surrogates) identified cases occurring in the community and patients hospitalized several weeks earlier. There was delayed detection at the community level. The earliest recognized case identified retrospectively was August 30, 2000, and it was first notified to the MoH on October 8, 2000.[2,18] The epidemic was not recognized until six weeks later. By this time, a number of cases had occurred and many were incubating the disease. The ages of presumptive case patients ranged from 3 days to 72 years (median 28 years); 269 (63%) were women. The mean time from symptom onset to death was eight days (95% confidence interval = ±5 days); 218 (51%) presumptive cases were laboratory confirmed. In Masindi the outbreak was limited to a single-family chain of transmission. One of the family members had traveled to Gulu for a burial and was infected. He passed on the infection to other members of his family. In Mbarara, a soldier who had traveled from

Gulu moved with the infection to Mbarara, where he passed on the infection to his attendant in the hospital.

This was recognized as the largest reported outbreak of EHF in the world. Control interventions were very successful in containing the epidemic. This was also the first outbreak response coordinate by the WHO under the Global Outbreak Alert and Response Network, a voluntary organization that had recently been created to coordinate technical and financial resources to developing countries during outbreaks.

Challenges to control efforts included inadequate and poor-quality protective materials, nosocomial transmission of EHF, deaths of health workers, numerous rumors, and the rejection of convalescent cases by members of the community.

20.2.2 *The 2007 Marburg Outbreaks in the Kamwemge District*

A 29-year-old male miner became sick on July 4, 2007, and was admitted to a Kampala hospital on July 7, 2007.[2] He presented with a three-day history of fever, chills, headache, and arthralgias. Three days after admission, he continued to be febrile and began vomiting. On July 13, he developed confusion, seizures, hematemesis, and hematochezia and died later that day. A blood sample taken on July 13 was sent to the Kenya Medical Research Institute (KEMRI) for analysis. All tests at KEMRI were negative. However, due to the symptoms really showing that the patient fitted the case definition for a viral hemorrhagic infection, the samples were sent on July 23 to the CDC for further testing. MARV was detected by RT-PCR on July 29 and subsequently by virus isolation and antigen-capture ELISA. The Uganda MoH, KEMRI, and WHO were notified of the results on July 30, 2007. The Uganda MoH quickly announced that the MARV (family *Filoviridae*) had been confirmed as the causative agent of a suspected case of hemorrhagic fever in a male mine worker from Kamwemge district, Western Uganda, and a Marburg VHF outbreak was declared.[32,33]

An international team, including experts from WHO, the CDC, Médecins sans Frontières (MSF), the Uganda Virus Research Institute (UVRI), the African Field Epidemiology Network (AFENET), and local nongovernmental organizations (NGOs) supporting the MoH was set up. This team started active surveillance, contact tracing, infection control, logistics, and social mobilization activities in an effort to contain this outbreak. A search for other infected persons was initiated, and control activities were started. A surveillance team was sent to the Kamwemge district, where the mine is situated, and to the nearest hospital located in the Ibanda district. Another team was dispatched to search for contacts within Kampala.

Initial contact tracing revealed that the deceased had provided care to a coworker, aged 23 years, who worked at the same mine and had developed hemorrhagic fever one week earlier. This mine worker first presented to a health center in Kamwemge district on June 21 with fever, headache, arthralgias, and vomiting and was transferred to Hospital B the next day when he developed what appeared to be generalized seizures. He was discharged on June 29 but had persistent weakness and dizziness and was transported overnight to another hospital in Kampala on July 1. The deceased and another coworker had both cared for him while he was being transported to the hospitals. He recovered and was discharged on July 9, 2007. A blood sample from this case patient was sent to the CDC and tested positive for antibodies against MARV.

Contacts of these two cases were followed for at least 21 days after last contact with the case patient in order to identify any secondary transmission. Overall, 267 contacts with both case patients were reported, of which 161 were located and followed.[2] Of the contacts followed, 15% had experienced fever and 2% had bleeding since the beginning of June. Three contacts had both symptoms in the review period, resulting in potential suspect cases (referred to as alert cases). None of the contacts developed illness compatible with MHF during this time period. Active surveillance for new cases was continued in the Ibanda and Kamwemge districts for an additional 21 days; during the second 21-day period, not a single patient meeting the suspected case definition was identified.

Blood samples were collected from 87 high-risk contacts and sent to the CDC for testing. Of the 17 mine workers tested, one male aged 22 years had evidence of prior exposure to MARV by IgG. This worker was also a contact, present in the vehicle that transported both the case patient with confirmed MARV infection and the second case patient from the hospital in the Ibanda district to the hospital in Kampala on July 1, 2007. He had developed fever, joint pains, and a headache in early June and had traveled from the Ibanda district to a health center in Kampala to seek treatment. He returned to find the first case patient ill at the mining camp. None of the two who tested positive for anti-MARV IgG had anti-MARV IgM, so the timing of infection can only be determined for the patient from whom MARV was isolated.

Interviews were conducted with both the IgG-positive contacts and other miners to learn more about these events and other practices related to the mining activities. Mine workers reported that local residents who worked at the mine in the early 1980s experienced a similar outbreak of hemorrhagic fever, in which approximately 20–40 mine workers became ill and 10 died. Interviews of seven former mine workers who survived this outbreak revealed that some, but not all, had signs and symptoms consistent with hemorrhagic fever; blood samples were obtained from six survivors for serologic testing for evidence of prior MARV infection.[2]

20.2.3 *The 2007–2008 Ebola Outbreak in the Bundibugyo District*

During September 2007, there had been rumors of a strange disease going on in the Bundibugyo district (Fig. 20.2) of Western Uganda at the border with the DRC. An initial investigative team composed of personnel from the MoH, the UVRI, Uganda Peoples Defense Forces, and WHO went in toward the end of the month (September 29, 2007) to investigate a number of reported deaths in Kikyo Health Center IV (HCIV) in one of the counties of the Bundibugyo district. A lady had died, and the child she left behind with the grandmother also died after a couple of days. Then the grandmother sickened and died. The initial investigation was not able to get any samples from the family members, as the homestead was quite far in the mountains and the rainy season did not allow easy movement in this terrain. However, samples of blood were collected from other patients in the district who exhibited hemorrhagic symptoms. These samples were sent to the CDC, Atlanta, and a negative report was sent back for these samples.

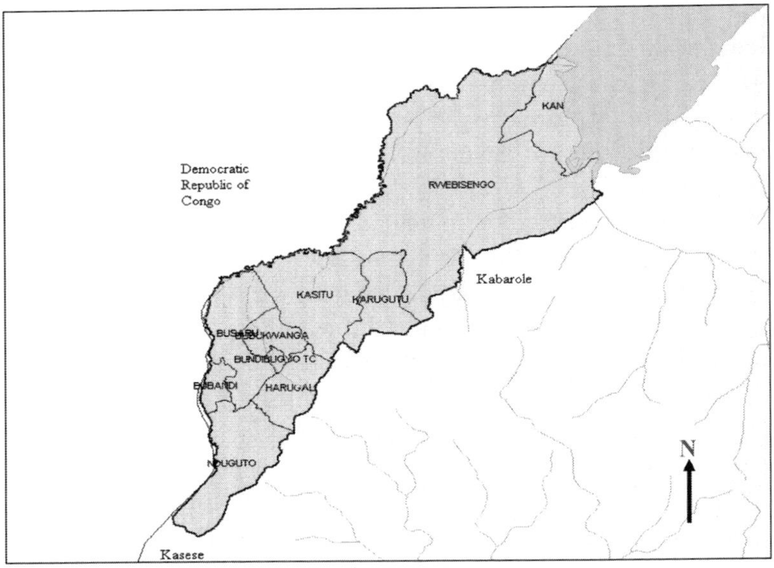

Figure 20.2 Map of the Bundibugyo district.

Despite that, the rumors of the strange disease continued, and several deaths were again reported. The MoH decided to send a team again, and this team was supposed to remain in the district and respond immediately to any reported cases so that it could investigate and collect specimens that could be used to diagnose the infection. The team commenced the investigations with a briefing from the District Health Team (DHT) about the reported outbreak and magnitude of the problem. The team then conducted a clinical evaluation of the suspected cases admitted, together with reviewing all the records in the registers, to identify and verify other cases that had presented earlier on. The team also traced to the communities, a selection of suspected cases, especially if the local conditions permitted, to establish the risk factors and provide sensitization to the communities.

The disease was variously attributed to witchcraft, eating infected goat, an encounter with diseased monkeys, a virus released by unscrupulous people, and more. In one homestead, there were 15 cases, of which 4 died; the initial cases were linked to eating the meat from a dead goat. The goat had died three weeks prior to the occurrence of cases in the home. The team visited the grazing area, found poisonous ferns, and therefore concluded that the goats probably passed away after feeding on a poisonous fern. The occurrence of cases following the goat deaths was likely to be a coincidence and not linked in any way. A sample of the meat from one of the goats was obtained for laboratory investigations and proved negative for hemorrhagic fever viruses. Meanwhile the team collected samples from all likely patients, and the samples were sent for diagnosis.

On November 26, 2007, the CDC received blood samples from the Uganda MoH, taken from 20 of the 49 patients involved in the unknown illness in the Bundibugyo district. These were the patients who had reported fever, enteritis, and bleeding. Of the 49, 14 had died. Genetic sequencing of a small segment of viral RNA from samples indicated the presence of a previously unknown strain of Ebola virus.[26,28] On November 30, 2007, the Uganda MoH declared an outbreak of Ebola in the Bundibugyo district in Western Uganda.[27]

At the invitation of the Uganda MoH, the CDC, WHO, the MSF, and other collaborators deployed field investigators to the affected region; additionally, a laboratory was set up in Entebbe at the UVRI. Definitions of the alert cases and for suspected cases were distributed throughout the district, and information on the disease was provided over the radio and charts and at community meetings. Guidelines on management of alert and suspected cases were also given out to health personnel. Two isolation wards were set up at Kikyo HCIV and at Bundibugyo District Hospital. Proper infection control was instituted and health education massively imparted to the communities.

As the outbreak neared conclusion in January 2008, the total number of suspected cases was 149, with 37 deaths. The epidemic came to an official end on February 20, 2008.[27] While the index case was not definitely identified, it was established that three branches of the infection were from people who were infected while at the hospital with the index case: the health worker, who attended the index case and two other patients, who passed on the infection to attendants, and the visitors, sparking off the spread of the infection.

During the investigations, it was noted that there was a lot of person-to-person transmission, the risk factors including the following:

i) Nursing a suspected case, with 38.8% giving a history of nursing a suspected case prior to falling sick.
ii) Staying in the same home with a suspected case, with 55.1% giving a history of staying in the same house with a case prior to onset of the illness.
iii) Visiting a suspected case: 53.1% had visited the hospital prior to falling sick.
iv) Handling bodies of suspected deaths: 10% reported handling the bodies prior to onset of the illness.
v) Overall, 69.4% reported a history of contact with either a suspected case or a dead person prior to falling sick.

20.3 The Challenges

20.3.1 *Time Spent on Receiving Information and Time Spent before a Response Is Made*

Outbreaks normally start in remote places where health facilities are usually poor and the medical personnel are often not suspecting serious diseases.[34–36] The symptoms and signs of VHFs are similar to those of the many other common infections found in Uganda. For example, malaria is a common infection and many fevers are routinely treated as such. It is only after a cluster of related people in a community have died and usually when some health personnel have been infected that an outbreak is recognized.

In the Ebola outbreak of 2000 in the Gulu district, the disease was recognized in October 2000 but the first cases were traced back to August or even July 2000.[24] This was a delay of one to two months. The Ebola outbreak in the Bundibugyo district in 2007[27] was confirmed at the end of November 2007. The first reports of a strange disease were in September 2007. The first cases could have been as early as August 2007. Again, there was a delay of more than one month. For the Marburg outbreak,[2] the patients moved to the city (Kampala) where the disease was quickly suspected. Within two weeks, barrier nursing was instituted, although confirmation of the outbreak came after another two weeks. Overall, such delays lead to more people being infected and higher numbers of deaths.

20.3.2 *Time Spent without Confirmation of an Outbreak*

Because of the severity of Ebola and Marburg viruses and their ability to be transmitted from person to person,[13,37] suspected cases must be diagnosed as soon as possible after the onset of the disease. Isolation of the virus from blood samples is the method of choice, but definite identification takes more than one week.[27,38,39] In many of the developing countries, the capacities and capabilities to handle highly infectious pathogens are limited. When such diseases are recognized, specimens are collected and sent to specialized laboratories in other countries for diagnoses.[40] For the Ebola outbreak in Gulu in 2000, specimens were collected and hand-carried to the National Institutes for Communicable Diseases (NICD) and the NIV, South Africa, where diagnoses were made within a week.

The specimens collected from the first suspected case of Marburg were sent to the Global Emerging Infections Surveillance and Response System (GEIS) laboratories at KEMRI in Nairobi, Kenya. A negative diagnosis was made, but aliquots of the specimens were sent to the CDC, Atlanta, for confirmation. A positive diagnosis was then made. The time spent to get a positive confirmation was 13 days from when the specimens were first sent out for diagnosis. The Biosafety Level 4 (BSL 4) laboratory at NICD, South Africa, had been down for a couple of years, and samples were not sent there.

The specimens collected during the Ebola outbreak in Bundibugyo were sent to the CDC, Atlanta. The initial samples sent first were probably from people who were not infected with the Ebola virus since they were collected in counties where the disease, at that time, had not been reported, even though they exhibited bleeding symptoms. For the second set of samples again no positive result was given. Finally for the third set of samples, a positive diagnosis was made, but it took time because the PCR test that was initially used could not detect the new strain of Ebola that caused the infection. It was only after adjustments were made to the ELISA tests that a positive diagnosis evolved. This delay in recognition was because the Ebola strain was different from the earlier-known strains. Thus, test adjustments were first needed to be made before a positive diagnosis could be obtained.

20.3.3 *Numbers of People Infected and Deaths*

When people start to die in a community, first there is witch hunting and there are rumors; and as more people die, there is disarray and there is migration of people, resulting because of the fear of death.[40] This results in the spread of infection. The longer the diagnosis of the infection takes, the worse the problem. There is a lot of speculation at the beginning, and if the diagnosis is not done early people will migrate from the affected regions. There is a great challenge to limit people from moving out of affected areas. The more the people die, the greater the migration.[41]

20.3.4 *Numbers of Health Workers Infected and Deaths*

In the recent Ebola outbreak in the Bundibugyo district, being a health worker carried a risk of only 10%, but all the cases were linked to a suspected case initially admitted in Kikyo HCIV.[28] This indicated that much of the initial transmission occurred in the ward. This particular health center has one ward were all the maternity, surgical, and medical cases are admitted. Due to shortage of staffing at the health unit it is very common for the health worker attending maternity cases to also treat medical cases. Another factor that could have fuelled nosocomial transmission was the shortage of basic supplies like gloves and disinfectants to observe standard precautions for infection control.

When health workers get infected and die, the people in the community avoid the health centers for fear of becoming infected and also the belief that if medical workers can be susceptible to infection and death, then it is useless to go to the hospitals. It is only when proper infection control and care for the infected have been instituted that a few people will start venturing to the health centers.[41]

20.3.5 *The Need for Quick Diagnostic Tests*

20.3.5.1 The needed capacities

A variety of tests are known to be specific and useful for Ebola and MARV identification.[42–47] Antigen-capture ELISA, for IgM and IgG antibodies, is a useful diagnostic tool.[42] IgM antibodies to these viruses appear between 2 and 9 days after symptoms begin and disappear between 30 and 168 days after.[42,43] IgG antibodies appear between 6 to 18 days after symptom onset and can persist for many months.[48,49] Viral antigen detection by ELISA and RT-PCR are particularly useful tests that are rapid and sensitive.[42,50–52] Ebola and Marburg viral antigens are detectable in the blood 3 to 6 days after symptom onset, but antigen positivity disappears 7 to 16 days after symptoms have begun. Indirect fluorescent antibody tests have demonstrated a lack of specificity and a tendency to produce false positives.[42,43,53–56] While the above tests are useful, they cannot be performed everywhere, as discussed later.[57]

20.3.5.2 Infrastructural capacities

Definitive laboratory diagnoses of Ebola and Marburg viruses require ELISA and RT-PCR capability.[57,58] These procedures are presently not practical in the field or other settings where outbreaks normally occur, as they require laboratories with the appropriate biosafety containment for safe handling of the viruses. The appropriate biosafety level is ideally BSL 4, which is designed for work with dangerous agents that pose a high risk of laboratory infection and life-threatening disease.[59,60] There is only one such laboratory on the African continent,[35] and this laboratory has been closed for renovation for the past four years. These are highly specialized laboratories, which require a lot of maintenance and well-trained personnel. While BSL 3 facilities have been used in certain instances, the CDC recommends administrative measures and/or additional personal protective equipment to reduce risk. Special guidelines and protective wear are required. These too are not commonly available.[61]

20.3.6 *Trained Personnel*

Ebola and Marburg are highly infectious viral diseases.[61,62] Just as there is a need to handle these viruses in high-level biosafety laboratories,[61,62] so also there is a need to have highly trained personnel to handle the diagnostics for these infections. It takes time and money to train these personnel. There should be a critical number of trained personnel to work in shifts of two or three to watch over the safety of each other.

20.3.7 *Availability of Funds*

Funds for the establishment and maintenance of appropriate laboratories, purchase of equipment and required supplies, and training of necessary personnel and their remuneration are very difficult for resource-poor countries where the majority of the serious outbreaks occur. Unless funds are available, reliance has to be made on collecting and sending specimens to laboratories where these facilities can be obtained. This of course means that delays in diagnoses will continue.

20.3.8 *The Complexity of the Present Tests and the Need for Simplification*

Good diagnostic tests are required to quickly identify the emergence of serious disease in a community.[63–65] Transmission of a dangerous infectious disease threatens not merely the local population but also the world at large as a result of immigration and increased and faster travel.[10] Any outbreak elicits considerable concern and demands that various precautionary methods be instituted and the disease be contained as quickly as possible. The detection should be rapid, reliable, and accurate.[66–68]

With the advance in biotechnology, electronics, and engineering, new advanced tools have been developed.[67,69,70] These tools, however, in many instances, require sophisticated know-how for their use, are very expensive, require inaccessible supplies, have elaborate protocols that are difficult to follow, and sometimes require advanced laboratories in which to function and thus cannot

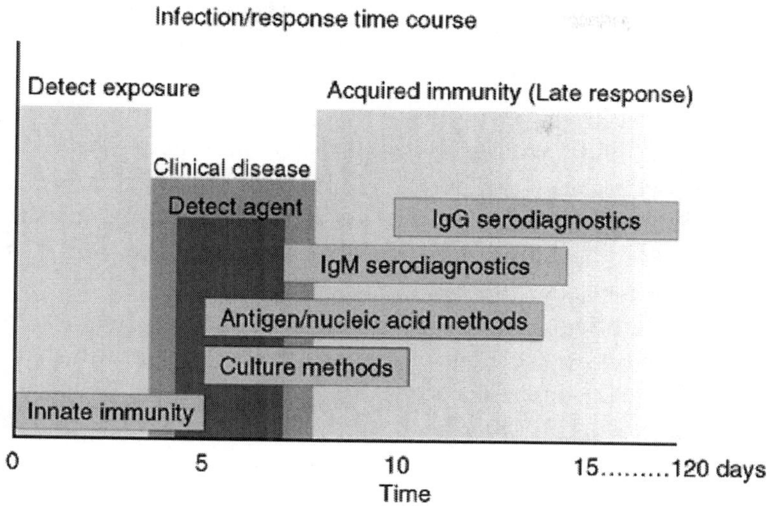

Figure 20.3 Infection and response time course.

easily be deployed for field investigations.[71] The RT-PCR for Ebola and Marburg viruses takes four to five hours to perform but requires expensive equipment and reagents and supplies, some of which cannot be sourced easily on the open market.[72] The antigen-capture IgM and IgG antibody ELISA for Ebola and Marburg takes at least 16 hours to perform. These technologies can only be deployed where there are relatively good laboratories.

Various detection methodologies have highly different entry points in their use in human disease (Fig. 20.3). Medical intervention has less success as points of time extend. The earlier the time, the more successful the prognosis for most diseases. Tools that can diagnose the infection early soon after exposure and during clinical disease are more useful than those that are limited to diagnose the infection later. Early diagnostic tools assist patients' care and reduce further infection. Current efforts in rapid diagnostics should shift the window of detection closer to the point at which clinical disease symptoms become evident. Ultimately, future rapid diagnostics should shift the window to a point soon after exposure, giving the clinician the greatest opportunity to intervene in the disease process.

The need for quick diagnostic tools increases every day. Ideally, new technologies should be developed with the following characteristics in mind:

(1) Portable and not necessarily requiring connections to electricity power lines.
(2) Requiring a few processing steps: The fewer the steps, usually the easier the test, and the fewer the chances of making mistakes in processing the samples.
(3) Not requiring rigorous safety practices but relatively safe: Because safety facilities may be unavailable in the rural areas where many outbreaks occur, the technologies need to be safe with minimal precautions.[66,67,73]

20.4 Way Forward for Uganda

The UVRI, through the Department of Arborvirology, Emerging and Re-emerging Viral Infections, has developed the capacity for diagnosis and confirmation of filoviral infections using RT-PCR and ELISA through the assistance of the CDC, Special Pathogens Division, Atlanta, USA. A BSL 3 laboratory was refurbished, and diagnostics for the highly infectious pathogens are carried out in this laboratory. Three personnel have been trained for working in the laboratory. A series of other laboratories has also been refurbished.

Through collaboration with the CDC, Fort Collins, USA, the capacity for diagnosis of many arborvirus infections has been developed in the Department of Arborvirology, Emerging and Re-emerging Viral Infections. The capacity for rapid response for infection investigations and support has been improved.

While Uganda experiences many outbreaks of infectious diseases, there is a general inability to diagnose many of the infections. There is a need for recruiting more staff, training them, and sustaining them on the job. There is also need for support in terms of infrastructure development to set up functional laboratories, and international collaboration.

There is a continued problem of sustained provisions of supplies and reagents. This requires funding and goodwill from collaborators.

The reagents are needed for a varied number of infections if the laboratory is going to be able to diagnose as many diseases as possible. There is also a problem of specimen referral within the country. There is no referral system, and samples sometimes arrive late and in poor condition for testing. If simple portable technologies were available then these would be dispatched in the field.

References

1. World Health Organization. Ebola haemorrhagic fever in Sudan, 1976. Report of a WHO/international study team. 1978. *Bull. World Health Organ.*, **56**(2), 247–270.
2. Outbreak of Marburg haemorrhagic fever: Uganda, June–August 2007. 2007. *Wkly. Epidemiol. Rec.*, **82**(43), 381–384.
3. Dick, G. W. Epidemiological notes on some viruses isolated in Uganda; yellow fever, Rift Valley fever, Bwamba fever, West Nile, Mengo, Semliki forest, Bunyamwera, Ntaya, Uganda S and Zika viruses. 1953. *Trans. R. Soc. Trop. Med. Hyg.*, **47**(1), 13–48.
4. Eisa, M., E. D. Kheir El Sid, A. M. Shomein, and J. M. Meegan. An outbreak of Rift Valley fever in the Sudan: 1976. 1980. *Trans. R. Soc. Trop. Med. Hyg.*, **74**(3), 417–419.
5. Henderson, B. E., D. Metselaar, K. Cahill, G. L. Timms, P. M. Tukei, and M. C. Williams. Yellow fever immunity surveys in northern Uganda and Kenya and eastern Somalia, 1966–67. 1968. *Bull. World Health Organ.*, **38**(2), 229–237.
6. Kornaszewski, W., and M. Kornaszewska. Africa as the source of 3 acute hemorrhagic fevers of viral origin (1967–1976). 1979. *Wiad Lek.*, **32**(6), 437–441.
7. Pourrut, X., M. Souris, J. S. Towner, P. E. Rollin, S. T. Nichol, J.-P. Gonzalez, and E. Leroy. Large serological survey showing cocirculation of Ebola and Marburg viruses in Gabonese bat populations, and a high seroprevalence of both viruses in *Rousettus aegyptiacus*. 2009. *BMC Infect. Dis.*, **9**, 159.
8. Bruce, J., and P. Brysiewicz. Ebola fever: the African emergency. 2002. *Int. J. Trauma Nurs.*, **8**(2), 36–41.
9. Colebunders, R., and M. Borchert. Ebola haemorrhagic fever: a review. 2000. *J. Infect.*, **40**(1), 16–20.

10. Feldmann, H., M. Czub, S. Jones, D. Dick, M. Garbutt, A. Grolla, and H. Artsob. Emerging and re-emerging infectious diseases. 2002. *Med. Microbiol. Immunol.*, **191**(2), 63–74.

11. Viral haemorrhagic fever/Marburg, Democratic Republic of the Congo. 1999. *Wkly. Epidemiol. Rec.*, **74**(20), 157–158.

12. Lawrence, J. J., and D. R. Hill. Largest ever Marburg haemorrhagic fever outbreak, Angola. 2005. *Eur. Surveill.*, **10**(14), E050407 2.

13. Pourrut, X., B. Kumulungui, T. Wittmann, G. Moussavou, A. Délicat, P. Yaba, D. Nkoghe, J. P. Gonzalez, and E. M. Leroy. The natural history of Ebola virus in Africa. 2005. *Microbes Infect.*, **7**(7–8), 1005–1014.

14. Takada, A., and Y. Kawaoka. The pathogenesis of Ebola hemorrhagic fever. 2001. *Trends Microbiol.*, **9**(10), 506–511.

15. Okware, S. I., F. G. Omaswa, S. Zaramba, A. Opio, J. J. Lutwama, J. Kamugisha, E. B. Rwaguma, P. Kagwa, and M. Lamunu. An outbreak of Ebola in Uganda. 2002. *Trop. Med. Int. Health*, **7**(12), 1068–1075.

16. Bowen, E. T., G. S. Platt, G. Lloyd, R. T. Raymond, and D. I. Simpson. A comparative study of strains of Ebola virus isolated from southern Sudan and northern Zaire in 1976. 1980. *J. Med. Virol.*, **6**(2), 129–138.

17. Malherbe, H., and M. Strickland-Cholmley. Human disease from monkeys (Marburg virus). 1968. *Lancet*, **291**(7557), 1434.

18. Siegert, R., H. L. Shu, and, W. Slenczka. Detection of the "Marburg virus" in patients. 1968. *Ger. Med. Mon.*, **13**(11), 521–524.

19. Shu, H. L., R. Siegert, and W. Slenczka. The pathogenesis and epidemiology of the "Marburg-virus" infection. 1969. *Ger. Med. Mon.*, **14**(1), 7–10.

20. Mohamadzadeh, M., L. Chen, G. G. Olinger, W. D. Pratt, and A. L. Schmaljohn. Filoviruses and the balance of innate, adaptive, and inflammatory responses. 2006. *Viral Immunol.*, **19**(4), 602–612.

21. Mohamadzadeh, M., L. Chen, and A. L. Schmaljohn. How Ebola and Marburg viruses battle the immune system. 2007. *Nat. Rev. Immunol.*, **7**(7), 556–567.

22. Feldmann, H., V. Wahl-Jensen, S. M. Jones, and U. Ströher. Ebola virus ecology: a continuing mystery. 2004. *Trends Microbiol.*, **12**(10), 433–437.

23. Groseth, A., H. Feldmann, and J. E. Strong. The ecology of Ebola virus. 2007. *Trends Microbiol.*, **15**(9), 408–416.

24. Outbreak of Ebola haemorrhagic fever, Uganda, August 2000–January 2001. 2001. *Wkly. Epidemiol. Rec.*, **76**(6), 41–46.

25. Outbreak of Ebola hemorrhagic fever, Uganda, August 2000–January 2001. 2001. *Can. Commun. Dis. Rep.*, **27**(6), 49–53.

26. Alsop, Z. Ebola outbreak in Uganda "atypical", say experts. 2007. *Lancet*, **370**(9605), 2085.
27. Outbreak news. Ebola haemorrhagic fever, Uganda: end of the outbreak. 2008. *Wkly. Epidemiol. Rec.*, **83**(10), 89–90.
28. Towner, J. S., T. K. Sealy, M. L. Khristova, C. G. Albariñ, S. Conlan, S. A. Reeder, P.-L. Quan, W. Ian Lipkin, R. Downing, J. W. Tappero, S. Okware, J. Lutwama, B. Bakamutumaho, J. Kayiwa, J. A. Comer; P. E. Rollin, T. G. Ksiazek, and S. T. Nichol. Newly discovered Ebola virus associated with hemorrhagic fever outbreak in Uganda. 2008. *PLoS Pathog.*, **4**(11), e1000212.
29. Maurice, J. The Uganda Ebola outbreak: not all negative. 2000. *Bull. World Health Organ.*, **78**(12), 1476–1477.
30. Bitekyerezo, M., C. Kyobutungi, R. Kizza, J. Mugeni, E. Munyarugero, F. Tirwomwe, E. Twongyeirwe, G. Muhindo, V. Nakibuuka, M. Nakate, L. John, A. Ruiz, K. Frame, G. Priotto, L. Pepper, J. Kabakyenga, S. Baingana, and D. Ledo. The outbreak and control of Ebola viral haemorrhagic fever in a Ugandan medical school. 2002. *Trop. Doct.*, **32**(1), 10–15.
31. Sanchez, A., and P. E. Rollin. Complete genome sequence of an Ebola virus (Sudan species) responsible for a 2000 outbreak of human disease in Uganda. 2005. *Virus Res.*, **113**(1), 16–25.
32. Marburg haemorrhagic fever, Uganda: update. 2007. *Wkly. Epidemiol. Rec.*, **82**(36), 317.
33. Outbreak news. Marburg haemorrhagic fever, Uganda. 2007. *Wkly. Epidemiol. Rec.*, **82**(33), 297–298.
34. Geisbert, T. W., and P. B. Jahrling. Exotic emerging viral diseases: progress and challenges. 2004. *Nat. Med.*, **10**(12 Suppl), S110–S121.
35. Feldmann, H., T. Geisbert, and Y. Kawaoka. Filoviruses: recent advances and future challenges. 2007. *J. Infect. Dis.*, **196**(Suppl 2), S129–S130.
36. Gotch, F., and J. Gilmour. Science, medicine and research in the developing world: a perspective. 2007. *Nat. Immunol.*, **8**(12), 1273–1276.
37. Peters, C. J., and J. W. LeDuc. An introduction to Ebola: the virus and the disease. 1999. *J. Infect. Dis.*, **179**(Suppl 1), ix–xvi.
38. Guimard, Y., M. A. Bwaka, R. Colebunders, P. Calain, M. Massamba, A. De Roo, K. D. Mupapa, K. Kibadi, K. J. Kuvula, D. E. Ndaberey, K. R. Katwiki, B. B. Mapanda, O. B. Nkuku, Y. Fleerackers, E. Van den Enden, and M. A. Kipasa. Organization of patient care during the Ebola hemorrhagic fever epidemic in Kikwit, Democratic Republic of the Congo, 1995. 1999. *J. Infect. Dis.*, **179**(Suppl 1), S268–S273.

39. Gilchrist, M. J. A national laboratory network for bioterrorism: evolution from a prototype network of laboratories performing routine surveillance. 2000. *Mil Med.*, **165**(7 Suppl 2), 28–31.

40. Bausch, D. G., A. G. Sprecher, B. Jeffs, and P. Boumandouki. Treatment of Marburg and Ebola hemorrhagic fevers: a strategy for testing new drugs and vaccines under outbreak conditions. 2008. *Antiviral Res.*, **78**(1), 150–161.

41. Lamunu, M., J. J. Lutwama, J. Kamugisha, A. Opio, J. Nambooze, N. Ndayimirije, and S. Okware. Containing a haemorrhagic fever epidemic: the Ebola experience in Uganda (October 2000–January 2001). 2004. *Int. J. Infect. Dis.*, **8**(1), 27–37.

42. Ksiazek, T. G., P. E. Rollin, A. J. Williams, D. S. Bressler, M. L. Martin, R. Swanepoel, F. J. Burt, P. A. Leman, A. S. Khan, A. K. Rowe, R. Mukunu, A. Sanchez, and C. J. Peters. Clinical virology of Ebola hemorrhagic fever (EHF): virus, virus antigen, and IgG and IgM antibody findings among EHF patients in Kikwit, Democratic Republic of the Congo, 1995. 1999. *J. Infect. Dis.*, **179**(Suppl 1), S177–S187.

43. Ksiazek, T. G., C. P. West, P. E. Rollin, P. B. Jahrling, and C. J. Peters. ELISA for the detection of antibodies to Ebola viruses. 1999. *J. Infect. Dis.*, **179**(Suppl 1), S192–S198.

44. Sanchez, A., T. G. Ksiazek, P. E. Rollin, M. E. G. Miranda, S. G. Trappier, A. S. Khan, C. J. Peters, and S. T. Nichol. Detection and molecular characterization of Ebola viruses causing disease in human and nonhuman primates. 1999. *J. Infect. Dis.*, **179**(Suppl 1), S164–S169.

45. Drosten, C., S. Göttig, S. Schilling, M. Asper, M. Panning, H. Schmitz, and S. Günther. Rapid detection and quantification of RNA of Ebola and Marburg viruses, Lassa virus, Crimean-Congo hemorrhagic fever virus, Rift Valley fever virus, dengue virus, and yellow fever virus by real-time reverse transcription-PCR. 2002. *J. Clin. Microbiol.*, **40**(7), 2323–2330.

46. Weidmann, M., E. Muhlberger, and F. T. Hufert. Rapid detection protocol for filoviruses. 2004. *J. Clin. Virol.*, **30**(1), 94–99.

47. Murray, P. R. 2003. *Manual of Clinical Microbiology*, ASM Press, Washington, DC.

48. Ksiazek, T. G. Laboratory diagnosis of filovirus infections in non human primates. 1991. *Lab Anim.*, **20**, 34–46.

49. Becker, S., H. Feldmann, C. Will, and W. Slenczka. Evidence for occurrence of filovirus antibodies in humans and imported monkeys: do subclinical filovirus infections occur worldwide? 1992. *Med. Microbiol. Immunol.*, **181**(1), 43–55.

50. Sanchez, A., and H. Feldmann. 1995. Detection of Marburg and Ebola virus infections by polymerase chain reaction assays, in *Protocols for Diagnosis of Human and Animal Virus Diseases* (G. Darai, Y. Becker, eds), pp. 411–419.

51. Niikura, M., T. Ikegami, M. Saijo, I. Kurane, M. E. Miranda, and S. Morikawa. Detection of Ebola viral antigen by enzyme-linked immunosorbent assay using a novel monoclonal antibody to nucleoprotein. 2001. *J. Clin. Microbiol.*, **39**(9), 3267–3271.

52. Towner, J. S., P. E. Rollin, D. G. Bausch, A. Sanchez, S. M. Crary, M. Vincent, W. F. Lee, C. F. Spiropoulou, T. G. Ksiazek, M. Lukwiya, F. Kaducu, R. Downing, and S. T. Nichol. Rapid diagnosis of Ebola hemorrhagic fever by reverse transcription-PCR in an outbreak setting and assessment of patient viral load as a predictor of outcome. 2004. *J. Virol.*, **78**(8), 4330–4341.

53. Rodriguez, L. L., A. De Roo, Y. Guimard, S. G. Trappier, A. Sanchez, D. Bressler, A. J. Williams, A. K. Rowe, J. Bertolli, A. S. Khan, T. G. Ksiazek, C. J. Peters and S. T. Nichol. Persistence and genetic stability of Ebola virus during the outbreak in Kikwit, Democratic Republic of the Congo, 1995. 1999. *J. Infect. Dis.*, **179**(Suppl 1), S170–S176.

54. Rowe, A. K., J. Bertolli, A. S. Khan, R. Mukunu, J. J. Muyembe-Tamfum, D. Bressler, A. J. Williams, C. J. Peters, L. Rodriguez, H. Feldmann, S. T. Nichol, P. E. Rollin, and T. G. Ksiazek. Clinical, virologic, and immunologic follow-up of convalescent Ebola hemorrhagic fever patients and their household contacts, Kikwit, Democratic Republic of the Congo. Commission de Lutte contre les Epidémies à Kikwit. 1999. *J. Infect. Dis.*, **179**(Suppl 1), S28–S35.

55. Casillas, A. M., A. M. Nyamathi, A. Sosa, C. L. Wilder, and H. Sands. A current review of Ebola virus: pathogenesis, clinical presentation, and diagnostic assessment. 2003. *Biol. Res. Nurs.*, **4**(4), 268–275.

56. Nyamathi, A. M., J. L. Fahey, H. Sands, and A. M. Casillas. Ebola virus: immune mechanisms of protection and vaccine development. 2003. *Biol. Res. Nurs.*, **4**(4), 276–281.

57. Grolla, A., A. Lucht, D. Dick, J. E. Strong, and H. Feldmann. Laboratory diagnosis of Ebola and Marburg hemorrhagic fever. 2005. *Bull. Soc. Pathol. Exot.*, **98**(3), 205–209.

58. Strong, J. E., A. Grolla, P. B. Jahrling, and H. Feldmann. 2006. Filoviruses and arenaviruses, in *Manual of Molecular and Clinical Laboratory Immunology* (B. Detrick, R. G. Hamilton, J. D. Folds, eds), ASM Press, Washington, DC, pp. 774–790.

59. Saijo, M., M. Niikura, S. Morikawa, T. G. Ksiazek, R. F. Meyer, C. J. Peters, and I. Kurane. Enzyme-linked immunosorbent assays for detection of antibodies to Ebola and Marburg viruses using recombinant nucleoproteins. 2001. *J. Clin. Microbiol.*, **39**(1), 1–7.
60. Niedrig, M., H. Schmitz, S. Becker, S. Gnther, J. ter Meulen, H. Meyer, H. Ellerbrok, A. Nitsche, H. R. Gelderblom, and C. Drosten. First international quality assurance study on the rapid detection of viral agents of bioterrorism. 2004. *J. Clin. Microbiol.*, **42**(4), 1753–1755.
61. Borio, L., T. Inglesby, C. J. Peters, A. L. Schmaljohn, J. M. Hughes, P. B. Jahrling, T. Ksiazek, K. M. Johnson, A. Meyerhoff, T. O'Toole, M. S. Ascher, J. Bartlett, J. G. Breman, E. M. Eitzen Jr, M. Hamburg, J. Hauer, D. A. Henderson, R. T. Johnson, G. Kwik, M. Layton, S. Lillibridge, G. J. Nabel, M. T. Osterholm, T. M. Perl, P. Russell, and K. Tonat; Working Group on Civilian Biodefense. Hemorrhagic fever viruses as biological weapons: medical and public health management. 2002. *JAMA*, **287**(18), 2391–2405.
62. Niiler, E. Bioterrorism: biotechnology to the rescue? 2002. *Nat. Biotechnol.*, **20**, 21–25.
63. Murphy, F. A. Emerging zoonoses. 1998. *Emerg. Infect. Dis.*, **4**, 429–435.
64. Chomel, B. B. Control and prevention of emerging zoonoses. 2003. *J. Vet. Med. Educ.*, **30**(2), 145–147.
65. Garcia-de-Lomas, J., and D. Navarro. New directions in diagnostics. 1997. *Pediatr. Infect. Dis. J.*, **16**(3 Suppl), S43–S48.
66. Lim, D. V., J. M. Simpson, E. A. Kearns, and M. F. Kramer. Current and developing technologies for monitoring agents of bioterrorism and biowarfare. 2005. *Clin. Microbiol. Rev.*, **18**(4), 583–607.
67. Petrosova, A., T. Konry, S. Cosnier, I. Trakht, J. Lutwama, E. Rwaguma, A. Chepurnov, E. Mühlberger, L. Lobel, and R. S. Marks. Development of a highly sensitive, field operable biosensor for serological studies of Ebola virus in central Africa. 2007. *Sens. Actuators, B*, **122**, 578–586.
68. Amano, Y., and Q. Cheng. Detection of influenza virus: traditional approaches and development of biosensors. 2005. *Anal. Bioanal. Chem.*, **381**(1), 156–164.
69. Palacios, G., T. Briese, V. Kapoor, O. Jabado, Z. Liu, M. Venter, J. Zhai, N. Renwick, A. Grolla, T. W. Geisbert, C. Drosten, J. S. Towner, J. Ju, J. Paweska, S. T. Nichol, R. Swanepoel, H. Feldmann, P. B. Jahrling, and W. Ian Lipkin. MassTag polymerase chain reaction for differential diagnosis of viral hemorrhagic fever. 2006. *Emerg. Infect. Dis.*, **12**(4), 692–695.

70. Yu, J. S., H. X. Liao, A. E. Gerdon, B. Huffman, R. M. Scearce, M. McAdams, S. M. Alam, P. M. Popernack, N. J. Sullivan, D. Wright, D. E. Cliffel, G. J. Nabel, and B. F. Haynes. Detection of Ebola virus envelope using monoclonal and polyclonal antibodies in ELISA, surface plasmon resonance and a quartz crystal microbalance immunosensor. 2006. *J. Virol. Methods.*, **137**(2), 219–228.

71. Lucht, A., P. Formenty, H. Feldmann, M. Gotz, E. Leroy, P. Bataboukila, A. Grolla, F. Feldmann, T. Wittmann, P. Campbell, C. Atsangandoko, P. Boumandoki, E. J. Finke, P. Miethe, S. Becker, and R. Grunow. Development of an immunofiltration-based antigen-detection assay for rapid diagnosis of Ebola virus infection. 2007. *J. Infect. Dis.*, **196**(Suppl 2), S184–S192.

72. Panning, M., T. Laue, S. Olschlager, M. Eickmann, S. Becker, S. Raith, M. C. Courbot, M. Nilsson, R. Gopal, A. Lundkvist, Ad Caro, D. Brown, H. Meyer, G. Lloyd, B. M. Kummerer, S. Gunther, and C. Drosten. Diagnostic reverse-transcription polymerase chain reaction kit for filoviruses based on the strain collections of all European biosafety level 4 laboratories. 2007. *J. Infect. Dis.*, **196**(Suppl 2) S199–S204.

73. Zhai, J., G. Palacios, J. S. Towner, O. Jabado, V. Kapoor, M. Venter, A. Grolla, T. Briese, J. Paweska, R. Swanepoel, H. Feldmann, S. T. Nichol, and W. Ian Lipkin. Rapid molecular strategy for filovirus detection and characterization. 2007. *J. Clin. Microbiol.*, **45**(1), 224–226.

Chapter 21

Amperometric Immuno- and DNA Sensors for Rapid and Specific Identification of Viruses

Rodica E. Ionescu,[a] **Serge Cosnier,**[b] **Vasile Magearu,**[c] **and Robert S. Marks**[d]

[a]*Université de Technologie de Troyes, Institut Charles Delaunay, UMR CNRS 6281, Laboratoire de Nanotechnologie et d'Instrumentation Optique, BP-42060, 10004 Troyes, Cedex, France*
[b]*Département de Chimie Moléculaire UMR CNRS 5250, Université Joseph Fourier, Grenoble, France*
[c]*Department of Analytical Chemistry, University of Bucharest, Bucharest, Romania*
[d]*Department of Biotechnology Engineering, Ben-Gurion University of the Negev, Beer-Sheva, Israel*
elena_rodica.ionescu@utt.fr, serge.cosnier@ujf-grenoble.fr, rsmarks@bgu.ac.il

Molecular and cellular sciences, as well as engineering efforts over the past decade, have given rise to numerous novel approaches for the rapid detection of viruses. One of such approach is amperometric transduction that has substantial potential in early detection of viruses, providing high specificity in immuno- and geno-sensing. We have included some basic theoretical aspects about electrochemistry for better understanding. We conclude with concrete examples of viruses that have been identified amperometrically.

Viral Diagnostics: Advances and Applications
Edited by Robert S. Marks, Leslie Lobel, and Amadou Alpha Sall

ISBN 978-981-4364-43-0 (Hardcover), 978-981-4364-44-7 (eBook)
www.panstanford.com

21.1 Introduction

Viruses are among the most important causes of human disease and are of increasing concern as possible agents of biowarfare and bioterrorism. Structurally, viruses are the simplest microorganisms that consist of a nucleocapsid protein coat containing genetic material, either ribonucleic acid (RNA) or deoxyribonucleic acid (DNA). Early viral detection is of great importance to differentiate the symptoms from those of bacterial infections so as to provide for an efficient cure.

Early detection of viral pathogens will lead to more efficient management of infected patients, play a key role in surveillance, and aid in limiting the spread of viruses through infection control practices. In this context, it has been estimated that 60% of all human infections are caused by viruses, and although they are the simplest, they are yet the least understood of all the microorganisms or small living particles. Nowadays, new and better techniques to study viruses continue to evolve since it is well-known that most viral infections are untreatable, and what "antiviral" drugs or vaccines are available, they are limited.

Because the control of viral infections is difficult and often impossible with the conventional methods, the best strategy requires the development of rapid and sensitive identification techniques, such as amperometric biosensors, to prevent the worldwide transmission of pathogenic viruses.[1] Examples of viral detections using amperometry are presented in Table 21.1.

Despite a vast diversity of the conventional methods for specific detection of viruses, there is still a tremendous need for their improvement in terms of sensitivity, specificity, and the test costs required. Thus, it appears that traditional molecular biology procedures like hemagglutination, inhibition, immunofluorescence, neutralization, single radial hemolysis, enzyme-linked immunosorbent assay (ELISA), etc., are either too long or semiquantitative or too expensive, and this may create problems in the medical field, where sensitive, rapid, and quantitative assays are required for detection of specific pathogens, the continuous monitoring of human health as well as water and foodstuffs. One of the most promising approaches described in the literature that can

Table 21.1 Amperometric virus detection

Viruses	Immobilization onto surfaces	Detection method	Amperometric setup/detection limit	Ref.
Variola virus (smallpox via bacteriophage MS2)	Paramagnetic microbeads STA-coated + biotinylated anti-MS2 IgG	• RDE amperometry • IDA array	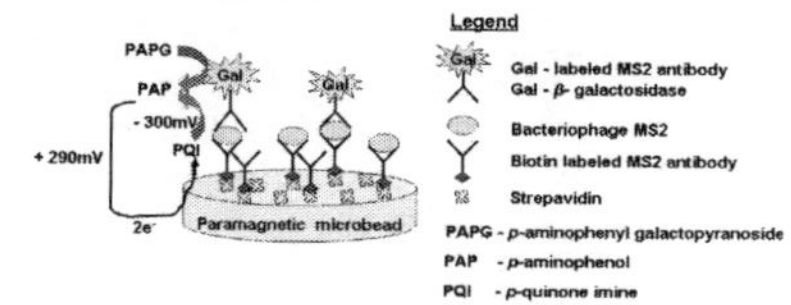 RDE: 200 ng/mL MS2 or 3.2 × 10^{10} MS2 particles/mL IDA: 90 ng/mL MS2 or 1.5 × 10^{10} MS2 particles/mL	60
BLV	ssDNA within a Ppy film BLV glycoprotein *gp51* entrapped into Ppy	Pulsed amperometry Pulsed amperometry	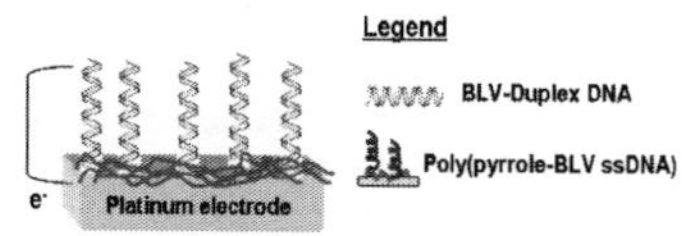 0.37 ng/mL target DNA 10 ng/mL glycoprotein *gp51*	62, 63
IfluV A/Para IfluV	Graphite powder particles + antibodies	Flow amperometry	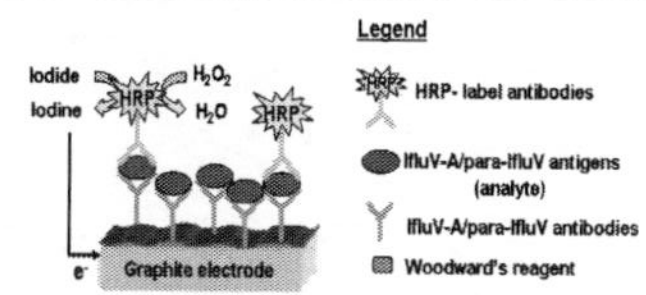 0.5 ng/mL (LOD) or 300 ng/mL (HOD) influenza 10^4 ng/mL para-influenza	66

(*Contd.*)

Table 21.1 (*Contd.*)

Viruses	Immobilization onto surfaces	Detection method	Amperometric setup/detection limit	Ref.
JEV	Antiserum on a bilayer nano-Au/ PDA polymer + Prussian blue	• Cyclic voltammetry • Amperometry	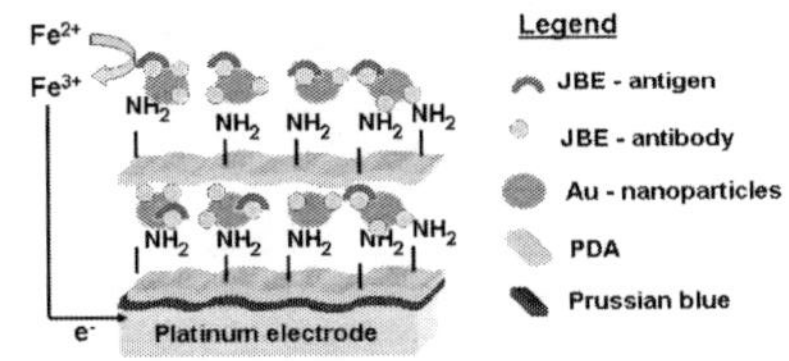 6 x 10^{-9} lg pfu/mL	69
Artificial WNV	cDNA-WNV probe onto poly(pyrrole-NHS)	Amperometry	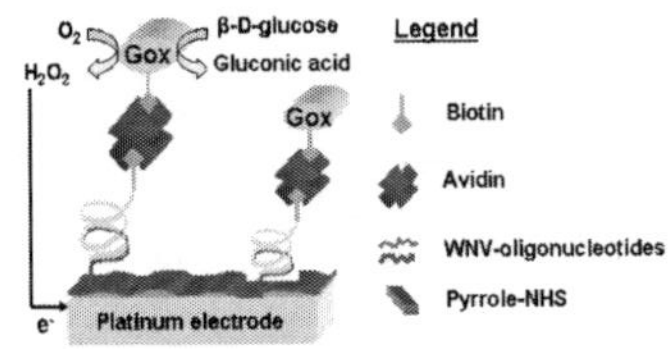 10^{-15}g/mL cDNA-WNV	20
WNV	• Duplex cDNA-WNV • Redox acridinic intercalator	• Cyclic voltammetry • Amperometry	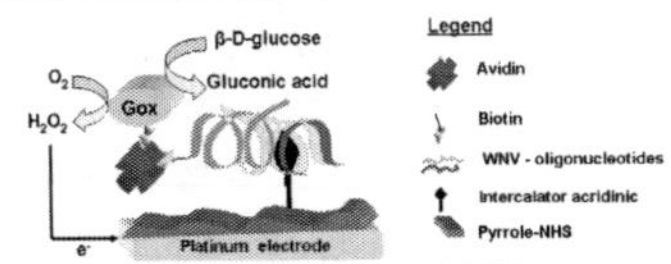 1 pg/mL cDNA-WNV	19

WNV	T7 bacteriophage with a specific WNV epitope photografted/entrapped onto (co)Ppy films	Amperometry	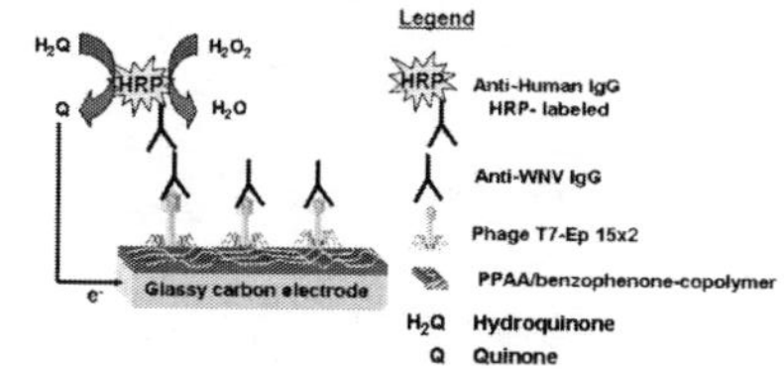 1:10^6 WNV antibody (93) 1:10^7 WNV antibody (94)	73, 74
BVDV	STA-siloxane hybrids + biotinylated cDNA	Amperometry	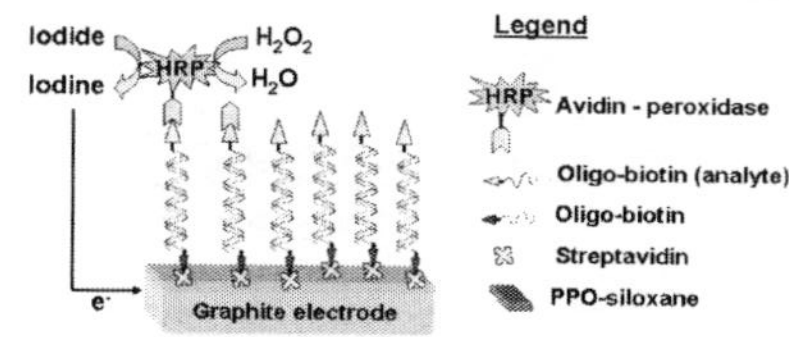 Patient sera (dilution 1:80)	75
BVDV	ITO-Pani + antibodies	Amperometry	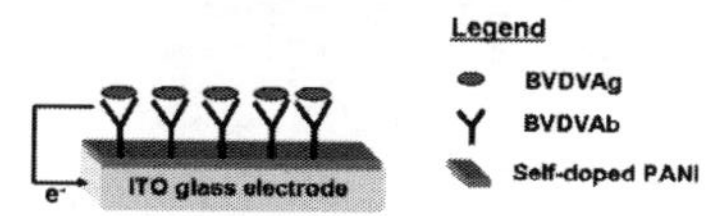 10^4 CCID mL BVDV	77

(*Contd.*)

Table 21.1 (*Contd.*)

Viruses	Immobilization onto surfaces	Detection method	Amperometric setup/detection limit	Ref.
HBV	Gold nanoparticles + hepatitis B antibody	• Amperometry • Potentiometry	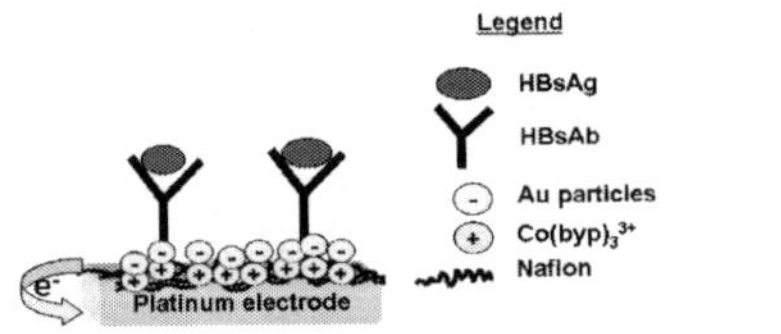 A 5 ng/mL hepatitis B antigen	80
HBV	Silanazed gold nanoparticles + hepatitis B antibody	Amperometry	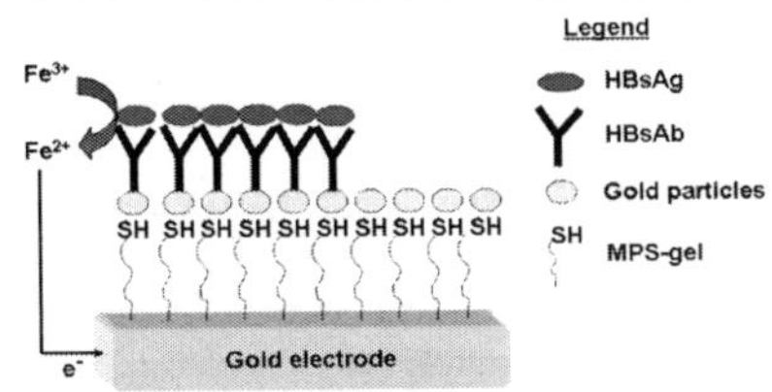 B 2 ng/mL hepatitis B antigen	81
HBV	Peroxidase-modified gold nanoparticles + hepatitis B antibody	Amperometry	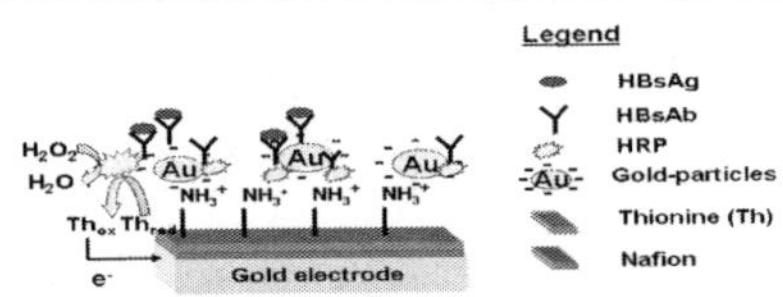 C 0.85 ng/mL hepatitis B antigen	82

HT	HT antigen adsorbed onto carbon particles	Flow amperometry	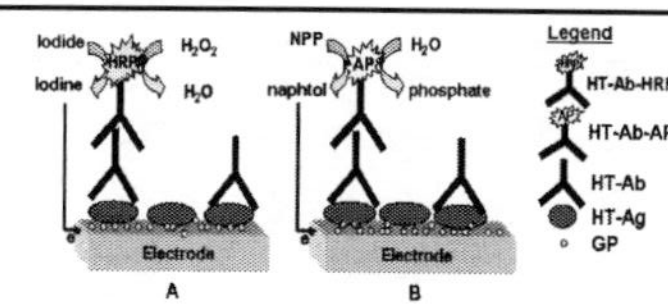 HT antibody	85, 86
NDV	NC antigen immobilized into graphite paste matrix	Amperometry	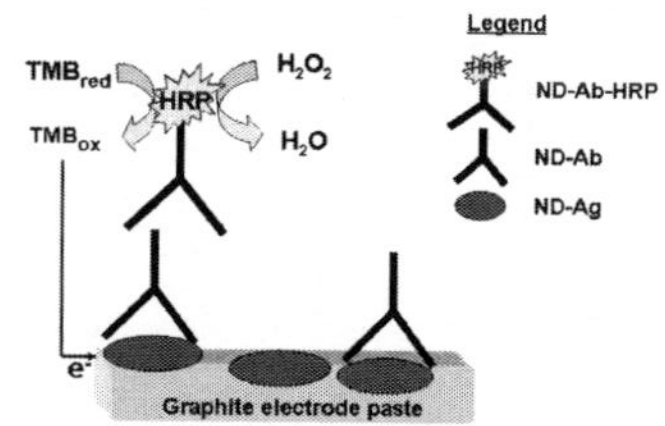 11.1 ng/mL NC antibody	88

Abbreviations: STA, streptavidin; RDE, rotating disk electrode; IDA, interdigital array electrode; Ppy, polypyrrole; BLV, bovine leukemia virus; IfluV, influenza virus; JEV, Japanese encephalitis virus; WNV, West Nile virus; HCV, hepatitis C virus; BVDV, bovine viral diarrhea virus; HBV, hepatitis B virus; HT, *Hantavirus*; NDV, Newcastle disease virus; ssDNA, single-stranded DNA; CCID, cell culture infectious dose; LOD, low limit of detection; HOD, higher limit of detection; pfu, plaque-forming unit; lg, common logarithm; PDA, *o*-phenylenediamine; NHS, *N*-hydroxysuccinimide; PPAA, poly(pyrrole-alkyl ammonium); HBsAb, hepatitis B surface antibody; HBsAg, hepatitis B surface antigen; MPS, mercaptopropyl-trimethoxysilane sol-gel solution; HRP, horseradish peroxidase; $Co(byp)_3^{3+}$, tris(2,2′-bipyridyl)cobalt(III); ND-Ab, Newcastle disease antibody; ND-Ag, Newcastle disease antigen; ND-Ag-HRP, Newcastle disease secondary antibody peroxidase labeled; Th_{red}/Th_{ox}, reduced and oxidized forms of thionine; TMB_{red}/TMB_{ox}, reduced and oxidized forms of tetramethylbenzidine; NC, Newcastle virus; ITO-Pani, indium tin oxide–polyaniline.

be applicable to study physiological fluids (e.g., blood, saliva) is provided by biosensor technologies that can be adapted for routine identification and characterization of bioanalytes, (e.g., viruses) using either antibodies/antigens or nucleic acid probes as the detection element.[2,3]

A biosensor according to the International Union of Pure and Applied Chemistry (IUPAC)[4–6] is "... a self-contained integrated device which is capable of providing specific quantitative or semi-quantitative analytical information using a biological recognition element (biochemical receptor) which is in direct spatial contact with a transducer element. A biosensor should be clearly distinguished from a bioanalytical system which requires additional processing steps, such as reagent addition. Furthermore, a biosensor should be distinguished from a bioprobe which is either disposable after one measurement, i.e. single use, or unable to continuously monitor the analyte concentration." Often the biological recognition element is an enzyme or an antibody, but it also can be a protein (e.g., a peptide or polypeptide), a nucleic acid, a microorganism (e.g., a bacterium), or even tissue material. The transducer is the device on which the biological recognition element is attached and that converts the biological recognition reaction into a quantifiable signal. When the dimensions of the transducer are minimized to a micrometer scale, biosensors are referred to as microsensors, microbiosensors, or microelectrode biosensors. The detection at the transducer can occur via different principles, for example, optically, calorimetrically, acoustically, or electrochemically, and the latter can be further subdivided into amperometric, conductometric, potentiometric, or impedance monitoring.[7–10] Although in theory many different types of biosensors can be constructed, by far the most frequently used type is the enzymatic amperometric biosensor.

Biosensors based on electrochemical transducers have the advantage of being economic and portable, with a fast response time and the possibility of automatization, which allows their application in a wide number of samples.[11]

Biosensors can be subdivided into two classes based on the type of biorecognition molecule: catalytic and affinity. In the case of catalytic biosensors, enzymes or microorganisms can act as the biorecognition entities that catalyzes a reaction involving the analyte to give a product. Common analytes for catalytic biosensors are small organic molecules like glucose, phenol, urea, and dopamine. The biorecognition molecules commonly used in the other main type of biosensors, affinity ones, include antibodies, DNA, peptides, and lectins. Affinity biosensors are characterized by a binding event between the biorecognition molecule and the analyte (the affinity reaction) often with no further reaction occurring. Hence the challenge then becomes transducing the biorecognition event. This class of biosensors is compatible with the detection of virtually all biological agents (including viruses), making possible the development of portable devices for field work. Transduction of affinity biosensors can be achieved using either labeled species or label-free approaches. If transduction is achieved using labeled species the principles are very similar to an immunoassay with the amount of analyte detected being inferred from the amount of label that binds to the interface.

This chapter presents an overview of amperometric affinity biosensors in the detection of viruses or immunoglobulins they elicited.

21.2 Theoretical Aspects for Amperometric Enzyme Biosensors

In general, diagnostic tools used for detecting or quantization of biological analytes (viruses, bacteria) rely on ligand-specific binding between a ligand and a receptor pair as antigen-antibody, hormone-receptor, drug-receptor, cell surface antigen–lectin, biotin-avidin, substrate/enzyme, and complementary nucleic acid strands. The analyte to be detected may be either a member of the binding pair or a ligand analog that competes with the ligand for binding to the complement receptor. The captured target (analyte) is then electrochemically detected.

21.2.1 *Introduction to Basic Electrochemical Principles*

In principle, the electrochemical techniques are based on monitoring changes in an electrical signal due to an electrochemical reaction at an electrode surface, usually as a result of an imposed potential or current. Between electrochemical techniques, amperometry is the branch of electrochemistry that deals with the addition (reduction) or the removal (oxidation) of electrons from a molecule or an atom that is qualified as an electroactive species.

The transfer of charge occurring in an electrochemical cell can be described by the following generalized equation:

$$\text{Oxidant} + n \Leftrightarrow \text{Reductant} \qquad (21.1)$$

where n is the number of electrons (e^-) transferred between oxidant and reductant. This equation describes a simple, chemically reversible charge transfer process, which takes place across the phase boundary between an electrode (which acts as a source or sink of electrons) and an ionically conducting medium that contains the electroactive analyte.

In a solution, the equilibrium concentrations of the reduced and oxidized forms of a redox couple are linked to the potential (E) via the Nernst equation[12]

$$E = E^0 + RT/n\text{F} \ln(C_{\text{ox}}/C_{\text{red}}) \qquad (21.2)$$

where E and E^0 are the applied and standard potentials, respectively; C_{ox} and C_{red} are the concentrations of oxidized and reduced forms, respectively; R is the molar gas constant; T is the temperature in Kelvin; n is the number of electrons transferred; and F is the Faraday constant (96,485 C mol^{-1}).

The movement of electrons to and from the electrode can be measured as an oxidation or reduction current, and this technique is referred to as *amperometry* (e.g., redox couple $[\text{Fe(CN)}_6]^{3-/4-}$). Oxidation or reduction processes cause a *Faradaic current* to flow in the electrochemical cell when a voltage is applied between the working and reference electrodes.

21.2.2 *Voltammetry*

Voltammetry is defined as an electrochemical measuring technique used for electrochemical analysis, for the determination of the kinetics and mechanism of electrode reactions, and for corrosion studies. In this respect, voltammetry applies a constant and/or varying potential (typically with a potentiostat) at an electrode's surface (working electrode) in contact with the analyte and measures the resulting current with a three-electrode system. Thus, voltammetry[13] can reveal the oxidation or reduction potential of an analyte and electrochemical reactivity. Voltammetry is nondestructive since only a very small amount of the analyte is consumed at the two-dimensional surface of the working electrode, as well as sometimes at the surface of the auxiliary electrode. In practice the analyte solutions are usually disposed of, since it is difficult to separate the analyte from the bulk electrolyte and the experiment requires only a small amount of analyte. An experiment typically involves an analyte concentration between 0.5 and 10 mM.

21.2.3 *Amperometry*

Amperometry is an electroanalytical technique based upon the measurement of the current flowing through the working electrode of an electrochemical cell. Moreover, amperometry encompasses coulometry, voltammetry, and constant potential techniques and is widely used to identify and quantify electroactive species in the liquid or gas phase. For liquid-phase analytes, the electrodes and analytes are immersed in a common electrolyte solution. In contrast, application of amperometry to gas-phase analytes involves a unique gas/liquid/solid boundary (analyte-electrolyte electrode) and an interfacial transport process that frequently controls the response characteristics and analytical performance of the amperometric gas sensor.

Practically, amperometry is based on the measurement of the current resulting from the electrochemical oxidation or reduction of an electroactive species at a constant applied potential. Amperometric techniques are linearly dependent on analyte concentration and give a normal range in the measurement of current. Almost all microorganisms can be sensed amperometrically by their enzyme-catalyzed electrooxidation/electroreduction or their involvement in a bioaffinity reaction.

On the other hand, by an appropriate choice of the applied potential, it is possible to discriminate between different redox couples. The range of potentials that can be applied in amperometric

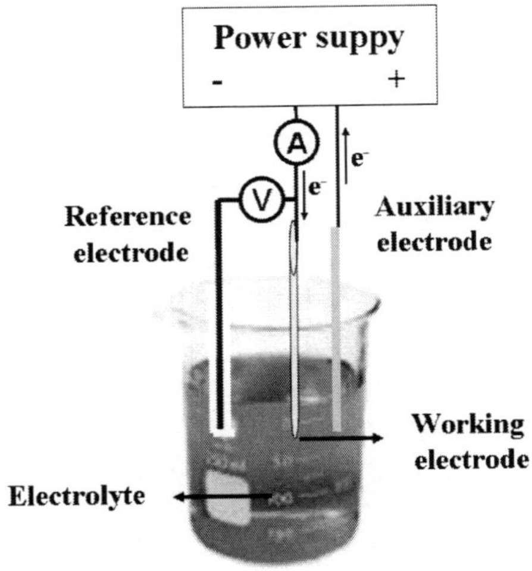

Figure 21.1 Electrochemical setup of a three-electrode amperometric/voltammetric cell used when current flowing through a reference electrode would change the potential of the reference electrode by changing the surface concentration of the species that controls the potential (e.g., Ag/AgCl : $AgCl + e^- \rightarrow Ag + Cl^-$); no current passes through the reference electrode while passing through working and auxiliary (counter) electrodes.

detection is, however, generally limited by redox processes involving the solvent, for example, the oxidative and reductive evolution of oxygen ($2H_2O \rightarrow O_2 + 4H^+ + 4e^-$) and hydrogen ($2H_2O + 2e^- \rightarrow H_2 + 2OH^-$) in water. A wide range of heavy metals, transition metals, and their complexes, as well as many physiologically and pharmacologically important substances, exhibit standard potentials within this accessible potential range.

Amperometry is based on electrochemical cells consisting of a working electrode and counter- and reference electrodes, which are in connection through an electrolyte phase (Fig. 21.1). The reference electrode is constructed so that its potential is constant regardless of the solution into which it is dipped. The potential is controlled between the indicator (or working) electrode and the reference electrode, but the current flows between the auxiliary electrode and the indicator electrode. The electrolyte phase has to carry the cell current by enabling the transport of charge carriers in the form of ions and often has to provide co-reactants (usually water, protons, or hydroxide ions) to the electrode as well as to allow the removal of ionic products from the reaction site. The analyte (the probe to be identified) is consumed at the working electrode. To balance the current flow, a counterreaction has to occur. Since the current flow is always in the same direction a reducible or oxidizable species is therefore consumed at the counterelectrode whenever measurements are performed. These redox species have to be either included in the sensor construction, even though the available amount may limit the lifetime of the sensor or presented in the electrolyte solution.

An important aspect in the amperometric detection is the choice of the working electrode material. This stems from the required electron transfer between the electrode and the analyte. Amperometry is usually performed at a platinum-, gold-, copper-, or carbon-based working electrode or an array of electrodes. Carbon-based electrode materials, such as glassy carbon and porous graphite, perform well the detection of catechols and phenolic acids, respectively.[24] Other suitable biologically important substances include amino acids and peptides, which can be detected using copper electrodes;[15] proteins (antibodies, antigens) or enzymes or nucleic acids can be detected on platinum electrodes.[16–20]

21.2.4 *Amperometric Biosensors' Classes*

The greatest challenge in constructing a proper functioning biosensor is to realize a fast and selective electron transfer between the prosthetic group of the enzyme and the electrode surface. Enzymes are generally very selective, but often the biosensor loses this enzyme-induced selectivity due to biological interferences in the electron transfer cascade. Enzyme-based biosensors most frequently use an enzyme of the class of oxidoreductases. Over the years, amperometric biosensors have been divided into three classes (Fig. 21.1).

21.2.4.1 First class of amperometric biosensors

The architecture of the first class type of biosensor consists of an enzyme attached to an electrode surface (Fig. 21.1A). The prosthetic group of the enzyme is recycled by freely diffusing oxygen (O_2), and either the decrease of O_2 or the increase of the reaction product, that is, hydrogen peroxide (H_2O_2), is monitored. This detection, which is based on freely diffusing redox species, is referred to as a *shuttle mechanism*. The first biosensors[21,22] belong to this class: glucose oxidase (GOX) was immobilized on top of a Clark oxygen electrode (an electrode that measures oxygen on a catalytic platinum surface using the reaction $O_2 + 2e^- + 2H_2O \rightarrow H_2O_2 + 2OH^-$), and the consumption of O_2 was detected by monitoring changes in the reduction of O_2. The enzyme's function is to selectively oxidize the analyte by the reduction of O_2 to H_2O_2; that is, GOX selectively catalyzes the following two reactions:

$$\text{Glucose} + \text{GOX} - \text{FAD} \leftrightarrow \text{Gluconolactone} + \text{GOX} - \text{FADH}_2$$

$$O_2 + \text{GOX} - \text{FADH}_2 \leftrightarrow \text{GOX} - \text{FAD} + H_2O_2$$

where GOX-FAD and GOX-FADH$_2$ represent the oxidized and reduced states, respectively, of the flavin active site of the GOX enzyme.

21.2.4.2 Second class of amperometric biosensors

In the second biosensor class, a Clark oxygen electrode for glucose analysis in connection with the GOX enzyme has been used. This type of biosensor was based on monitoring the oxygen consumption where the formation of the product or consumption of reagent could be monitored and directly related to the measure of the analyte concentration.

The disadvantages of the first class of biosensors have stimulated the development of the second class type of biosensors (Fig. 21.2B). In this approach artificial redox mediators are used instead of O_2. Second-class biosensors involve a two-step procedure in which the substrate is converted by the enzyme in the first redox reaction, whereas in the second part the prosthetic group of the enzyme is reoxidized by the redox mediator.

Mediators (ferrocene, quinones, quinoid-like dyes, organic conducting salts, and viologens) are redox active molecules that facilitate the electron transfer between the enzyme and the electrode.

Most enzymes (e.g., GOX), in particular oxidases, used the same cofactor, namely, oxygen, as the oxidizing agent. The latter may be

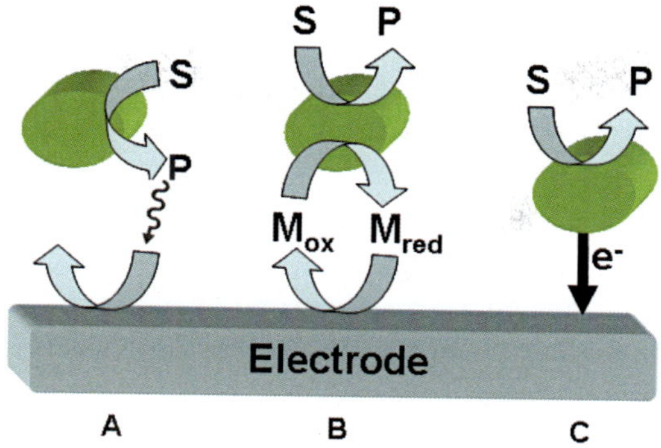

Figure 21.2 Amperometric biosensors classes. (A) first, (B) second, and (C) third. The biocatalyst is shown schematically by the green circles; S, substrate; P, product; M_{ox}, mediator oxidized form; M_{red}, mediator reduced form.

replaced by a variety of artificial oxidizing agents able to exchange electrons with the prosthetic site of the enzyme without steric constraints. These compounds must also exhibit a redox potential that is similar to or slightly more positive than that of the active site in the enzyme, as in the following reaction:

$$\mathrm{GOX-FADH_2+Mediator_{ox}\leftrightarrow GOX-FAD+Mediator_{red}}$$

In this way, so-called *reagentless* biosensors are fabricated: all components that are required for the electrochemical reaction are immobilized on the electrode surface, and only the concentration of the substrate of interest is a variable parameter. For proper functioning of the biosensor, the redox mediator has to display stable oxidized and reduced forms, a fast reaction rate with both the enzyme and the electrode surface, and the possibility of a tight attachment into the architecture of the biosensor.[10] Mainly soluble, low-molecular-weight metal complexes are used for this purpose, for example, ferrocene and osmium derivatives, but other materials are also used, for example, quinones, tetrathialfulvalene (TTF), tetracyanoquinodimethane (TCNQ), conducting salts (e.g., TTF-TCNQ), and organic dyes (e.g., methylene blue, Prussian blue, and thionin).[7,8,10]

21.2.4.3 Third class of amperometric biosensors

The direct enzyme-electrode coupling or mediatorless biosensors based on direct electron transfer mechanisms belong to the third class. In this case, the electron is directly transferred from the electrode to enzyme and to the substrate molecule or vice versa (Fig. 21.2C). In this mechanism the electron acts as a second substrate for the enzymatic reactions and results in the generation of a catalytic current. The substrate transformation (electrode process) is essentially a catalytic process.[23,24] Third-class biosensors offer all the benefits of second-class sensors and some new ones as well. The latter arise from the self-contained nature of the sensor. Since neither a mediator nor an enzyme must be added, this design facilitates repeated measurements. Since the sensor can be of use for multiple analyses this minimizes cost pressures on sensor design. Theoretically, this has many advantages, as direct electron transfer avoids intermediate electron transfer steps via self-exchange reactions, which are prone to interference. However, direct electron transfer is only possible for relatively small enzymes with an easily accessible prosthetic group. Connection with the enzyme occurs at the outer protein shell of the enzyme,[25] at the inner surface of the enzyme in close proximity to its active center,[55] or directly at the prosthetic group of the enzyme.[26] With the third biosensor approach, the redox center of the enzyme is *plugged* directly to an *electrical cable*. The connection between the enzyme and the electrode surface is mediated via conducting polymers,[28,29] conducting polymers modified with redox mediators,[30,31] or entrapment of enzymes in conducting materials, such as sol-gel composites with graphite particles.[32]

21.3 Classification of Amperometric Biosensors

21.3.1 *Immunosensors*

Immunoassays have spread rapidly in clinical analysis, while immunosensors are of great interest because of their potential utility being specific, simple, and rapid, with small sample requirements compared to conventional immunoassay techniques. They are designed to detect the highly binding event between antibody and antigen without a need for steps of separation and washing; thus, they require an immobilizing antibody or antigen by either covalent binding or physical adsorption on the electrode surfaces.[33] There are four types of immunosensor detection devices, depending on the transducer employed: electrochemical (amperometric, potentiometric, capacitative, or impedance), optical (fluorescence, luminescence, electrochemiluminescence, surface plasmon resonance, refractive index), microgravimetric, and thermometric devices.[34–36]

The immobilized sensing element can be either an antibody to monitor the presence of a viral antigen or an antigen (whole virus, chemically neutralized virus, phage-displayed epitopic antigen, synthetic peptides derived from viral polypeptides, or even purified viral macromolecules) to measure the elicited antibodies.

Indeed, whole viruses, which are molecular genetic parasites that use cellular systems for their own replication,[37] are not considered living as they require a host habitat for metabolic activities.[38] It is the surface coat proteins that are the viral antigen that triggers the immune response and antibody production that will be measured. Viruses (and bacteria, fungi, and parasites) have a large number of antigenic determinants on their surfaces, and therefore each organism can bind a number of antibody units. Another possibility would be the use of synthetic peptides derived from epitopic viral antigen or cloned epitopes within phage displays and of course purified antigenic viral macromolecules. In the case of phages,[39] selected from a *phage display library* "a collection of engineered phages, where each one of them contains a short foreign coding sequence spliced into the major coat protein gene enables their use so that the altered amino acids are displayed on every coat protein subunit." These can be constructed utilizing a number of peptides having random or partially random amino acid sequences. Phages may also be created and selected after multiple rounds of sequence mutagenesis and affinity selection.[40,41] The peptide specified by the foreign coding sequence is displayed on the surface of the phage or *virion*. Each phage clone displays many copies of a single foreign peptide, thus increasing sensitivity to the particular viral epitope of interest. Because the viral carrier is infective, phages can be propagated indefinitely, helping to reduce cost.[42]

To measure the presence of an immunogenic reaction secondary antibodies are required labeled with enzyme markers such as HRP.

21.3.2 *DNA Sensors*

Electrochemical detection of DNA can be achieved by direct or catalyzed oxidation of DNA bases, by the electrochemical response generated, by enzyme or other redox markers, or by a specific binding event with the target DNA.

Typically, electrochemical biosensors (DNA sensors) use DNA probe molecules—well-defined sequences of single strands as a biological receptor—are immobilized to an electrically active solid surface. If the probe hybridizes (combines) with the unknown nucleic acid because of pairing of complementary base recognition, detection and identification are possible, for example, by measuring current (amperometry) or resistance changes (impedance) caused by hybridization of target DNA.[20,43] Also, by coupling hybridization to a redox-active double-stranded DNA (dsDNA) intercalator with a catalytic mediator, the sensitivity is increased to make detection possible without amplification.[19]

Moreover, recent experiments designed to improve probe immobilization using modified self-assembled monolayers[44] or conducting polymers[45] demonstrate detection of single-nucleotide mismatches. Enzymatic discrimination methods have improved detection of single-nucleotide mismatches to the picomolar range.[46,47] Thus, recently an electrochemical method to sense a single-base mutation in an oligonucleotide was demonstrated using an oligonucleotide-tagged redox enzyme and an oligonucleotide-associated redox polymer as a sensing and electrical contacting matrix.[48] The effectiveness of electrical contacting and the resulting amperometric response were found to be controlled by the number of base mismatches between the oligonucleotide probe and the gene (analyte).

Moreover, directly measuring target DNA oxidation using an electrochemical mediator with a probe-coated electrode results also in sensitive detection. In addition, DNA-based analytical methods such as amperometry seem to be a very specific method for detecting genetic modifications, being one of the most sensitive approaches for detecting microorganisms such as viruses with either single-stranded or double-stranded RNA or DNA.[49]

21.4 Viral Detection Using Amperometry

21.4.1 *Variola Virus*

Smallpox was first experienced in Europe in the fifteenth century[50] and is an infectious disease unique to humans, caused by either of two virus variants named *Variola major* (*V. major*) and *Variola minor* (*V. minor*), which belong to the family *Poxviridae*.

Smallpox localizes in small blood vessels of the skin and in the mouth and throat. *V. major* produces a more serious disease and has an overall mortality rate of 30%–35%. *V. minor* causes a milder form of the disease (also known as alastrim, cottonpox, milkpox, whitepox, or Cuban itch), which kills $\sim$ 1% of its victims.[51]

Smallpox killed an estimated 400,000 Europeans each year during the eighteenth century (including five reigning monarchs) and was responsible for a third of all blindness.[52] After successful vaccination campaigns throughout the nineteenth and twentieth centuries, the World Health Organization (WHO) certified the eradication of smallpox in 1979. To this day, smallpox is the only human infectious disease to have been completely eradicated from nature.[53]

However, smallpox detection using rapid technologies is still attracting world research groups. And one of a potential route is using paramagnetic microbeads as powerful tools for numerous biochemical assays, such as immunoassays[54–56] and DNA assays.[57] These beads often have a STA coating on the surface that binds with various biotinylated molecules with a dissociation constant of ca. 10^{-14} M. Paramagnetic microbeads can be used as mobile solid phases onto which enzyme immunoassay sandwiches can be assembled,[54,55] being good candidates for use in a fluidic system, as demonstrated previously for an electrochemical immunoassay with mice IgG[55] and bacteriophage MS2.[56] Bacteriophage MS2 is a well-characterized virus with a 26 nm diameter[81] and a known nucleic acid sequence and crystal structure and is a recognized stimulant for viruses, such as smallpox.[59]

In 2004 the Thomas Group developed an amperometric immunoassay for detecting bacteriophage MS2—the stimulant for smallpox—using paramagnetic beads of 2.8 μm diameter.[60] The immunoassay sandwich is described in Table 21.1 and is based on attaching a biotinylated rabbit anti-MS2 IgG to a STA-coated bead, capturing the virus, and then attaching a rabbit anti-MS2 IgG-β-galactosidase conjugate to another site on the virus. β-galactosidase converts *p*-aminophenyl galactopyranoside (PAPG) to *p*-aminophenol (PAP). PAPG is electroinactive at the potential at which PAP is oxidized to *p*-quinone imine (PQI), so the current that results from the oxidation of PAP to PQI is directly proportional to the concentration of antigen in the sample. The immunoassay was detected with rotating disk electrode (RDE) amperometry and an interdigitated array (IDA) electrode. With an applied potential of +290 mV versus Ag/AgCl, the detection limit was 200 ng/mL MS2 or 3.2×10^{10} viral particles/mL with RDE amperometry. The authors also fabricated a trench of an IDA electrode incorporated into a poly(dimethyl siloxane) channel, within which beads were collected, incubated with PAPG, and PAP generation monitored. With the two working electrodes holding it at +290 and −300 mV versus Ag/AgCl, and electrochemical recycling of the PAP/PQI couple, the limit of detection was lowered to 90 ng/mL MS2, or 1.5×10^{10} MS2 particles/mL.

21.4.2 Retroviridae *Family*

21.4.2.1 Bovine leukemia virus

BLV is a bovine virus widely distributed and endemic in many cattle herds being closely related to human tumor T-lymphotropic virus. BLV is a retrovirus (family *Retroviridae*), which integrates a DNA intermediate as a provirus into the DNA of B-lymphocytes of blood and milk. It contains an oncogene coding for a protein called *Tax*. Most cattle infected with BLV do not exhibit clinical signs. BLV infection is lifelong in cattle, so demonstration of serum antibodies to BLV indicates persistent infection. Persistent (or fluctuating) lymphocytosis demonstrated in peripheral blood films or complete blood counts develops in approximately 30% of

BLV-infected cattle. Lymphoma (or lymphosarcoma) develops in approximately 3% of BLV-infected cattle but usually not until they are at least six years of age.

Nevertheless in its natural host cattle leukemia is rare. Because the oncogenic properties of the virus were discovered early, a search for evidence of pathogenicity in humans started soon after discovery. Many potential routes of BLV transmission exist. Transmission through procedures that transmit blood between animals, such as gouge dehorning and vaccination and ear tagging with instruments or needles that are not changed or disinfected between animals, is a significant means of BLV spread.[61]

Due to the importance of rapid BLV identification, Ramanaviciene and Ramanavicius reported in 2004 on the pulsed amperometric detection of a bovine leukemia provirus DNA target using platinum electrodes modified by single-stranded DNA (ssDNA) 20-mers entrapped within polypyrrole (Ppy) films (Table 21.1). The authors mentioned that even though the detection limit of the DNA biosensor was 0.37 ngmL^{-1}/target DNA, the sensitivity of this assay is still insufficient to enable a direct assay without the polymerase chain reaction (PCR) amplification step.[62]

Moreover, the same researcher group also reports for the first time on the preparation and basic characterization of a Ppy-based molecularly imprinted polymer (MIP) for label-free amperometric detection of BLV glycoprotein *gp*51 (*gp*51). Ppy was used as a matrix for preparation of MIP. Ppy doped by *gp*51 was prepared by electrochemical deposition of this polymer on the surface of a platinum-black electrode. Then, molecules of *gp*51 were removed from the polymeric backbone and molecularly imprinted polypyrrole (mPpy) was ready for recognition of *gp*51 in the aqueous solution. Pulsed amperometric detection was applied for label-free detection of *gp*51 in the samples. Anti-*gp*51 antibodies and secondary antibodies labeled with HRP were involved as markers for the control of mPpy preparation procedures. Afterward, catalytic activity of HRP was tested spectrophotometrically. Application of anti-*gp*51 and HRP-labeled secondary antibodies confirmed that the generation of an analytical signal was based on redoping of mPpy by 10 ng/mL *gp*51. The authors noticed during experiments that only a few mPpy redoping/dedoping cycles were effective, and they think that this method may be very effective for the future development of mPpy-based MIPs.[63]

21.4.3 *The* Orthomyxoviridae *Family*

21.4.3.1 Parainfluenza and influenza A viruses

IfluV A is a genus of the *Orthomyxoviridae* family of viruses. IfluV A includes only one species—*Influenza A virus*, which causes influenza in birds and some mammals. Strains of all subtypes of influenza A virus have been isolated from wild birds, although disease is uncommon. Some isolates of influenza A virus cause severe disease both in domestic poultry and, rarely, in humans. Occasionally viruses are transmitted from wild aquatic birds to domestic poultry, and this may cause an outbreak or give rise to human influenza pandemics.[64]

IfluVs afflict millions of people and animals every year. IfluVs belong to *Orthomyxoviridae*, which can be divided into two different protein components, identified as antigens. They are composed of spike-like features called hemagglutinin (H) and neuraminidase (N) components. In total there are 15 subtypes of H and 9 subtypes of N. Unfortunately, all of the H and N subtypes can infect humans, animals, and birds.

Influenza is caused by the H type component, an antibody that causes the clumping of red blood cells. Influenza killed 240 people between 2003 and 2008, but although strains have been found in the United States, there is no evidence that it is fatal to humans.[65]

Various researchers used the enzyme-linked immunosorbent assay (ELISA) as an efficient test for influenza diagnosis. But ELISA requires a purified antigen, which cannot be available without some delay. At any rate, the development of rapid, sensitive, and simple technologies, such as amperometric biosensors, for the detection of IfluVs is still needed.

Therefore, Sitdikov and Wilkins, in 2007, were focusing on testing a miniature biosensor that measures amperometric response using a flow-amperometric immunoassay device for fast, qualitative detection of low concentration (0.5 ng/mL) of influenza A and para IfluVs. The authors used Woordward's reagent K (*N*-ethyl-

5-phenyliso-xazolium-3′-sulfonate) to immobilize covalently IfluV antibodies to the surface of the carbon electrode (covalently linked immunoagent solid-phase conjugates). The sandwich scheme of the IfluV amperometric detection is presented in Table 21.1. It was found that the higher limit of detection can be 300 ng/mL and more than 10^4 ng/mL for influenza A and parainfluenza, respectively.[66]

21.4.4 *The* Flaviviridae *Family*

21.4.4.1 Japanese encephalitis virus

JEV is a mosquito-borne flavivirus that induces severe encephalitic and neurologic disease manifestations and is the cause of many cases of acute fatal encephalitis each year in Asia. Typically, the symptoms of JEV infection appear following an incubation period of 4 to 20 days. Patients may arrive at the hospital with evidence of severe central nervous system infection, including high fever, paralysis, and coma. There is no effective treatment.[67]

Approximately one-third of hospitalized patients will die within 10 days, and another one-third will recover with severe, permanent neuropsychiatric deficits. Each year more than 10,000 JEV-induced deaths are officially reported, although the actual number is estimated to be much higher.[68]

The Yuan team[69] fabricated in 2005 a sensitive, fast, and label-free amperometric immunosensor for the direct detection of Japanese B encephalitis vaccine (Table 21.1). Antiserums of Japanese B encephalitis were immobilized on a bilayer nano-Au/*o*-phenylenediamine polymer film with Prussian blue deposited as an electronic mediator on the platinum electrode. The electrochemical behavior of the biosensor was studied with $Fe^{2+/3+}$ as probe on the platinum surface using the cyclic voltammetry technique. The variation of amperometric response to the concentration of the Japanese B encephalitis vaccine, the target antigen, was evaluated by cyclic voltammetry in phosphate buffer saline. The immunosensor showed a specific response to the Japanese B encephalitis vaccine in the range 1.1×10^{-8} to 1.9×10^{-6} lg pfu/mL with a detection limit of 6×10^{-9} lg pfu/mL.

21.4.4.2 West Nile virus

WNV is closely related to other flaviviruses of global importance, including dengue, yellow fever, and tick-borne encephalitis viruses, among others.[70] These pathogens are typically transmitted to vertebrates by mosquitoes or ticks and are responsible for severe morbidity and mortality in both humans and animals.

Most infections are asymptomatic and remain undetected, except by later serologic evidence of exposure to the virus.[71] Approximately 1 in 150 infected individuals, however, develops neurologic symptoms, such as meningitis, encephalitis, or flaccid paralysis. The elderly and patients with compromised immune systems are particularly at risk for the more severe forms of the disease. For example, in the United States the total number of human cases collectively reported from 1999 to 2005 was 19,655, with 782 fatalities.[72] Unfortunately, current treatment for WNV infection primarily consists of supportive measures.

Knowing the importance of developing rapid and sensitive WNV detection devices, Ionescu et al. reported in 2006 on the construction of an amperometric cDNA-WNV sensor using an amino-21-mer oligonucleotide probe covalently anchored to a poly(pyrrole-NHS), previously electropolymerized on a platinum electrode (Table 21.1). After incubation with a target model of the cDNA-WNV, the hybridization events on the sensor surface were detected by an additional hybridization process with a complementary biotinylated 15-mer WNV cDNA followed by the specific attachment of biotinylated GOX via an avidin bridge. The hybridization event was then monitored at 0.6 V versus Ag/AgCl by amperometric detection of H_2O_2, generated by the enzyme marker in the presence of glucose. A relatively short (two hours) hybridization period allows quantification of the WNV-DNA target in the range 10^{-10}–10^{-15} g/mL.[20]

The same group of researchers evolved a genuine cDNA-WNV amperometric sensor using a grafted acridinic redox intercalator derivate onto a Ppy film functionalized by NHS groups that self-insert into the dsDNA-WNV solution (Table 21.1). This modified electrode was used for the detection of ssDNA derived from a WNV sequence. The latter was amperometrically detected after its hybridization in solution with a biotinylated complementary oligonucleotide, followed by its anchoring and labeling by GOX at 1 pg/mL.[29]

Moreover, the same research group was involved in the fabrication of two types of WNV immunosensors. One biosensor configuration was using an electrogenerated copolymeric film of poly(pyrrole-benzophenone and tris(bipyridine pyrrole) ruthenium) units for the photochemical grafting of T7 bacteriophages, displaying a specific WNV epitope[73] (Table 21.1). The resulting electrodes were applied to the amperometric detection of the WNV antibody at 0 V in an aqueous electrolyte via a secondary peroxidase-labeled antibody. The addition of H_2O_2 and hydroquinone as the enzyme substrate induces the enzymatic formation of quinone, which is detected at the electrode surface through its reduction, the response time (30 sec) being relatively short. The intensity of the cathodic current is thus proportional to the immobilized amount of the WNV antibody, and a calibration curve based on different WNV antibody dilutions ranging from 10 to 10^6 was created.

The second amperometric biosensor configuration[74] was based on the immobilization via entrapment of T7 phages (previously modified by an additional peptide WNV sequence) into a Ppy film electrogenerated from an adsorbed coating constituted of electropolymerizable amphiphilic pyrrole ammonium and phage molecules (Table 21.1). After successive incubation with a WNV antibody solution and then a secondary peroxidase-labeled antibody, the immunosensors were applied to the quantitative amperometric determination of the WNV antibody at 0 V versus Ag/AgCl via the reduction of the enzymatically generated quinone in the presence of hydroquinone and H_2O_2. This immunosensor configuration was more sensitive than the first WNV immunosensor setup and detected the lowest WNV antibody dilution, namely, a titer of $1:10^7$.

21.4.4.3 Hepatitis C virus

Hepatitis C is a blood-borne infectious disease that is caused by HCV, infecting the liver. HCV is the only known member of the *Hepacivirus* genus in the family *Flaviviridae*. There are six major genotypes of HCV, which are indicated numerically (e.g., genotype 1, genotype

2). HCV infection can cause liver inflammation (hepatitis) that is often asymptomatic, but ensuing chronic hepatitis can result later in cirrhosis (fibrotic scarring of the liver) and liver cancer.

HCV is spread by blood-to-blood contact. No vaccine against hepatitis C is available. The symptoms of infection can be medically managed, and a proportion of patients can be cleared of the virus by a long course of antiviral medicines. Although early medical intervention is helpful, people with HCV infection often experience mild symptoms and consequently do not seek treatment. An estimated 150–200 million people worldwide are infected with hepatitis C.

Rapid diagnosis and genotyping of HCV (genotypes 1, 2A/C, 2B, and 3) was performed in 2006 by the Riccadi Group[75] using an amperometric DNA biosensor (Table 21.1). The biosensor used STA encapsulated in thin films of siloxane-poly(propylene oxide) hybrids prepared by the sol-gel method and deposited on the graphite electrode surface by the dip-coating process, while biotinylated 18-mer probes were immobilized through STA. The HCV RNA from serum was submitted to reverse transcriptase polymerase chain reaction (RT-PCR), and biotin-labeled cDNA was obtained. Thus, the cDNA was hybridized to the target-specific oligonucleotide probe immobilized on the graphite electrode surface, followed by the addition of the avidin and peroxidase (HRP) conjugate. The HRP response was monitored by using H_2O_2 and KI, where the amount of I_2 generated was detected by constant potential amperometry at −0.45 V versus Ag/AgCl. The authors believe that their methodology will be a suitable and convenient tool for STA immobilization and diagnosis of HCV disease.

21.4.4.4 Bovine viral diarrhea virus

Bovine viral diarrhea is a viral disease of cattle and other ruminants, caused by BVDV, a member of the *Pestivirus* genus and the family *Flaviviridae*. There are four recognized species within the *Pestivirus* genus. These species are BVDV-1, BVDV-2, border disease virus of sheep, and classical swine fever virus, previously known as hog cholera virus. Clinical signs of mucosal erosions and diarrhea, which occur in the acute form of BVDV, have a significant effect on the animals infected, but much more costly is the persistent/chronic infection of animals.

Humans are not susceptible to BVDV. However, immunosuppression is one of the consequences of BVDV infection. Immunosuppression may increase the susceptibility of an infected animal to other viruses and bacteria, and some of these viruses and bacteria may be detrimental to human health. Humans may become infected due to some species of *E. coli* and *Salmonella*.[76]

Recently, Muhammad et al.[77] reported on the construction of an ITO-Pani amperometric biosensor for the detection of BVDV (Table 21.1). The Pani biosensor design was based upon the specific nature of antibodies to capture the target virus and the conductive properties of self-doped Pani to translate the antibody-antigen binding into a quantifying signal. The authors found that the ITO-aniline biosensor configuration is versatile for the detection of other pathogens of interest by changing the specificity of the antibodies.

21.4.5 *The* Hepadnaviridae *Family*

21.4.5.1 Hepatitis B virus

HBV infects the liver of entities from the *Hominoidae* family, including humans. The disease was originally known as "serum hepatitis"[78] and has caused epidemics in parts of Asia and Africa. Hepatitis B is endemic in China and various other parts of Asia.[79] Symptoms of the acute illness caused by the virus include liver inflammation, vomiting, jaundice, and, occasionally, death. Chronic hepatitis B may eventually cause liver cirrhosis and liver cancer, a fatal disease with very poor response to current chemotherapy. But HBV infection could be preventable by vaccination.

Since assays for detection of HBV infection involve serum or blood tests that detect either viral antigens (proteins produced by the virus) or antibodies produced by the host, the whole interpretation of results is a complex process. HBsAg is most frequently used to screen for the presence of HBV disease because it is the first detectable viral antigen that appears during infection.

To simplify the hepatitis diagnosis process, biosensor technology is one attractive option due to its advantages of simplicity and rapidity.

One example of HBsAg detection was reported by the Tang Group in 2005 (Table 21.1) concerning the elaboration of amperometric and potentiometric immunosensors using the following configuration: a layer of plasma-polymerized Nafion film (PPF) deposited on the platinum electrode surface and then positively charged with an electron mediator system such as tris(2,2′-bipyridyl)cobalt(III) [$Co(bpy)_3^{3+}$] and negatively charged gold nanoparticles assembled on the PPF-modified platinum electrode using a layer-by-layer technique.[80] Finally, HBsAb was electrostatically adsorbed on the gold nanoparticles surface. When HBsAg was bound with antibodies, the antigen-antibody complex coating on the surface of the electrode inhibited the electron transfer. Therefore, the current responses decrease after the formation of immunocomplexes. The multilayer film-modified immunosensor was used for amperometric and potentiometric HBsAg determination, and both systems provided the same linear ranges from 0.05 to 4.5 μg/mL with different detection limits for the amperometric system 0.005 μg/mL (at −0.35V vs. saturated calomel electrode [SCE]) and for the potentiometric system 0.015 μg/mL.

$$[Co(bpy)_3{}^{3+}] \Leftrightarrow [Co(bpy)_3{}^{2+}] + 1e^-$$

An HBsAg amperometric immumosensor[81] has also been developed by self-assembling gold nanoparticles to a thiol containing sol-gel network where the linearity of HBsAg is in the range of 2–360 ng/mL (Table 21.1). Initially, a cleaned gold electrode was first immersed in a hydrolyzed mercapto-propyltrimethoxysilane (MPS) sol-gel solution to assemble a three-dimensional silica gel, and then gold nanoparticles were chemisorbed onto the thiol groups of the sol-gel network. Finally, HBsAb was adsorbed onto the surface of the gold nanoparticles. Thus, an interfacial design of a bare gold electrode (BGE)/MPS/Au[HBsAb] was prepared to detect HBsAg in human serum on the basis of the specific reaction of HBsAb and HBsAg. The electrochemistry of the ferricyanide redox reaction was used as a marker to probe the interface and as a redox probe to determinate HBsAg. The authors compared their approach with the glutaraldehyde-binding approach, and they found that the antibodies immobilized by their method present larger amounts and higher immunoactivity.

Even more, an amperometric immunosensor based on colloidal gold nanoparticles and an HRP-modified gold electrode[82] for the determination of HBsAg has been developed (Table 21.1). The authors report on the first use for the HRP, instead of bovine serum albumin (BSA), to block the possible remaining active sites of the gold nanoparticles monolayer, to avoid the nonspecific adsorption and amplify the response of the antigen-antibody reaction at the same time. The HBsAg immunosensor was initially constructed using a negatively charged Nafion polymeric film (that increases electrode stability) to which phenothiazine dye molecules with amino groups (such as thionine) were absorbed via the ion exchange process. On the formed matrix, negatively charged colloidal gold nanoparticles were further attached and used subsequently to immobilize HBsAb. The amperometric current responses were recorded in phosphate saline buffer (pH: 6.5) containing 0.27 mM H_2O_2 after the immunosensor was incubated with a different HbsAg solution for five minutes. The calibration curve for the detection of hepatitis B antigens using HRP-blocked immunosensors emphasizes a good linear range from 2.56 to 563.2 ng/mL with a sensitive detection limit of 0.85 ng/mL.

21.4.6 *The* Bunyaviridae *Family*

21.4.6.1 Hantaviruses

HV is a genus of the family *Bunyaviridae*, which presently consists of approximately 25 viruses. Like all of the *Bunyaviridae*, HVs are maintained in natural reservoirs, such as in small mammals (rodents and possibly in shrews).[83]

HVs are distributed worldwide, including Asia, Europe, and the United States. However, there are regions, namely, Africa and Australia, where only very limited information about HV-infected rodents and human HV infections is available so far. Thus, only very few studies report the presence of HV-specific antibodies in humans in Central and East African countries.

Four pathogenic types for humans from the *Bunyaviridae* family are hosted by rodents: the *Sin Nombre virus* (SNV) responsible for severe respiratory syndrome in the United States; the *Hantaan* and *Seoul* viruses responsible for hemorrhagic fevers with renal syndrome (HFRS) of severe-to-moderate expression in Asia and the Balkans; and the *Puumala* virus, which produces HFRS of moderate expression, or the nephropatia epidemica in Europe.[84]

The determination of specific antibodies against HV in blood plasma has been developed by Krishnan et al. in 1996 (Table 21.1). Their assay technique is based on a flow-through amperometric immunoelectrode where highly dispersed carbon material serves as an immunosorbent and as an electrode material. The analysis employed the "sandwich" assay scheme with the recombinant protein of HV immobilization on the surface of the immunosorbent. The target analyte captured by the immunosorbent at the first stage of incubation interacts with the peroxidase-labeled antihuman antibodies at the second stage of incubation. The detection of the peroxidase label was conducted amperometrically by electroreduction of iodine formed as a product of the peroxidase-catalyzed reaction. The technique permits determination of specific antibodies against HV in blood plasma using a 22-minute assay time.[85]

Sitdikov et al. report also on the fabrication of an automated electrochemical immunosensor device of blood for the detection in mice of specific anti-HV antibodies (Table 21.1). The assay technique is also based on a flow-through amperometric immunosensor with a flow rate of reagents of 120 μl/min and using the dispersed carbon particles as immunoelectrodes, which were modified with recombinant SNV nucleocapsid protein. Further, a *sandwich* immunoassay was performed, where naphthol was formed as a result of enzymatic dephosphorylation of naphtyl phosphate in the presence of alkaline phosphatase. The overall time of measurements was 22 minutes.[86]

21.4.7 *The* Paramyxoviridae *Family*

21.4.7.1 Newcastle disease virus

NDV is an RNA virus of the family *Paramyxovidae*, causing a highly contagious disease transmissible to poultry and over 250 other species of birds. Mortality from infection with virulent strains of NDV can reach 100% in poultry flocks unprotected by vaccination,[87] and thus vaccination is relied upon in most countries with commercial and noncommercial poultry production. NDV effects are most notable in domestic poultry due to their high susceptibility and the potential for severe impacts of an epidemic on the poultry industry. NDV is endemic to many countries.

NDV transmission occurs by exposure to fecal and other excretions from infected birds and through contact with contaminated feed, water, equipment, and clothing.

Exposure of humans to infected birds can cause mild conjunctivitis and influenza-like symptoms, but NDV otherwise poses no hazard to human health.

An amperometric immunosensor (Table 21.1) for detection of NDV poultry infectious disease through a Newcastle disease antibody (ND-Ab) was developed in 2003 by the Gong Group.[88] The Newcastle disease antigen (ND-Ag) was immobilized in a graphite paste matrix. The assay procedure is based on the bioreaction of the analyte ND-Ab and enzyme-labeled HRP-ND-Ab competing for the ND-Ag sites at the newly regenerated biocomposite surface. The determination is accomplished by amperometry using tetramethylbenzidine (TMB) and H_2O_2 as the substrates. Two different graphite pastes were compared using either sol-gel or solid paraffin as binders. The sol-gel biocomposite showed better response characteristics and higher reproducibility. The ND-Ab can be determined up to 443.24 ng/mL with a detection limit of 11.1 ng/mL.

21.5 Future Directions

Infectious diseases are responsible for almost half the mortality in developing countries. These deaths occur primarily among the poorest people because they do not have access to the diagnostics tools, drugs, and commodities necessary for rapid detection, prevention, or cure. Approximately half of infectious disease mortality can be attributed to just three diseases: human immunodeficiency virus (HIV), tuberculosis (TB), and malaria.

To combat these diseases, rapid, simple, cost-effective medical devices for screening medical diseases and infectious pathogens are essential for early diagnosis and improved treatments of many illnesses.

For such purposes, amperometric biosensors devices are attractive because they have the advantages of being highly sensitive, rapid, and inexpensive to detect pathogens (e.g., bacteria and viruses) at field locations located far from sophisticated laboratories and they do not require skilled workers.[89]

Moreover, to achieve the required level of sensitivity and specificity in detection, it is often necessary to use a device that is capable of identifying and differentiating a large number of biochemical constituents in complex environmental samples. And this is why, in the near future, amperometric multisensor biochip devices will combine automated sample collection systems and multichannel sensing capability, allowing simultaneous detection of multiple pathogens (e.g., viruses) present in complex samples. Therefore, amperometric biochip technology will provide an important tool to warn of exposure to pathogenic viruses and to be used in health protection and disease prevention for animals and humans.

Acknowledgements

This research is conducted with NTU-HUJ-BGU Nanomaterials for Energy and Water Management Programme under Campus for Research Excellence and Technological Enterprise (CREATE) that is supported by the National Research Foundation, Prime Minister's Office, Singapore. The authors also thank NRF-POC project 002-026 (2014–2015) title *"Electrochemical lateral flow biosensor for detection and quantification of Dengue virus in blood"*.

References

1. http://www.vnaa.org/vnaa/g/?h=html/germ_protection_center_viruses.

2. Iqbal, S. S., M. W. Mayo, J. G. Bruno, B. V. Bronk, C. A. Batt, and J. P. Chambers. A review of molecular recognition technologies for detection of biological threat agents. 2000. *Biosens. Bioelectr.*, **15**, 549–578.
3. Wang, J. From DNA biosensors to gene chips. 2000. *Nucleic Acids Res.*, **28**, 3011–3016.
4. Thévenot, D. R., K. Toth, R. A. Durst, and G. S. Wilson. (1999) Electrochemical biosensors: recommended definitions and classification. 2000. *Pure Appl. Chem.*, **71**, 2333–2348.
5. Thévenot, D. R., K. Toth, R. A. Durst, and G. S. Wilson. Electrochemical biosensors: recommended definitions and classification. 2001. *Biosens. Bioelectr.*, **16**, 121–131.
6. Schuhmann, W. Amperometric enzyme biosensors based on optimized electron-transfer pathways and non-manual immobilization procedures. 2002. *Mol. Biotechnol.*, **82**, 425–441.
7. Habermüller, K., M. Mosbach, and W. Schuhmann. Electron-transfer mechanisms in amperometric biosensors. 2000. *Fresenius J. Anal. Chem.*, **366**, 560–568.
8. Palmisano, F., P. G. Zambonin, and D. Centonze. Amperometric biosensors based on electrosynthesised polymeric films. 2000. *Fresenius J. Anal. Chem.*, **366**, 586–601.
9. Pearson, J. E., A. Gill, and P. Vagdama. Analytical aspects of biosensors. 2000. *Annu. Clin. Biochem.*, **37**, 119–145.
10. Chaubey, A., and B. D. Malhotra. Mediated biosensors. 2002. *Biosens. Bioelectr.*, **17**, 441–456.
11. Luong, J. H. T., A. Mulchandani, and G. G. Guilbault. Developments and applications of biosensors. 1988. *Trends Biotechnol*, **6**, 310–316.
12. Nernst, W., and E. Brunner. Theorie der reaktionsgeschwindigkeit in heterogenen systemen. 1904. *Z. Phys. Chem.*, **47**, 52–57.
13. http://en.wikipedia.org/wiki/Electrochemical_reaction_mechanism.
14. Achilli, G., G. P. Cellerino, G. V.M dEril, and S. Bird. Simultaneous determination of 27 phenols and herbicides in water by high-performance liquid chromatography with multielectrode electrochemical detection. 1995. *J. Chromatogr. A*, **697**, 357–362.
15. Ye, J., and R. P. Baldwin. Determination of amino acids and peptides by capillary electrophoresis and electrochemical detection at a copper electrode. 1994. *Anal. Chem.*, **66**, 2669–2674.
16. Ionescu, R. E., C. Gondran, S. Cosnier, L.A. Gheber, and R. S. Marks. Construction of amperometric immunosensors based on the electrogen-

eration of a permeable biotinylated polypyrrole. 2004. *Anal. Chem.*, **76**, 6808–6813.

17. Ionescu, R. E., C. Gondran, S. Cosnier, L.A. Gheber and R. S. Marks. Comparison between the performances of amperometric immunosensors for cholera antitoxin based on three enzyme markers. 2005. *Talanta*, **66**, 15–20.
18. Ionescu, R. E., R. S. Marks, and S. Cosnier. Protease amperometric sensor. 2006. *Anal. Chem.*, **78**, 6327–6331.
19. Cosnier, S., R. E. Ionescu, S. Herrmann, L. Bouffier, M. Demeunynck, and R. S. Marks. Electroenzymatic polypyrrole-intercalator sensor for the determination of West Nile virus cDNA. 2006. *Anal. Chem.*, **78**, 7054–7057.
20. Ionescu, R. E., S. Herrmann, S. Cosnier, and R. S. Marks. A polypyrrole cDNA electrode for the amperometric detection of the West Nile virus. 2006. *Electrochem. Commun.*, **8**, 1741–1748.
21. Clark, L. C., and C. Lyons. Electrode systems for continuous monitoring in cardiovascular surgery. 1962. *Annu. N. Y. Acad. Sci.*, **102**, 29–45.
22. Updike, S. J., and G. P. Hicks. The enzyme electrode. 1967. *Nature*, **214**, 986–988.
23. Ghindilis, A. L., P. Atanasov, and E. Wilkins. Enzyme-catalyzed direct electron transfer: fundamentals and analytical applications, 1997. *Electroanalysis*, **9**, 661–674.
24. Gorton, L., A. Lindgren, T. Larsson, F. D. Munteanu, T. Ruzgas, and I. Gazaryan. Direct electron transfer between heme-containing enzymes and electrodes as basis for third generation biosensors. 1999. *Anal. Chim. Acta*, **400**, 91–108.
25. Schuhmann, W., T. J. Ohara, H. L. Schmidt, and A. Heller. Electron transfer between glucose oxidase and electrodes via redox mediators bound with flexible chains to the electrode surface. 1991. *J. Am. Chem. Soc.*, **113**, 1394–1397.
26. Dequaire, M., and A. Heller. Screen printing of nucleic acid detecting carbon electrodes. 2002. *Anal. Chem.*, **74**, 4370–4377.
27. Katz, E., A. Riklin, V. Helegshabtai, I. Willner, and A. F. Bückmann. Glucose oxidase electrodes via reconstitution of the apo-enzyme: tailoring of novel glucose biosensors. 1999. *Anal. Chim. Acta*, **385**, 45–58.
28. Cosnier, S. Biomolecule immobilization on the electrode surface by entrapment or attachment to electrochemically polymerized films. 1999. *Biosens. Bioelectr.*, **14**, 443–456.

29. Koopal, C. G. J., A. A. C. M. Bos, and R. J. M. Nolte. Third-generation glucose biosensor incorporated in a conducting printing ink. 1994. *Sens. Actuators, B*, **18**, 166–170.

30. Schuhmann, W. 1998. Immobilisation using electrogenerated polymers, in *Immobilized Biomolecules in Analysis. A Practical Approach* (T. Cass, F. S. Ligler, eds), Oxford University Press, New York, 187210.

31. Willner, I., and E. Katz. Integration of layered redox proteins and conductive supports for bioelectronic applications. 2000. *Angew. Chem., Int. Ed.*, **39**, 1180–1218.

32. Coche-Guerente, L., S. Cosnier, and L. Labbe. Sol-gel derived composite materials for the construction of oxidase/peroxidase mediatorless biosensors. 1997. *Chem. Mater.*, **9**, 1348–1352.

33. Ghindilis, A. L., R. Krishnan, P. Atanasov, and E. Wilkins. Flow-through amperometric immunosensor: fast "sandwich" scheme immunoassay. 1997. *Biosens. Bioelectr.*, **12**, 415–423.

34. Luppa, P. B., L. J. Sokoll, and D. W. Chan. Immunosensor: principles and applications to clinical chemistry. 2001. *Clin. Chim. Acta*, **314**, 1–26

35. Shah, J., and E. Wilkins. Electrochemical biosensors for detection of biological warfare agents. 2003. *Electroanalysis*, **15**, 157–167.

36. Hock, B. Advances in immunochemical detection of microorganisms. 1996. *Annu. Biol. Clin.*, **54**, 243–252.

37. Villarreal, L. P. 2005. *Viruses and the Evolution of Life*, ASM Press, Washington, DC, 395.

38. Mahner, M., and M. Bunge. 1997. *Foundations of Biophilosophy*, Springer-Verlag, Berlin.

39. Barry, M. A., W. J. Dower, and S. A. Johnston. Towards cell-targeting gene therapy vectors: selection of cell-binding peptides from random peptide-presenting phage libraries. 1996. *Nat. Med.*, **2**, 299–305.

40. Tuckey, C. D., and C. J. Noren. Selection for mutants improving expression of an anti-MAP kinase monoclonal antibody by filamentous phage display. 2002. *J. Immunol. Methods*, **270**, 247–257.

41. Chu, R., J. Takei, J. R. Knowlton, M. Andrykovitch, W. Pei, A. V. Kajava, P. J. Steinbach, X. Ji, and Y. Bai. Redesign of a four-helix bundle protein by phage display coupled with proteolysis and structural characterization by NMR and X-ray crystallography. 2002. *J. Mol. Biol.*, **323**, 253–262.

42. Atias, D., L. Lobel, M. Virta, R. S. Marks, D. Cullen, and C. Lowe. 2002. In *Handbook of Biosensors and Biochips* (H. H. Weetall, I. Karube, eds), Wiley-Interscience, New York.

43. Palecek, E. Past, present and future of nucleic acids electrochemistry. 2002. *Talanta*, **56**, 809–819.
44. Kerman, K., D. Ozkan, P. Kara, B. Meric, J. J. Gooding, and M.Ozsoz. Voltammetric determination of DNA hybridization using methylene blue and self-assembled alkanethiol monolayer on gold electrodes. 2002. *Anal. Chim. Acta*, **462**, 39–47.
45. Lee, T. Y., and Y. B. Shim. Direct DNA hybridization detection based on the oligonucleotide-functionalized conductive polymer. 2001. *Anal. Chem.*, **73**, 5629–5632.
46. Campbell, C. N., D. Gal, N. Cristler, C. Banditrat and A. Heller. Enzyme amplified amperometric sandwich test for RNA and DNA. 2002. *Anal. Chem.*, **74**, 158–162.
47. Patolsky, F., A. Lichtenstein, and I. Willner. Detection of single-base DNA mutations by enzyme-amplified electronic transduction. 2001. *Nat. Biotechnol.*, **19**, 253–257.
48. Caruana, D. J., and A. Heller. Enzyme amplified amperometric detection of hybridization and of a single base pair mutation in an 18-base oligonucleotide on a 7-μm diameter microelectrode. 1999. *J. Am. Chem. Soc.*, **121**, 769–774
49. http://www.ncbi.nlm.nih.gov/books/bv.fcgi?rid=mmed.figgrp.2270.
50. Barquet, N., and P. Domingo. Smallpox: the triumph over the most terrible of the ministers of death. 1997. *Annu. Inter. Med.*, **127**, 635–642.
51. Jezek, Z., W. Hardjotanojo, and A. G. Rangaraj. Facial scarring after varicella: a comparison with variola major and variola minor. 1981. *Am. J. Epidemiol.*, **114**, 798–803.
52. Behbehani, A. M. The smallpox story: life and death of an old disease. 1983. *Microbiol. Rev.*, **47**, 455–509.
53. De Cock, K. M. The eradication of smallpox: Edward Jenner and the first and only eradication of a human infectious disease. 2001. *Nat. Med.*, **7**, 15–16.
54. Wijayawardhana, C. A., S. Purushothama, M. A. Cousino, H. B. Halsall, and W. R. Heineman. Rotating disk electrode amperometric detection for a bead-based immunoassay. 1999. *J. Electroanal. Chem.*, **468**, 2–8.
55. Heineman, W. R., J. H. Thomas, C. A. Wijayawardhana, H. B. Halsall, T. H. Ridgway, J. W. Choi, K. W. Oh, C. Ahn, S. Dharmatilleke, P. Medis, and T. H. Henderson. BioMEMS: electrochemical immunoassay with microfluidic systems. 2001. *Anal. Sci.*, **17**, i281–i284.

56. Wilson, R., C. Clavering, and A. Hutchinson. Paramagnetic bead based enzyme electrochemiluminescence immunoassay for TNT. 2003. *J. Electroanal. Chem.*, **557**, 109–118.

57. Wang, J., A.-N. Kawde, M. Musameh, and G. Rivas. Dual enzyme electrochemical coding for detecting DNA hybridization. 2002. *Analyst*, **127**, 1279–1282.

58. Strauss, J. H., Jr., and R. L. Sinsheimer. Purification and properties of bacteriophage MS2 and of its ribonucleic acid. 1963. *J. Mol. Biol.*, **7**, 43–54.

59. Rowe, C. A., L. M. Tender, M. J. Feldstein, J. P. Golden, S. B. Scruggs, B. D. MacCraith, J. J. Cras, and F. S. Ligler. Array biosensor for simultaneous identification of bacterial, viral, and protein analytes. 1999. *Anal. Chem.*, **71**, 3846–3852.

60. Thomas, J. H., S. K. Kim, P. J. Hesketh, H. B. Halsall, and W. R. Heineman. Bead-based electrochemical immunoassay for bacteriophage MS2. 2004. *Anal. Chem.*, **76**, 2700–2707.

61. http://en.wikipedia.org/wiki/Bovine_leukemia_virus.

62. Ramanaviciene, A., and A. Ramanavicius Pulsed amperometric detection of DNA with an ssDNA/polypyrrole-modified electrode. 2004. *Anal. Bioanal. Chem.*, **379**, 287–293.

63. Ramanaviciene, A., and A. Ramanavicius. Moleculary imprinted polypyrrole-based synthetic receptor for direct detection of bovine leukaemia virus glycoproteins. 2004. *Biosens. Bioelectr.*, **20**, 1076–1082.

64. Eccles, R. Understanding the symptoms of the common cold and influenza. 2005. *Lancet Infect. Dis.*, **5**, 718–725.

65. http://www.who.int/csr/disease/avian_influenza. Accessed April 22, 2008.

66. Sitdikov, R., and E. Wilkins. Detection of para influenza and influenza A viruses using flow-injection amperometric immunosensor. 2007. *J. Appl. Res.*, **7**, 108–126.

67. Hoke, C. H., Jr., D. W.Vaughn, A. Nisalak, P. Intralawan, S. Poolsuppasit, V.Jongsawas, Titsyakorn, and R. T. Johnson. Effect of high-dose dexamethasone on the outcome of acute encephalitis due to Japanese encephalitis virus. 1992. *J. Infect. Dis.*, **165**, 631–637.

68. Burke, D. S., and C. J. Leake. 1988. Japanese encephalitis, in *The Arboviruses: Epidemiology and Ecology* (T. P. Monath, ed), CRC Press, Boca Raton, Florida, 63–92.

69. Yuan, R., L. Zhang, Q. Li, Y. Chai, and S. Cao. A label free amperometric immunosensor based on multi-layer assembly of polymerized

o-phenylenediamine and gold nanoparticles for determination of Japanese B encephalitis vaccine. 2005. *Anal. Chim. Acta*, **531**, 1–5.

70. Brinton, M. A. The molecular biology of West Nile virus: a new invader of the estern hemisphere. 2002. *Annu. Rev. Microbiol.*, **56**, 371–402.
71. Hayes, E. B., N. Komar, R. S. Nasci, S. P. Montgomery, D. R. O'Leary, and G. L. Campbell. Epidemiology and transmission dynamics of West Nile virus disease. 2005. *Emerg. Infect. Dis.*, **11**, 1167–1173.
72. http://www.cdc.gov /ncidod/dvbid/westnile/index.htm.
73. Ionescu, R. E., S. Cosnier, G. Herzog, K.Gorgy, B. Leshem, S. Herrmann, and R. S. Marks. Amperometric immunosensor for the detection of anti-West Nile virus IgG using a photoactive copolymer. 2007. *Enzyme Microb. Technol.*, **40**, 403–408.
74. Ionescu, R. E., S. Cosnier, S. Herrmann, and R. S. Marks. Amperometric immunosensor for the detection of anti-West Nile IgG. 2007. *Anal. Chem.*, **79**, 8662–8668.
75. Riccardi, C. D., K. Dahmouche, C. V. Santilli, P. I. da Costa, and H. Yamanaka. Immobilization of streptavidin in sol-gel films: application on the diagnosis of hepatitis C virus. 2006. *Talanta*, **70**, 637–643.
76. http://www.aphis.usda.gov/vs/ceah/cei/bi/bvdinfosheet.pdf.
77. Muhammad Tahir, Z. M., E. C. Alocilja, and D. L. Grooms. Indium tin oxide-polyaniline biosensor: fabrication and characterization. 2007. *Sensor*, **7**, 1123–1140.
78. Barker, L. F., N. R. Shulman, R. Murray, R. J. Hirschman, F. Ratner, W. C. Diefenbach, and H. M. Geller. Transmission of serum hepatitis. 1970. 1996. *J. Am. Med. Assoc.*, **276**, 841–844.
79. Williams, R. Global challenges in liver disease. 2006. *Hepatology*, **44**, 521–526.
80. Tang, D., R. Yuan, Y. Chai, Y. Fu, J. Dai, L. Yan, and X. Zhong. New amperometric and potentiometric immunosensors based on gold nanoparticles/tris(2,2'-bipyridyl)cobalt (III) multilayer for hepatitis B surface antigen determinations. 2005. *Biosens. Bioelectr.*, **21**, 539–548.
81. Liang, R. P., H. D. Qiu, and P. X. Cai. A novel amperometric immunosensor based on three-dimensional sol-gel network and nanoparticle self-assemble technique. 2005. *Anal. Chim. Acta*, **534**, 223–229.
82. Zhuo, Y., R. Yuan, Y. Chai, Y. Zhang, X. L. Li, Q. Zhu, and N. Wang. An amperometric immunosensor based on immobilization of hepatitis B surface antibody on gold electrode modified gold nanoparticles and horseradish peroxidase. 2005. *Anal. Chim. Acta*, **548**, 205–210.

83. Schmaljohn, C., and B. Hjelle. Hantaviruses: a global disease problem. 1997. *Emerg. Infect. Dis.*, **3**, 95–104.

84. Montes, A. L., J. P. Martinez, I. Lorenzo, J. J. Blanch, F. de Ory, C. G. Roldán, E. L. Rubio, F. Llamas, E. G. Valcárce, and E. Olivas. Nephropathia caused by Hantavirus Puumala: a case report. 2005. *Nefrologia*, **25**, 195–199.

85. Krishnan, R., A. L. Ghindilis, P. Atanasov, E. Wilkins, J. Montoya, and F. T. Koster. Fact amperometric immunoassay for hantavirus infection. 1996. *Electroanalysis*, **8**, 1131–1134.

86. Sitdikov, R. A., E. S. Wilkins, T. Yates, and B. Hjelle. Detection of Hantavirus using a new miniaturized biosensor device. 2007. *J. Appl. Res.*, **7**, 86–107.

87. Alexander, D. J. Newcastle disease and other avian paramyxoviruses. 2000. *Revue Sci. Tech. (Int. Office Epizoot.)*, **19**, 443–462.

88. Gong, J. L., F. Chun-Gong, G. Ming-Zeng, G. Li-Shen, and R. Qin-Yu. An amperometric immunosensor for the Newcastle disease antibody assay. 2003. *Anal. Lett.*, **36**, 287–302.

89. Ghindilis, A. L., P. Atanasov, M. Wilkins, and E. Wlikins. Immunosensors: electrochemical sensing and other engineering approaches. 1998. *Biosens. Bioelectr.*, **13**, 113–131.

Index